Swarm Studies and Inelastic Electron–Molecule Collisions

Swarm Studies and Inelastic Electron–Molecule Collisions

Proceedings of the Meeting of the Fourth International Swarm Seminar
and the Inelastic Electron–Molecule Collisions Symposium
July 19–23, 1985, Tahoe City, California, USA

Edited by
L.C. Pitchford B.V. McKoy
A. Chutjian S. Trajmar

With 170 Figures

Springer-Verlag
New York Berlin Heidelberg
London Paris Tokyo

Leanne C. Pitchford
Discharge Physics Department
GTE Laboratories
Waltham, Massachusetts 02254, USA

B. Vincent McKoy
Noyes Laboratory of Chemical Physics
California Institute of Technology
Pasadena, California 91125, USA

Ara Chutjian
California Institute of Technology
Jet Propulsion Laboratory
Pasadena, California 91109, USA

Sandor Trajmar
California Institute of Technology
Jet Propulsion Laboratory
Pasadena, California 91109, USA

Library of Congress Cataloging in Publication Data
International Swarm Seminar (4th : 1985 : Tahoe City, Calif.)
Swarm studies and inelastic electron-molecule collisions.
1. Electron-molecule collisions—Congresses.
2. Electron swarms—Congresses. I. Pitchford, L.C. (Leanne C.) II. Inelastic Electron-Molecule Collisions Symposium (1985 : Tahoe City, Calif.) III. Title.
QC794.6.C6I587 1985 539.7'54 86-20343

9 8 7 6 5 4 3 2 1

ISBN-13: 978-1-4612-9104-6 e-ISBN-13: 978-1-4612-4662-6
DOI: 10.1007/978-1-4612-4662-6

Preface

This volume presents the contributions of participants in the Symposium on Swarm Studies and Inelastic Electron–Molecule Collisions, held on July 19–23, 1985, in Tahoe City, California. This was a joint meeting of the Fourth International Swarm Seminar and the Electron–Molecule Collisions Symposium which have been traditionally separate satellite symposia to the International Conference on the Physics of Electronic and Atomic Collisions (ICPEAC). In the early stages of planning for these two satellite symposia to the XIVth ICPEAC, a group of us recognized the significant scientific merit and advantages of having a joint symposium. This idea was particularly appealing due to a large mutual interest in important advances (theoretical, experimental, and modeling) in both fields, and because it provides a forum to bring together a single-collision point of view with a multiple-collision one. For example, studies of multiple-term solutions to Boltzmann's equation and their application to swarm systems are intrinsically coupled to the availability of both integral and differential cross-sections for electron–molecule collisions. In turn, experimental and theoretical studies of these electron–molecule scattering cross-sections are becoming quite sophisticated, accurate, and comprehensive. Furthermore, in swarm studies, computational and experimental methods have advanced to the point where detailed and meaningful comparison with, and use of, single-collision beam data is now possible. Moreover, recent experimental advances in the study of single-collision electron attachment phenomena have provided a significant overlap with swarm data and extension to subthermal energies.

With such objectives in mind, this symposium brought together a selected group of speakers who are centrally involved in the important areas of swarm modeling, swarm measurements, and studies of elastic and inelastic electron–molecule collisions. The first day of the symposium was devoted exclusively to traditional swarm (multiple-collision) topics. However, on the following two days, the presentation by the speakers in the single- and multiple-collision area was organized and intertwined into a single, coherent program. The goals of these presentations were as follows:

- to explain applications of current electron–molecule inelastic collision data
- to indicate where more data is needed for both near- and far-term swarm applications in particular devices
- to compare swarm- and beam-acquired electron scattering and attachment data
- to delineate the state of the art in our experimental and theoretical understanding of electron–molecule inelastic scattering.

We would like to thank our speakers for the special effort they made to coordinate their presentations and to make them relevant to the special objectives of this joint symposium. We also want to thank the following agencies and organizations for their financial support of the conference: the National Science Foundation, Air Force Office of Scientific Research, Office of Naval Research, Naval Surface Weapons Center, Los Alamos National Laboratories, and GTE Laboratories. Also, we wish to express our thanks and appreciation to the staff of the Granlibakken Conference Center in Tahoe City; particularly Bill and Henk Parson, Norma Parson, Ann Page, Mary Brown, and Dennise Connors. Their help and support contributed greatly to the success of this symposium.

Contents

Part I. Selected Reviews in Swarm Studies

Part II. Kinetic Theory

Part III. Abstracts of Contributed Swarm Papers

ORGANIZERS/EDITORS

L. Pitchford, GTE Laboratories

V. McKoy, California Institute of Technology

A. Chutjian, California Institute of Technology, Jet Propulsion Laboratory

S. Trajmar, California Institute of Technology, Jet Propulsion Laboratory

PARTICIPANTS

Aladzhadzhyan, A.	Jet Propulsion Laboratory, California Institute of Technology
Antoni, T.	Jet Propulsion Laboratory, California Institute of Technology
Boeuf, J.-P.	CNRS, Ecole Superieure d'Electricité, France
Brescansin, L.	California Institute of Technology
Broad, J.	Universität Bielefeld
Burke, P.	The Queen's University of Belfast
Cartwright, D.	Los Alamos National Laboratory
Cavalleri, G.	Universita Cattolica, Bresica, Italy
Christodoulides, A.	University of Tennessee
Christophorou, L.	Oak Ridge National Laboratory
Chutjian, A.	Jet Propulsion Laboratory, California Institute of Technology
Comer, J.	Manchester University
Cooper, R.	Texas Tech University
Crompton, R.	Australian National University
De Urquijo, J.	National University of Mexico
Domcke, W.	Universitat Heidelberg
Ehrhardt, H.	Universität Kaiserslautern
Eisele, F.	Georgia Institute of Technology
Elford, M.	Australian National University
Freund, R.	AT&T Bell Laboratories
Furst, J.	University of Oklahoma

Gallagher, J.	Joint Institute for Laboratory Astrophysics
Garscadden, A.	Wright-Patterson Air Force Base
Gianturco, F.	Universita di Roma
Green, C.	Louisiana State University
Hatano, Y.	Tokyo Institute of Technology
Hayashi, M.	Nagoya Institute of Technology
Hays, G.	Sandia National Laboratories
Hazi, A.	Lawrence Livermore National Laboratory
Hiskes, J.	Lawrence Livermore National Laboratory
Hotop, H.	Universität Kaiserslautern
Huo, W.	NASA-Ames Research Center
Hunter, S.	Oak Ridge National Laboratory
Jain, A.	Joint Institute for Laboratory Astrophysics
Jelenkovic, B.	Joint Institute for Laboratory Astrophysics
Johnsen, R.	University of Pittsburgh
Kano, S.	University of Electro-Communications, Tokyo
Keesee, R.	Pennsylvania State University
Kline, L.	Westinghouse Research and Development
Kondo, L.	Anan College of Technology, Tokyo
Kumar, K.	Australian National University
Li, Yan Ming	GTE Laboratories, Inc.
Lima, M.	California Institute of Technology
Mahgerefteh, M.	University of Oklahoma
Massaro, P.	Universita Degli Studi di Bari, Italy
McCorkle, D.	University of Tennessee
McKoy, V.	California Institute of Technology
Morgan, L.	Royal Holloway College

Morgan, W.	Lawrence Livermore National Laboratory
Morrison, M.	University of Oklahoma
Nakamura, Y.	Keio University
Ness, K.	University of British Columbia
Newell, W.	University College
Norcross, D.	Joint Institute for Laboratory Astrophysics/National Bureau of Standards
Noble, C.	Daresbury Laboratory
O'Malley, T.	General Electric Company
Phelps, A.	Joint Institute for Laboratory Astrophysics
Pitchford, L.	GTE Laboratories, Inc.
Raseev, G.	Université de Paris
Reid, I.	University of British Columbia
Schneider, B.	Los Alamos National Laboratory
Ségur, P.	Université Paul Sabatier
Shimamura, I.	Institute of Physics, RIKEN, Japan
Shizgal, B.	University of British Columbia
Skullerud, H.	The Norwegian Institute of Technology/University of Trondheim
Tagashira, H.	Hokkaido University
Torzo, G.	Padova, Italy
Trajmar, S.	Jet Propulsion Laboratory, California Institute of Technology
Tronc, M.	Universite Pierre et Marie Curie II
Van Brunt, R.	National Bureau of Standards, Gaithersburg, Maryland
Varracchio, E.	Universita Degli Studi di Bari, Italy
Wedding, A.	Joint Institute for Laboratory Astrophysics
Winters, H.	IBM Research Laboratories

PART I

SELECTED REVIEWS IN SWARM STUDIES

A COMPARATIVE STUDY OF THE COMPUTING METHODS ACTUALLY IN USE FOR ACCURATE DETERMINATION OF SWARM PARAMETERS

Pierre Ségur, M.-C. Bordage and M. Yousfi

Centre de Physique Atomique,
Unité de Recherche Associée au C.N.R.S. n° 277,
118 route de Narbonne, 31062 Toulouse Cedex, France

Introduction

At the present time, measurement of most swarm parameters can generally be made with a high level of accuracy. Consequently, in order to achieve meaningful comparisons between measured and calculated values of these swarm parameters, it is necessary not only to build up an accurate theory describing the motion of a swarm (this is the well-known theory of hydrodynamic regime; cf. Kumar et al [1]), but also to develop powerful and accurate methods of solution for the equations employed in this theory.

Since a lot of work has recently been done in order to try to obtain new and more accurate methods of solutions, it appears interesting to carry out a comparative review of most of the methods existing in this field. This is one of the aims of this work.

First, we shall very briefly give the equations used in the theory of hydrodynamic regime. Then, we shall introduce the main methods adapted to the solution of these equations. We shall distinguish between three classes of methods. These are, respectively, integral methods, integrodifferential methods, and stochastic (Monte Carlo) methods.

We shall emphasize the second class of methods, especially for the so-called half-range methods which have been used for a long time in traditional transport theory and which have been recently adapted by ourselves to the calculation of swarm parameters. In our opinion, these methods are, at the present time, the most powerful and the simplest ones.

We shall then give some typical results in the case of model gases for conservative and non-conservative situations.

Note that we limit our investigations here to the methods appropriate for the study of electron motion.

Brief Outline of the Theory of Hydrodynamic Regime

The accurate theory of hydrodynamic regime emerged after the discovery, around 1970, of some inconsistency between measured and calculated diffusion coefficients. According to the theory admitted at that time, the diffusion coefficient was a scalar quantity. Since experiments showed large differences between longitudinal (along electric field) and transversal components of the diffusion coefficients, a revision of this old theory was necessary.

In the new theory, (mainly developed by Kumar, Skullerud, Robson [1] and Tagashira's group [2]), the diffusion coefficient becomes not only a tensorial quantity, but furthermore new definitions are given for the swarm parameters, which become dependent on the type of experiments considered and which are different according to the kind of collision processes involved (conservative or not).

There are roughly two classes of swarm experiments. The first one is characterized by an exponential growth of the total density of electrons in time, and the second one by an exponential growth of the density with position.

The time-independent coefficients measured in the first class of experiments (time-of-flight or time-resolved experiments) are defined by the following relationships:

$$\underline{W} = \int \underline{dv}\ \underline{v}\ F^{(o)}(\underline{v}) + \int \underline{dv}(\nu_{ion}(v) - \nu_{at}(v))\underline{F}^{(1)}(\underline{v}) \quad (1)$$

$$\underline{D} = \int \underline{dv}\ \underline{v}\ \frac{\underline{v} - \underline{W}}{\nu(v)+\omega^\circ} F^{(o)}(\underline{v}) - \int \underline{dv}\ \underline{v}\ \frac{\underline{a}.\nabla v}{\nu(v)+\omega^\circ}\ \underline{F}^{(1)}(\underline{v}) + \int \underline{dv}\ (\nu_{ion}(v) - \nu_{at}(v))\underline{F}^{(2)}(\underline{v}) \quad (2)$$

In the relations above, $\underline{W}$ is the drift velocity (this is the velocity of the center of mass of the swarm), $\underline{D}$ is the diffusion tensor, $\underline{v}$ is the velocity vector, $\nu(\overline{v})$ is the total collision frequency, $\nu_{ion}(v)(\nu_{at}(v))$ is the ionization (attachment) frequency. ω° is the mean value of the effective ionization frequency and $\underline{a}$ is the acceleration vector equal to $-|e|\ \underline{E}/m$ where $\underline{E}$ is the electric field vector and m the electron mass.

We may first note that $\underline{W}$ and $\underline{D}$ depend on three functions $F^{(o)}(\underline{v})$, $\underline{F}^{(1)}(\underline{v})$, $\underline{F}^{(2)}(\underline{v})$ which are scalar, vectorial, and tensorial quantities, respectively. The functions $F^{(o)}(\underline{v})$, $\underline{F}^{(1)}(\underline{v})$, $\underline{\underline{F}}^{(2)}(\underline{v})$ are solutions of the three following equations:

$$L \; F^{(o)}(\underline{v}) = -\omega° \; F^{(o)}(\underline{v})$$

$$L \; \underline{F}^{(1)}(\underline{v}) = \underline{v} \; F^{(o)}(\underline{v}) - \underline{W} \; F^{(o)}(\underline{v}) - \omega° \; \underline{F}^{(1)}(\underline{v}) \qquad (3)$$

$$L \; \underline{\underline{F}}^{(2)}(\underline{v}) = \underline{v} \; \underline{F}^{(1)}(\underline{v}) - \underline{W} \; \underline{F}^{(1)}(\underline{v}) - \underline{\underline{D}} \; F^{(o)}(\underline{v}) - \omega° \; \underline{\underline{F}}^{(2)}(\underline{v})$$

Where L is an operator defined by the following relationship:

$$L = \underline{a} \cdot \nabla_{\underline{v}} + \nu(v) - J$$

J is the scattering operator, which includes all the collisional processes to be taken into account (elastic and inelastic).

To these equations, we must add the two normalization conditions:

$$\int \underline{dv} \; F^{(o)}(\underline{v}) = 1 \qquad (4)$$

$$\int \underline{dv} \; \underline{F}^{(k)}(\underline{v}) = 0 \quad , \quad k \geq 1$$

Furthermore, the $\omega°$ frequency is defined by the following relationship:

$$\omega° = \int \underline{dv}(\nu_{ion}(v) - \nu_{at}(v))F^{(o)}(\underline{v}) \qquad (5)$$

Note that:

1. In non-conservative situations, equations (3) are non-linear (through $\omega°$).

2. Relationships (1) and (2) are very different in conservative or non-conservative cases, since $\omega°$ and the terms depending on ionization and attachment cross-sections vanish when there is neither ionization nor attachment.

3. The first term, in the right-hand side of (1) is the mean velocity of all the electrons in the swarm.

4. The first term in the right-hand side of (2) is made up of two parts. The first one is the thermal part of the diffusion coefficient (it was the only term used in the old theory) and the second one is connected to the drift energy

of electrons in the swarm.

The position-independent parameters measured in the second class of experiments (steady state Townsend experiments) are the ionization and attachment coefficients defined by the following relationships:

$$\alpha = \int \underline{dv}\ \nu_{ion}(v) F_{SST}(\underline{v}) / \int \underline{dv}\ v_z\ F_{SST}(\underline{v}) \tag{6}$$

$$\eta = \int \underline{dv}\ \nu_{at}(v) F_{SST}(\underline{v}) / \int \underline{dv}\ v_z\ F_{SST}(\underline{v}) \tag{7}$$

The distribution function $F_{SST}(\underline{v})$ is solution of the following equation:

$$L\ F_{SST}(\underline{v}) + v_z(\alpha-\eta) F_{SST}(\underline{v}) = 0 \tag{8}$$

Here v_z is the longitudinal component of $\underline{v}$. As the set of equations (3), the equation above is non-linear in non-conservative situations.

Equations (3) and (8) have roughly a similar shape and they can be written under the following form:

$$M\ \ H\ (\underline{v}) = S\ (\underline{v}) \tag{9}$$

Where $S\ (\underline{v})$ is a known source term and M is an operator equal to

$$M = L + \omega^{\circ} \tag{10}$$

for the first class of experiments and to

$$M = L + v_z(\alpha-\eta) \tag{11}$$

for the second class of experiments.

The aim of all the current methods is to solve equations similar to equation (9).

Methods of Solution for Equations Similar to Eq. (9)

The most natural way to attack the solution of (9) (assuming that the scattering term J is known and noting that, in this case, eq. (9) becomes a partial differential equation) is to integrate formally this equation. We obtain the integral equation below, equivalent to eq. (9).

$$H(\underline{v}) = \int_{-\infty}^{v_z} \frac{dv_{z'}}{|\underline{a}|} \exp\left(-\Gamma \frac{(v_z - v_{z'})}{|\underline{a}|}\right) \left[\underline{a}.\nabla_{\underline{v}} + \Gamma - M) H + S\right] \quad (12)$$

Since the scattering term J is not known, the solution of (12) is generally made by iterations, starting from an initial guess for $H(\underline{v})$. Γ is a constant parameter chosen in order to speed up convergence.

Although integral methods have been known for a long time in traditional transport theory, the integral form above was used for the first time by Rees [3] in the field of solid state physics, and then adapted by Kleban and Davies [4] to the calculation of swarm parameters in model CH_4. A different form (without the Γ constant) has been recently used by Skullerud and Kuhn [5].

Although integral methods can give good results, they are generally too time consuming with respect to other methods and in our opinion must be avoided.

Another way to attack the solution of eq. (9) is, starting from the initial form of (9), to expand the function H into a series of spherical harmonics. An infinite set of differential equations is then obtained for the various components of the sperical harmonic expansion, and some additional assumptions must be made to solve this set.

The most current approximation employed is the so-called two-term approximation, where the expansion into spherical harmonics is truncated after the first two terms. This approximation has been used in a lot of codes [6]. However, recent works have shown that this approximation is no longer valid in some situations where inelastic collisions are important and for high values of electric field-density ratio.

A slight improvement to the two-term approximation is given by the three-term approximation codes. The best known are those of Wilhelm and Winkler (1969) [7], Makabe and Mori (1978) [8]. Usefulness of these codes is limited (because they are very complicated and they cannot give an answer to the validity of the two-term approximation), so they are now replaced by multiterm approximation (where more than three terms are used in the spherical harmonics expansion).

The first attempt to obtain an accurate multiterm solution of the Boltzmann equation was made by Lin et al [9] who used a moment method where the distribution function was expanded into a series of Sonine polynomials. For a long time, this code was only applied to conservative situations, and it is only recently (see [10]) that it has been modified to allow the treatment of non-conservative cases.

This method is interesting primarily because it allows the determination of most swarm parameters without the complete knowledge of the distribution function. However, there may be some drawback due to the lack of convergence of the Sonine polynomial expansion.

In the code of Pitchford et al [11], a Galerkin method is used where the terms of the Legendre series are expanded into B-spline functions. This code is only very powerful if it is run on a vectorial computer, but it is now one of the most popular and one of the most convenient among the full-range codes.

There are often some difficulties in full-range multiterm codes due to the existence of singular solutions close to zero energy and at infinity. This problem has been investigated and conveniently solved using different techniques in the works of Hammar (1973) [12], in the field of solid state physics, Winkler et al (1983) [13] and Mac Mahon (1983, 1984) [14].

A strictly numerical approach (called shooting method) was used to solve this problem by Canosa and Penafiel (1972) [15], in radiative transfer theory.

The accuracy of all these multiterm codes is generally good, but they need sophisticated techniques to overcome the problem of singular solutions and of numerical instabilities. Furthermore, they are not well adapted to the treatment of non-conservative situations and they often need very huge computers in order to solve the large linear system of equations which is generally needed.

Most of these difficulties can be avoided by using what we call half-range methods.

In order to understand the concept of half-range methods, we shall come back, first, to the traditional full-range Legendre expansion. In this case, the function $H(\underline{v})$ is generally expanded into the following series (μ is the cosine of the angle θ between $\underline{E}$ and $\underline{v}$, and $P_\ell(\cos\theta)$ is a Legendre polynomial):

$$H(\underline{v}) = H(v,\cos\theta) = \sum_{\ell=0}^{\infty} \frac{2}{2\ell+1} H_\ell(v) P_\ell(\cos\theta) \tag{13}$$

where

$$H_\ell(v) = \int_{-1}^{+1} d\mu \; H(v,\mu) P_\ell(\mu) = \int_0^1 d\mu (H^+(v,\mu) P_\ell(\mu) + H^-(v,\mu) P_\ell(-\mu))$$

we have set : (14)

$$H^+(v,\mu) = H\,(v,\mu) \qquad \mu > 0 \tag{15a}$$

$$H^-(v,\mu) = H\,(v,-\mu) \tag{15b}$$

$H^{\pm}(v,\mu)$ is the restriction of $H\,(v,\mu)$ to the positive (negative) values of μ.

It is clear that anisotropies $H_\ell(v)$ are made up of two terms depending on H^+ and H^-. Given a value of the velocity $\underline{v}$, a full-range method demands the knowledge of both J and H functions. In a half-range method, all the functions $\underline{H}^-$ are obtained before the functions H^+ and the values of H^- functions are closely connected to the values of H^+ functions.

Equation (9) must then be split in two parts according to the relationships:

$$\{ M+J\} \, H^{-(n+1)} = J \left[H_\ell\right]^{(n)} + S\,(v,\mu) \qquad \mu < 0 \tag{16a}$$

$$\{ M+J\} \, H^{+(n+1)} = J \left[H_\ell\right]^{(n)} + S\,(v,\mu) \qquad \mu > 0 \tag{16b}$$

We see that, since the scattering term J depends on a full-range quantity (i.e. H_ℓ), the solution of the equations above is necessarily iterative (in order to avoid using a huge system of equations). At the beginning, a guess of $J[H_\ell]$ is made and the calculation of H^- functions is first carried out by marching backward ($\mu < 0$, eq. (16b)). Then H^+ functions are obtained by marching forward ($\mu > 0$, eq. (16a)) and by using as initial conditions the values of H^- calculated at turning points ($\mu = 0$).

At the moment, we have developed three different half-range methods: a path differential method, a discrete ordinate method (SN method), and a half-range expansion method.

To understand what we call a path differential method, we have to look at figure 1 where we have plotted, in the

($\rho = \sqrt{2m}$ v, μ) plane, the variations of the characteristic lines $\mu(\rho)$. If there is no collision, an electron starting from point A "moves backward" along the characteristic line, turns back for $\mu = 0$ and then "moves" forward.

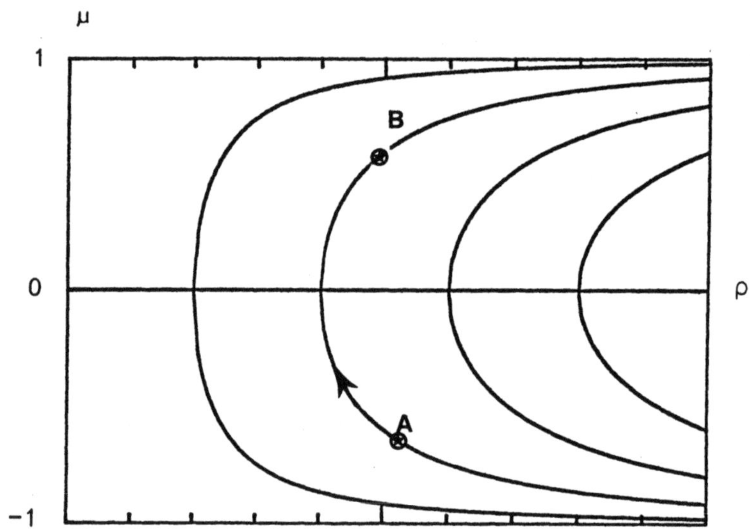

Figure 1. Motion of electrons in the (ρ,μ) plane.

Clearly, the state of an electron at B depends on its past history along this characteristic line and H^+ can be obtained only if H^- is known. In the path differential method, equation (9) is written along the various characteristic lines and it becomes (if we assume that the scattering term is known) a differential equation. The discrete scheme is built up by taking as discrete points the intersection between the characteristic lines and the parallels to the μ axis (cf. Ségur et al [16]).

It can be shown that the numerical scheme obtained is stable if marching is made in the direction of the motion of electrons (first backward, then forward). On the contrary, the scheme is unstable if marching is made in the opposite direction of electron motion.

The same conclusion about the direction of marching can be used for the SN method. The procedure is exactly the same as in the path differential method, except that, now, a set of rectangular cells is used in the (ρ, μ) plane. Furthermore, the location of the discrete points ρ_i and μ_j is chosen so that some conservative properties of equation (9) are verified by the new discrete equations [17]. Since the accuracy of this method is higher than that of the path differential method, it is generally used in our calculations.

The third method we have developed is called half-range expansion method, because it uses an expansion of the distribution functions $H^{\pm}$ into a series of Legendre polynomials defined upon the half spaces $\mu \in [-1,0]$ and $\mu \in [0,1]$ respectively, according to the relationship:

$$H^{\pm}(v,\mu) = \sum_{\ell=0}^{\infty} H_{\ell}^{\pm}(v) P_{\ell}(2\mu \mp 1) \tag{17}$$

As previously, $H_{\ell}^{-}(v)$ is obtained before $H_{\ell}^{+}(v)$ and using an iterative process.

No problem of stability occurs with half-range methods and, since they are iterative, non-conservative situations are treated in the same way as conservative situations. The only drawback may occur from a prohibitive number of iterations. In order to decrease this number, we have developed an acceleration process. Details about this acceleration method can be found in [16] and [17].

The Monte-Carlo method may be used to solve the Boltzmann equation. However, in the case of hydrodynamic regime, it cannot compete with direct methods of solution since, in order to achieve the same accuracy, a very long computational time is needed. However, better efficiency has been reached in modern Monte Carlo codes by using the null collision method (first introduced by Skullerud in 1968, [18]), and sampling before collision [19].

Recently, a new improvement has been made (Penetrante et al, [20]) and the calculation of diffusion coefficients can be carried out with a very good accuracy.

In the case of hydrodynamic regime, Monte Carlo methods are mainly used in order to check the accuracy of the direct methods of solution of the Boltzmann equation.

Results and comparisons

First, we want to show the magnitude of errors involved by the two-term approximation. To that end, we shall use some of the model gases whose cross-sections are given in figure 2. The first two have been used for the first time by Reid [21] and the third one approximates the cross-sections in CH_4.

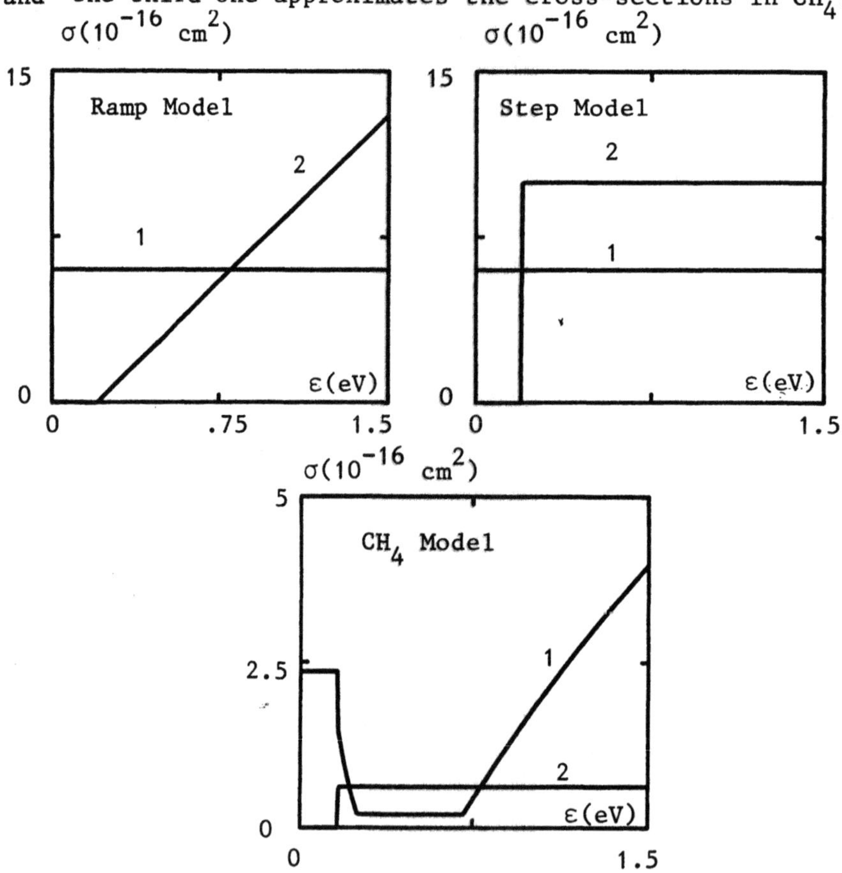

Figure 2. Collision cross-sections of model gases.
1 - Elastic momentum transfer cross-section.
2 - Inelastic cross-section.

The first two models are called "Ramp" and "Step", since the inelastic cross-sections are a ramp and a step respectively. These three gas models have only two collisional processes (elastic and inelastic) and they have been chosen in order to induce large anisotropies in the distribution function.

The worst situation may be found in the step model, where very strong anisotropies are likely to be induced by an abrupt change in the inelastic cross-section. In figure 3, we have plotted, as a function of E/N, the errors obtained by using the two-term approximation in the calculation of the drift velocity and the longitudinal and transversal characteristic energy. Clearly, errors may be very large for the characteristic energy (close to 100 % for DT/μ). But, in the drift velocity case, errors are lower than 20 %.

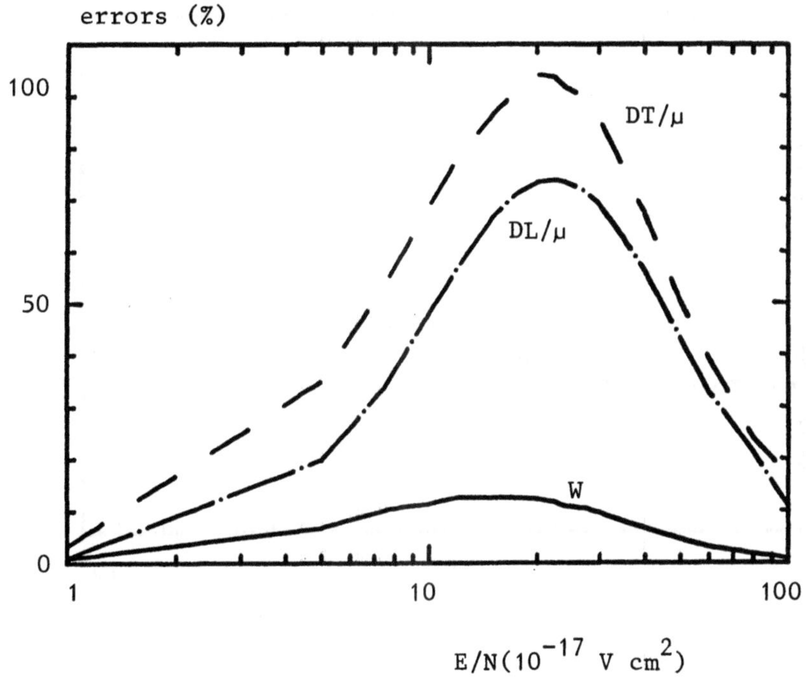

Figure 3. Relative errors in the calculation of swarm parameters between two-term and multiterm approximations: step model.

Similar results are obtained for the CH_4 model. But, in this case, errors are never so large.

Another point which must be stressed now is that not only the two-term approximation can cause significant errors but also that some physical processes can be masked. We will illustrate this point for the CH_4 model. In figure 4, we have plotted the variation as a function of E/N, of DT/μ, and DL/μ. Generally, DL/μ is lower than DT/μ but when the total momentum cross-section decreases very quickly DL/μ may be higher. As we can see, this is the case for values of E/N between 0.6 and 2 Td.

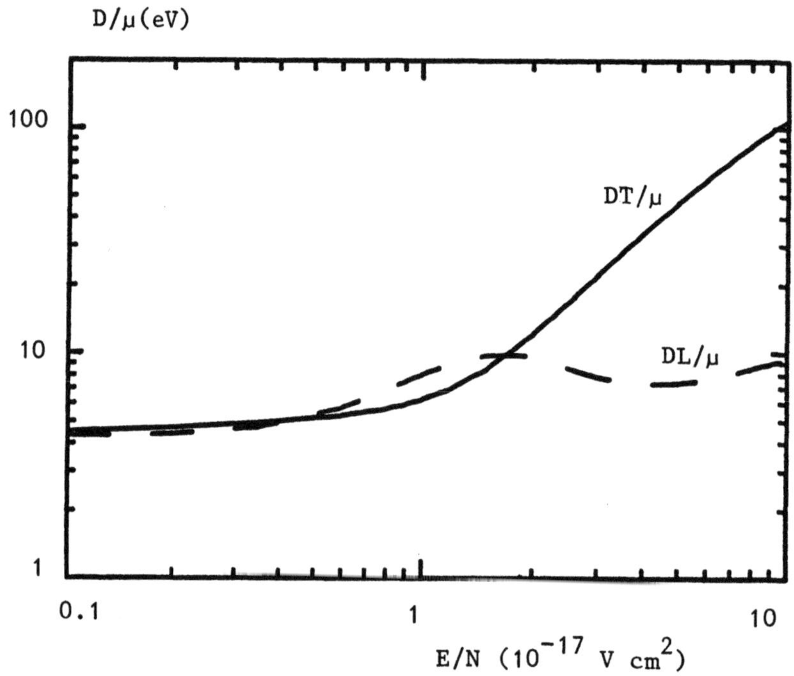

Figure 4. Longitudinal and transversal characteristic energies calculated by using the SN method.

In the next figure (figure 5), we have plotted the same quantities but, now, calculated with the two-term approximation. We see that now DL/μ is everywhere lower than DT/μ, which is an unphysical result.

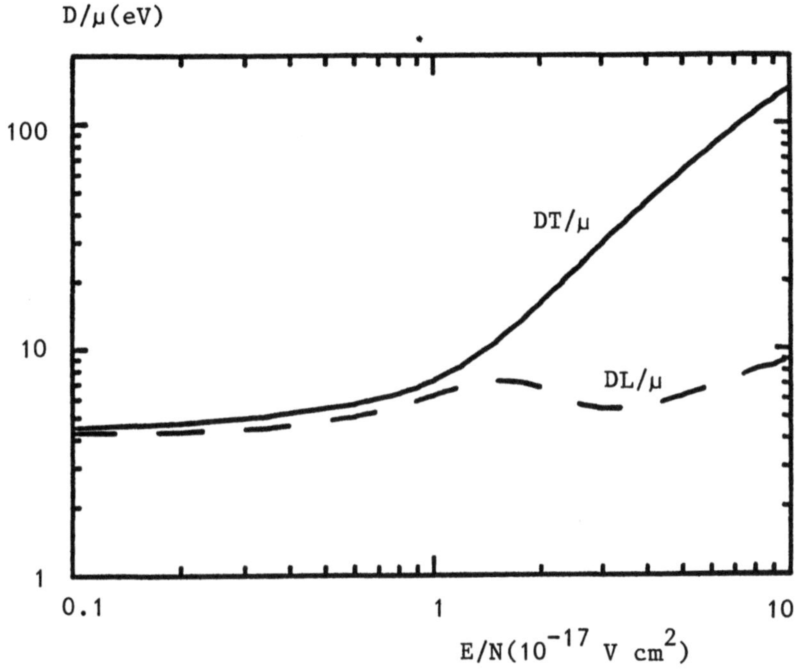

Figure 5. Longitudinal and transversal characteristic energies calculated by using the two-term approximation.

In order to check the accuracy of our half-range methods, we may carry out comparisons, for the model gases given above, between our results and some others, available in the literature.

In table 1, comparisons are made between our SN method

Table 1. Comparison of our results with those of Braglia, Mac Mahon (M.C.M.), Skullerud, and Kuhn for the case of the CH_4 model.

E/N (Td)	Source	W (10^6 cm/s)	NDT (10^{22} $cm^{-1}s^{-1}$)	NDL (10^{22} $cm^{-1}s^{-1}$)	$<\varepsilon>$ (eV)
1	M.C.M.(6 terms)	2.75	-	-	-
	Braglia (M.C.)	2.80	-	-	0.0759
	Method 1	2.74	1.319	3.164	0.0739
	Method 2	2.741	-	3.169	0.0739
2	M.C.M.(6 terms)	7.03	3.88	-	0.188
	Braglia (M.C.)	7.04	3.87	5.39	0.188
	Method 1	7.0	3.99	5.37	0.1885
	Method 2	7.025	-	5.367	0.1885
3.9	M.C.M.(6 terms)	9.14	7.59	-	0.479
	Braglia (M.C.)	9.16	7.57	1.87	0.476
	Method 1	9.15	7.58	2.029	0.4769
	Method 2	9.141	-	2.038	0.477
6	M.C.M.(6 terms)	8.12	-	1.25	0.731
	Braglia (M.C.)	8.07	7.88	1.09	-
	S.K.	7.8	7.8	1.2	-
	Method 1	8.0	7.89	1.093	0.7213
	Method 2	8.012	-	1.0835	0.72124

(Method I) and the half-range expansion method (Method II). Other results are those of Pitchford et al [11], Mac Mahon [14], Braglia [19], Reid [21] and Skullerud [18]. As we can see, agreement is generally good between all the results, and our two methods give very similar results.

In table 2, we give the values that we have obtained with the half-range expansion method. The three model gases have been checked. Convergence is very impressive. A two-term half-range expansion is generally sufficient and a three-term expansion gives in any case very accurate results. R is the number of points on the energy axis. Note that in this half-range expansion method, very good results are obtained with a small number of discrete energy points. In table 3, we give the calculation time of our two methods in the case of the CH_4 model. Clearly, the half-range expansion method is two times faster than the SN method.

Table 3. Comparison of the calculation time (t) between the half-range expansion method (H.R.E.M.) and the discrete ordinate method (D.O.M.) for the CH_4 model.

E/N (Td)	$t_{D.O.M.}$ (mn)	$t_{H.R.E.M.}$ (mn)	$t_{D.O.M.}/t_{H.R.E.M.}$
2	1.7	0.85	2
3.9	1.7	0.9	1.9
8	2.1	1.1	1.9

In the foregoing, all the calculations have been made in the case of conservative situations where all the equations are linear. It would now be interesting to check the accuracy of our methods when there is ionization and/or attachment. In order to do that, we shall use a model gas introduced for the first time by Lucas and Saelee [22]. In this model gas, three collisional processes are used (elastic, excitation, and ionization or attachment). This model is interesting because, besides the Monte Carlo calculations of Lucas and Saelee, some other calculations have been made in this case by Taniguchi et al [23] with a two-term code.

In figure 6, we may compare the results of Lucas and Saelee, Taniguchi et al with ours (SN method) for the longitudinal diffusion coefficient. Agreement is very good

Table 2. Convergence of the results as a function of the order of the anisotropy L_a for different model gases.

L_a	"RAMP", R = 181, E/N = 24 Td			"STEP", R = 201, E/N = 24 Td			"CH4", R = 251, E/N = 3.9 Td		
	$<\varepsilon>$ (eV)	W (10^6 cm/s)	NDL (10^{22} cm^{-1} s^{-1})	$<\varepsilon>$ (eV)	W (10^6 cm/s)	NDL (10^{22} cm^{-1} s^{-1})	$<\varepsilon>$ (eV)	W (10^6 cm/s)	NDL (10^{22} cm^{-1} s^{-1})
2	0.4087	8.8899	0.4519	0.19463	10.635	0.22193	0.48025	9.1965	0.20010
3	0.40799	8.8827	0.461	0.19417	10.534	0.24099	0.47728	9.1385	0.20404
4	0.408	8.8824	0.461	0.19414	10.545	0.24003	0.47744	9.1415	0.20380
5	0.408	8.8824	0.461	0.19415	10.544	0.24014	0.47743	9.1413	0.20381
6				0.19415	10.544	0.24013	0.47743	9.1413	0.20381
7				0.19415	10.544	0.24013			

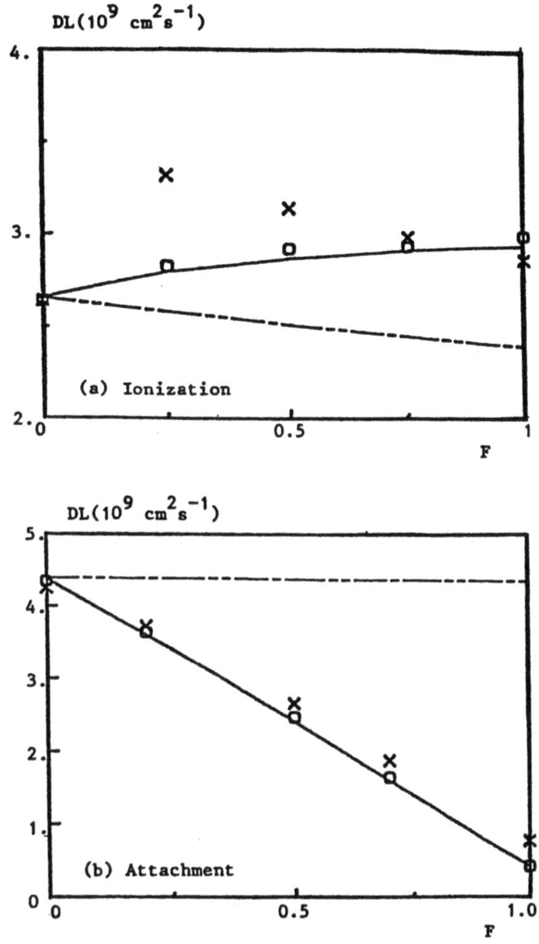

Figure 6. Variation of the longitudinal diffusion coefficient as a function of the level F of ionization (a) or attachment (b) in the model gas or Lucas and Saelee.
x Lucas and Saelee [22] ; o Taniguchi et al. [23] ; —— our calculations ; — – — our calculations by neglecting contribution of ionization and attachment.

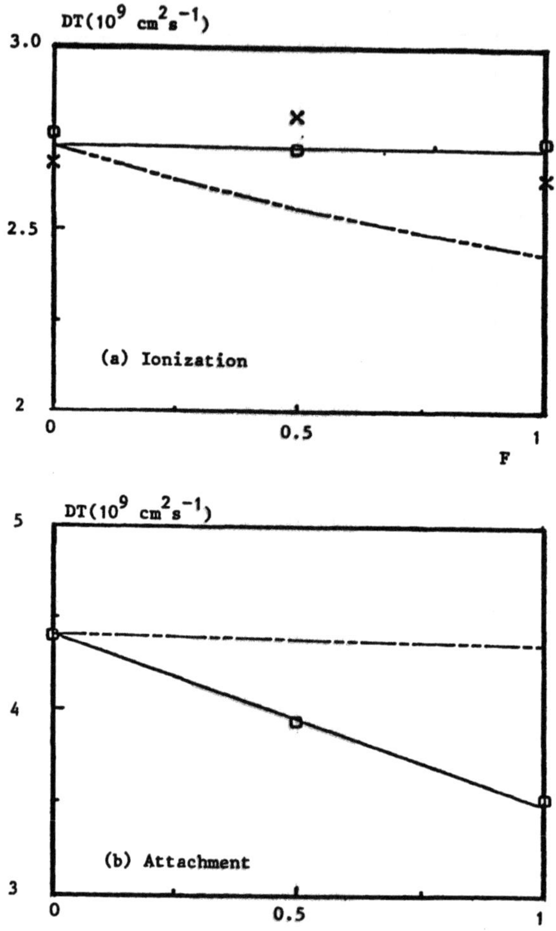

Figure 7. Transverse diffusion coefficient as a function of the level F of ionization (a) or attachment (b) in the model gas of Lucas and Saelee [22].
x Lucas and Saelee ; o Taniguchi ; —— our results ;
— — — our results by neglecting contribution of ionization and attachment.

between all Boltzmann results. On the contrary, Lucas' results differ greatly in some case. Agreement between our results and those of Taniguchi show that the two-term approximation is in this case valid. In dotted lines, we give the values obtained by neglecting contribution of ionization and attachment in eq. (1), and (2). Importance of these terms is clearly seen. In figure 7, the same comparisons are given for the transversal diffusion coefficient.

Conclusion

As we have shown in the foregoing, an accurate calculation of swarm parameters needs the use of sophisticated methods of solution of the Boltzmann equation.

A lot of accurate multiterm methods are now available in the literature. It is, however, difficult to choose among them.

In the conservative case, all the methods work well and most of the full-range methods are very convenient.

The half-range methods seem the best adapted to the non-conservative case. Among them, the half-range expansion method developed in Toulouse is, in our opinion, very promising and, once extended to the calculation of transversal diffusion coefficients, it is likely that it will be one of the more powerful methods for the calculation of swarm parameters.

References

[1] Kumar, K.; Skullerud, H.R.; Robson, R.E. Aust. J. Phys. 1980, 33, 343.
[2] Sakai, Y.; Tagashira, H.; Sakamoto, S. J. Phys. D: Appl. Phys. 1977, 10, 1035; Tagashira, H.; Sakai, Y.; Sakamoto, S. J. Phys. D: Appl. Phys. 1977, 10, 1051.
[3] Rees, H.D. J. Phys. Chem. Solids 1969, 30, 643.
[4] Kleban, P.; Davies, H.T. Phys. Rev. Let. 1977, 39, 456; Kleban, P.; Davies, H.T. J. Chem. Phys. 1978, 68, 2999.
[5] Skullerud, H.R.; Khun, S. J. Phys. D: Appl. Phys. 1983, 16, 1225.
[6] Frost, L.S.; Phelps, A.V. Phys. Rev. 1962, A127, 1621. Morgan, W.L.; JILA Information Center Report 1979, 19, Boulder Co.
[7] Wilhelm, J.; Winkler, R. Ann. Phys. 1969, 23, 28.
[8] Makabe, T.; Mori, T. J. Phys. Colloque C7 1979, 40, 43.

[9] Lin, S.L.; Robson, R.E.; Mason, E.A. J. Chem. Phys. 1979, 71, 3483.
[10] Ness, K.F.; Robson, R.E. These Proceedings.
[11] Pitchford, L.C.; O'Neil, S.V.; Rumble, J.R. Phys. Rev. 1981, A25, 540.
[12] Hammar, Cl. J. Phys. C: Solid State Phys. 1973, 6, 70.
[13] Winkler, R.; Wilhelm, J. Beitr. Plasma. Phys. 1983, 23, 485.
[14] Mac Mahon, D.R.A. I.D.U. Internal Rept., Australian National University 1983.
[15] Canosa J.; Penafiel, H.R. J. Comput. Phys. 1973, 13, 380.
[16] Ségur, P.; Bordage, M.C.; Balaguer, J.P.; Yousfi, M. J. Comput. Phys. 1983, 50, 116.
[17] Ségur, P.; Yousfi, M.; Bordage, M.C. J. Phys. D: Appl. Phys. 1984, 17, 2199.
[18] Skullerud, H.R. J. Phys. D: Appl. Phys. 1968, 1, 1567. Skullerud, H.R. Private communication 1981.
[19] Friedland, L. Phys. Fluid 1977, 20, 1461. Braglia, G.L. J. Chem. Phys. 1981, 74, 2990.
[20] Penetrante, B.M.; Bardsley, J.N.; Lin, J.L. Lett. Nuovo Cimento 1982, 34, 57.
[21] Reid, I.D. Aust. J. Phys. 1979, 32, 231.
[22] Lucas, J.; Saelee, H.T. J. Phys. D: Appl. Phys. 1975, 8, 640.
[23] Taniguchi, T.; Tagashira, H.; Sakai, Y. J. Phys. D: Appl. Phys. 1977, 10, 2301.

RECENT ADVANCES IN HIGH-PRESSURE SWARMS

R. Johnsen and H. S. Lee

Department of Physics and Astronomy
University of Pittsburgh
Pittsburgh, PA 15260

The paper describes the development and applications of an afterglow technique for measurements of deionization processes in plasmas of atmospheric neutral density.

The technique makes use of spark arrays to create the initial ionization. A newly developed rf-probe method and mass spectrometric analysis of the plasma ions are used to observe the decay of the charged-particle concentrations during the afterglow phase.

Measurements in atmospheric-density helium afterglows with several gaseous admixtures indicate that the experimental method is suitable for studies of electron attachment, electron-ion recombination, and recombination of positive with negative ions.

Introduction

Deionization processes in high-pressure plasmas are a subject of considerable applied interest, but in reviewing the literature one discovers rather quickly that there are extensive gaps in our knowledge concerning such basic processes as electron-ion and ion-ion recombination at high pressures. This is particularly true for ion-ion recombination (see the review by Flannery [1]), which has received much attention by theorists, but there is little experimental work for comparison.

The situation is somewhat better in the case of electron-ion recombination (see also the paper by Morgan [2]), which has been studied in several molecular and atomic gases [3, 4, 5, 6, 7]. Theory and experiment (see Morgan [2]), appear to be in agreement in at least some cases. It must be said, however, that the ionic species undergoing recombination in the experiments were not identified, and that there is no clear evidence that the measurements refer to a single species of ions over the range of pressures covered by the experiments. Perhaps the nature of the ion becomes unimportant in the high-pressure limit, but this may not be so at somewhat lower pressures.

The basic experimental methods used to study deionization phenomena at high pressures are similar to the familiar low-pressure plasma afterglow techniques, but the diagnostic and initial ionization methods differ substantially. If one has access to an electron-beam accelerator, e-beam ionization [8] or indirect x-ray ionization provide a powerful, though elaborate, method of producing a uniform plasma for such studies. Measurements of the electron-density decay can be carried out by microwave interferometry [3] and in some cases by optical spectroscopy [4]. Langmuir probes do not seem to be suitable for electron-density measurements at high pressures. Measurements of the ion composition of the plasma by mass spectrometry are subject to some complications arising from the extraction process, but one has little other choice, if one wishes to identify the ions.

In this paper we will describe the development of a relatively simple high-pressure afterglow technique. A radio-frequency (rf) probe method has been developed for measurements of electron densities, which can in some cases also be used to measure absolute ion densities. Achievement of a wide dynamic range for electron-density measurements was one goal in the development of this method. The physical principles of this technique are described in detail.

Using spark arrays for the purpose of uv-photoionization is not a new idea. The technique was employed here because of its great simplicity, and the results indicate that it is quite suitable for this type of measurement. Mass spectrometry of the plasma ions has been carried out also. The complications seem to be less severe than had been anticipated. The remainder of this paper presents the results of initial measurements of attachment, electron-ion and ion-ion recombination measurements. These measurements were largely intended to test the experimental methods, and to explore the range of phenomena accessible to measurement.

RF Probes for Measurements of Electron and Ion Densities

The basic idea of the rf-probe technique is quite simple.

The electrical conductivity of the plasma is measured by passing an rf-current between the two electrodes and the electron density is then inferred from the measured conductivity. At high gas density, low frequency, and small values of E/N the conductivity of a weakly ionized plasma is given by:

$$\sigma = e(n_e\mu_e + n_+\mu_+ + n_-\mu_-) \tag{1}$$

where n_e, n_+, n_- denote the densities of electrons, positive, and negative ions, the μ_i represent the corresponding mobilities, and e is the electronic charge. In the following, the ionic contribution to the conductivity will be neglected, because in most plasmas of interest $n_e \mathrel{\overline{\overline{\sim}}} n_+$ and $\mu_e >> \mu_+$. In plasmas where $n_e \mathrel{\overline{\overline{\sim}}} n_-$, the full expression should be substituted.

In using Eq. 1 to relate σ and n_e it is tacitly assumed that the electron-neutral collision frequency greatly exceeds the frequency of the applied electric field. At the neutral densities of interest here, the collision frequency is typically on the order of 10^{10} sec, while the applied frequencies are only 10^7/sec. Thus, use of the simple "ohmic" conductivity formula is justified, and current and field will be nearly in phase. A simple method estimating the residual, small phase shifts will be discussed later.

Obviously, the mobility of electrons in the gas of interest has to be known to infer n_e from the measured conductivity. For most gases accurate values of μ_e are available from either experiment or theory.

Let us now turn our attention to the problem of measuring the plasma conductivity. As is well known, the simple idea of passing a dc current between metal electrodes immersed in the plasma fails. The plasma at the electrodes will be depleted of charge carriers and the resulting "sheaths" act as large series resistance in the measuring circuit (the "finer points" of the sheath problem are not important here). The situation is depicted schematically in Fig. 1a, where the plasma is assumed to fill the space between planar electrodes (distance d, area a^2) uniformly except for the sheaths, here taken to be totally devoid of electrons. If we apply a dc voltage to the upper electrode, the charge carriers migrate to the surface of the plasmas until the resulting space-charge field (polarization field) equals the applied field, at which time the current in the measuring circuit vanishes.

It takes some time, however, for the polarization of the plasma to develop, suggesting that it can be prevented by applying ac fields of sufficiently high freqency.

An equivalent electrical circuit is shown in Fig. 1b where the sheaths are represented by their capacitance only. The plasma is viewed as a parallel circuit of its resistance R_P and a capacitance C_P with the values:

$$C_P = a^2 \varepsilon_o/d, \; R_P = d/\mu_e \, e \, n_e \, a^2 \tag{2}$$

It is assumed that the sheaths are very thin compared to the electrode spacing. In order to infer R_P from the measured rf current, the sheath impedance $1/\omega C_\delta$ should be small compared to R_P, which can be achieved by using a sufficiently large ω.

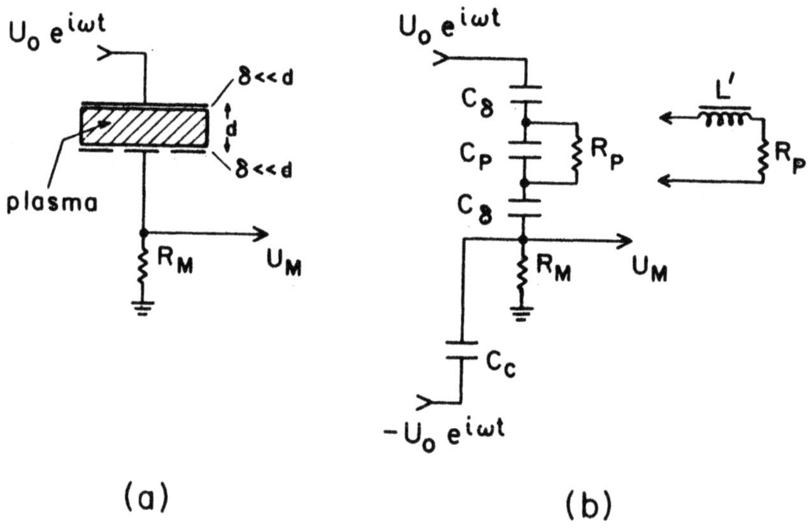

Fig. 1. Illustration of rf-probe principle and equivalent circuit.

As is immediately clear from the equivalent circuit, a current will also be detected in the absence of the plasma due to the finite capacitance between electrodes. As indicated in the figure, this displacement current has to be cancelled by using a bridge arrangement, i.e., applying an rf signal of opposite phase but of the same magnitude to the measuring resistor.

A further refinement may be added to the equivalent circuit, although it is not very important for the high-pressure plasmas of interest here. As is well known, a plasma exhibits resonant behavior at the plasma frequency

$$\omega_p = (e^2 n_e/m_e \varepsilon_o)^{1/2} \tag{3}$$

Adding a "pseudoinductance" L' in series with R_p (see Fig. 1b) duplicates the resonant behavior of L' is given the value:

$$L' = d\, m_e/e^2 a^2 n_e \tag{4}$$

As is easily estimated, the effect of L' is entirely negligible for an electron-conducting plasma at atmospheric pressure ($\omega L' << R_p$). In the case of ionic charge carriers (substitute the electron mass by the ionic mass in Eq. 4) it may have to be considered, and is mentioned only for that reason.

Theory for Cylindrical Probes

The probes that were actually built and tested had a cylindrical configuration, as shown in Fig. 2. The outer electrode consisted of fine screen to permit entry of the ionizing uv photons, to be discussed later. The inner electrode has grounded guard electrodes on either side to reduce the fringing fields.

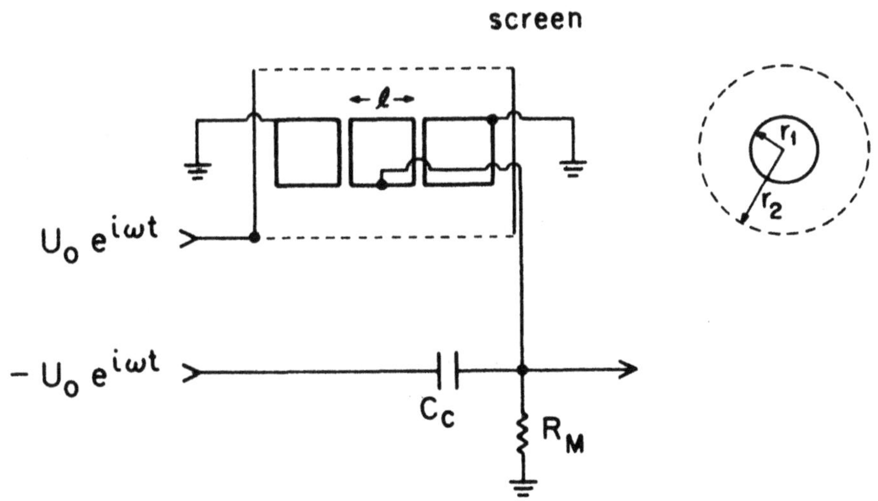

Fig. 2. Cylindrical probes. Dimensions r_1 = 0.635 cm, r_2 = 1.59 cm, ℓ = 1.29 cm.

The impedance of the probe is easily calculated by integration. One divides the volume into cylindrical shells and considers each shell to be a parallel circuit of a capacitance and the plasma resistance. In complex notation one obtains for the impedance Z:

$$Z = (A/2\pi\ell\ e\ \mu_e) \int_{r_1}^{r_2} (A\ n_e - i)(A^2\ n_e^2 + 1)^{-1}\ dr/_r \tag{5}$$

where

$$A = e\ \mu_e/\varepsilon_o\ \omega \tag{6}$$

The electron density is assumed to be a function of radius only. The notion of discrete, non-conducting sheaths used in the illustration of the principle is no longer invoked.

Inspection of Eq. 5 shows that Re(Z) will be proportional to n_e, if the quantity $A^2\ n_e^2$ is small compared to unity, in which case it can be neglected in the denominator of the integrand. The condition imposes a lower limit on the operating frequency, if one wishes to have a probe response linear in the electron density over a given range. Physically, the condition means that the plasma does not change the electric field appreciably from the "empty" probe field.

The "linearity condition" ($A^2\ n_e^2 << 1$) is similar to the perturbation approximation employed in analyzing the effect of plasmas on microwave-cavity resonances. The same approach can be taken in analyzing the rf-probe problem to obtain approximate solutions, but one loses physical insight into the limitations of the method.

A further analysis of the problem requires knowledge of the electron-density distribution $n_e(r)$, which in general may not be known accurately. Numerical integrations of Eq. 5 have been carried out for assumed distributions in order to gain some "feel" for the probe response for different $n_e(r)$. The model calculations included the circuit analysis of the rf-bridge. It is found that for most "reasonable" distributions $n_e(r)$, i.e., those fairly uniform in most of the volume but zero near the electrodes, the probe response U_M is linear over a much wider range in n_e than expected from the criterion $A^2\ n_e^2 << 1$. By "linear response" we mean that the signal U_M detected by the rf-bridge depends on a scaling factor c in the form $U_M(cn_e) = c\ U_M(n_e)$.

The response of the probes can be calculated very simply under the following assumptions: $n_e(r)$ is constant except for negligibly thin sheaths and the measuring resistor R_M is small compared to the plasma resistance. If the bridge is balanced in the absence of the plasma, the probe response U_M is given by:

$$U_M = U_O \; R_M \; 2\pi\ell \; [\ln(r_2/r_1)]^{-1} \; \mu_e \; e \; n_e \tag{7}$$

i.e., the signal is proportional to n_e and all factors are known from the probe dimensions.

The choice of the driving frequency will depend somewhat on the application. For the work described here a frequency $\omega/2\pi$ of 15 MHz was used for helium plasmas of atmospheric pressure. The quantity $A^2 \; n_e{}^2$ under these conditions would be about 0.03 for $n_e = 10^9 \; cm^{-3}$ (the reduced electron mobility in He at 300 K is about $9 \times 10^3 \; cm^2/V$ sec) [9].

There are several factors that limit the sensitivity of the probes for detecting small electron densities. Consideration of thermal noise and detection bandwidth needed to reproduce the time dependence of n_e in decaying plasmas indicated lower limits of about $10^5 \; cm^{-3}$ (for He gas at 1 at). Practical problems, such as stability of the rf-bridge, begin to be severe at about the same level of sensitivity.

Little will be said about technical details of the rf-bridge circuitry and rf-detectors. High-precision balancing of the bridge is necessary only for measurements of very low n_e. In the present work, capacitive balance to a precision of 10^{-5} pF and phase balance to within 10^{-3} degrees were attained without excessive difficulty. The rf-detectors and demodulators used were fashioned after standard circuits used in radio receivers. The detection bandwidth was typically 300 KHz, but could be reduced by filtering the low-frequency demodulated signal.

A reference resistor of 30 kOhm may be inserted in the compensating branch of the bridge. When the resistance of the plasma is equal to the reference resistance, the bridge signal approaches zero and this easily detected condition provides a convenient reference point for the electron density (in helium at 1 at. this occurs at $n_e = 2.6 \times 10^9$ cm^{-3}).

Much of the foregoing discussion remains valid for measurements of ion densities in plasmas containing only positive and negative ions but very few free electrons. The sensitivity for detecting ions is obviously lower by the ratio of ionic to electronic mobilities, but such measurements have been carried out. The optimal driving frequency for probes designed for ion detection may be considerably lower than that for electrons.

The rf voltages applied to the probes were typically on the order of 1 to 2 V (peak-to-peak). Perturbations of the plasma by the probe field (about 1V/cm) should be entirely negligible. Experimental tests were carried out with rf amplitudes a factor of thousand lower, but no change in the electron-density decay curves was observed.

UV Photoionization by Spark Arrays

The idea of using spark arrays to produce a weakly ionized plasma is not new, of course. It is frequently employed to pre-ionize laser discharges, but the physical processes leading to ionization at large distances from the sparks are far from clear (see, for instance, Seguin et al. [10]). The method was adopted here because of its simplicity compared to ionization by electron beams or x-rays.

The experimental arrangement is shown in Fig. 3. Eight spark gaps, each equipped with its own storage condenser, were arranged on a circle surrounding the rf-probes. The sparks were operated in the simple "relaxation oscillator" mode without external triggering. Synchronous firing of the sparks in such an arrangement occurs automatically, since the burst of uv light from one spark initiates breakdown in the others after a very short time (typically less than 1 μsec to several μsec). This very simple technique proved to be quite reliable and yielded shot-to-shot reproducibility of initial electron densities ($\sim 10^9$-10^{10} cm^{-3}) of several percent under good conditions.

In later work, after the mass spectrometer had been added to the apparatus, an additional spark gap was mounted near the mass spectrometer sampling orifice. Its purpose was to examine the ion composition at a distance from the sparks similar to that to the rf-probes. The mass spectra were found to be essentially the same as those obtained using the eight sparks.

As mentioned earlier, the mechanism of ionization in the vicinity of the sparks is not well understood. Faced with the difficulty of explaining the anomalously large absorption length for ionizing radiation (in helium at 1 atmosphere, for

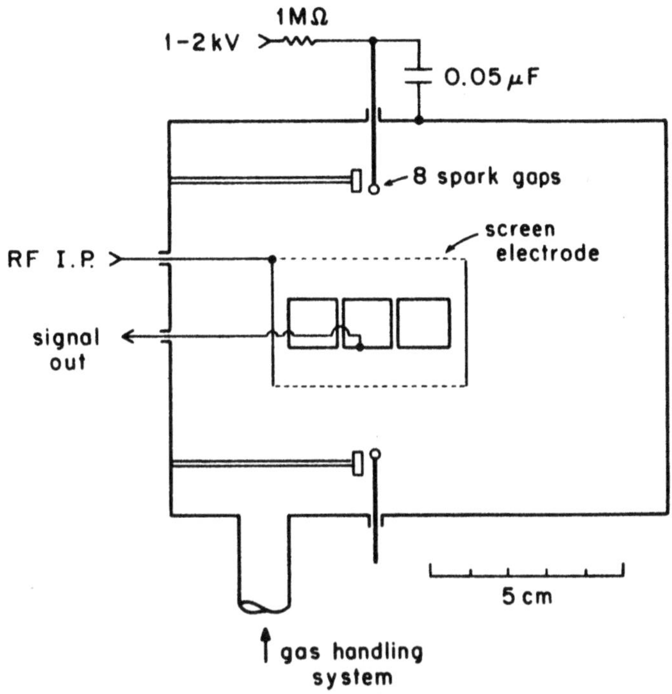

Figure 3. Experimental arrangement of spark array and rf-probes.

instance), some authors suggested [10] that ionization of impurities of low ionization potential by longer-wavelength, less absorbable radiation provides the principal ionization. The observations made in this work indicated that He_2^+ ions are the principal initial ions in pure helium irradiated by sparks in helium, rather than impurities. One might speculate that absorption of uv light by two helium atoms in the state of collision, i.e.,

$$h\nu + He + He \rightarrow He_2^+ + e$$

may lead directly to He_2^+ ions. Since the uv emission of the sparks is likely to consist largely of He molecular bands or collision-broadened lines, this explanation may not be too farfetched. Other mechanisms, possibly involving metastable-metastable ionizing collisions, may provide less exotic explanations. In the presence of additives or background impurities, direct photoionization, Penning ionization, and ion-molecule reactions are expected to be the dominant ionization mechanisms.

In the remainder of this paper, the afterglow phase of the plasma is the principal subject and the initial ionization processes are largely irrelevant.

Mass Spectrometry of the Plasma Ions

It is often said that mass spectrometric sampling of ions from high-pressure plasmas is almost useless because of the unpredictable changes in the mass spectrum due to effects of ion break-up and "condensation" (clustering) in the extraction process. This point of view seems overly pessimistic and it is difficult to find viable alternatives to mass spectrometry. Optical absorption or fluorescence have the advantage of "in-situ" detection but are not universally applicable and often, spectroscopic information on the species of interest is not available. Thus, it seems better to accept the shortcomings of mass spectrometry than not to have any knowledge of the ion composition of the plasma.

The mass spectrometer used to sample ions from the photoionized afterglows is shown in Fig. 4. Ions are extracted through a small, laser-drilled 50 μm orifice, traverse a first differentially pumped stage, and then enter a quadrupole mass filter. Detection of the ions is done by a channel multiplier. The ion countrate was quite adequate (about 10^3 ions of all masses per afterglow) for both recording mass spectra or determining the decay of the number density of a given ion species. As expected, the ion composition of the plasma was extremely sensitive to even minute impurities in the gases used, and the nature of the impurities seemed to be altered by the plasma environment. It was noticed, for instance, that in nominally-pure helium afterglows at atmospheric pressure the mass spectrum became dominated by ammonium cluster ions ($NH_4^+ (NH_3)_n$) after a short time of operation. Other ions, possibly nitric-acid clusters, appeared also. It appears likely that these impurities were synthesized from common impurities (N_2, H_2O, CO, H_2) by a chain of neutral reactions initiated by uv photons. The example may serve to illustrate the point that it is difficult to predict the ion composition, even if one has some knowledge of the initial impurities.

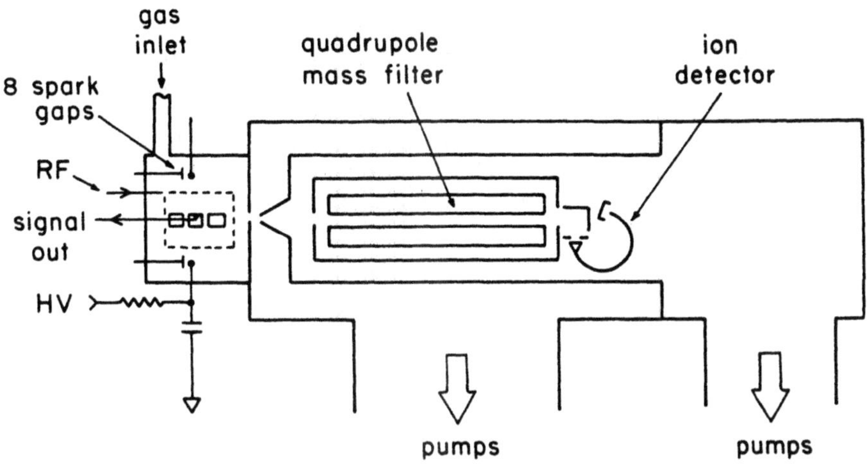

Fig. 4. Experimental apparatus including mass spectrometer.

In purified helium (Ultra-high-purity grade passed through zeolite at LN_2 temperature) the mass spectrum consisted largely of He_2^+ and He_3^+ ions at early afterglow times (up to several 100 μsec), but other ions were still present at later times. The abundance of He_3^+ relative to that of He_2^+ was higher by a factor of four than that calculated from equilibrium constants measured previously [11]. Assuming that the published equilibrium constants are correct, this finding may indicate that additional He_3^+ ions were formed in the cold gas jet emanating from the sampling orifice. No other ions clustered to helium were observed, however. There was no conclusive evidence that break-up of clusters (shedding of solvent molecules) during ion extraction was important in the case of hydronium-water clusters or ammonia clusters.

Measurements

Most of the measurements that have been carried out so far were "exploratory" in nature, i.e., they were intended to gain familiarity with the new methods, to investigate the range of validity of assumptions, and to show that data obtained with this technique are compatible with those obtained using other methods. The first set of measurements was designed to test the suitability of the rf- probes for measuring electron and ion densities. The mass spectrometer was not available in the initial phase.

In the second phase of the experiments, recombination of mass identified ions (Ar_2^+) with electrons was studied for helium pressures from 0.5 to 1 atmosphere. One does not expect pronounced third-body effects on electron ion recombination in ambient helium: The recombination rate coefficient measured should be close to that observed in low-pressure afterglows and show little variation with pressure. It seemed advisable to test this experimentally before studying recombination in molecular gases, which will exhibit stronger pressure dependences.

Electron Attachment to SF_6

A plasma decaying by electron attachment provides a good test of the experimental method for two reasons: The electron density should decay exponentially, i.e.,

$$n_e(t) = n_e(0)\ \exp(-k\ N_a\ t) \qquad (8)$$

where k is the rate coefficient for attachment and N_a the density of the attaching gas. Observation of a simple, exponential decay over several orders of magnitude in n_e would confirm a linear response of the rf-probe. Secondly, the late afterglow should contain mainly positive and negative ions, and one should be able to observe the transition from an electron-conducting to an ion-conducting plasma.

In the experiment, very small amounts of SF_6 were added to helium (on the order of 10 ppb) and the conductivity of the plasma was measured as a function of afterglow time. The results are shown in Fig. 5. The conductivity, normalized to that at t = 0, ($n_e(0)$ was about 10^{10} cm^{-3}), decays exponentially over more than two orders of magnitude, but the subsequent, slower decay of the conductivity follows a different decay law. The interpretation is simple: The initial, exponential decay reflects the loss of electrons by attachment. At later times, very few free electrons remain, and one observes the much smaller conductivity of the (+/-)-ion plasma, which decays by the slower process of ion-ion recombination.

A quantitative model of the plasma decay, which included attachment and ion-ion recombination, was found to be capable of reproducing the measured decay of the conductivity. Since the SF_6 density was not well known, the attachment rate coefficient could not be determined. There were also indications that the SF_6 density slowly declined, or that the SF_6

was chemically altered after many afterglow cycles, an effect which could be prevented by operating in a single-shot mode or replenishing the gas mixture. The preliminary results obtained indicate that attachment coefficients could be measured by this technique, that the negative product ions are detectable, and that one might be able to measure rates for ion-ion recombination as well.

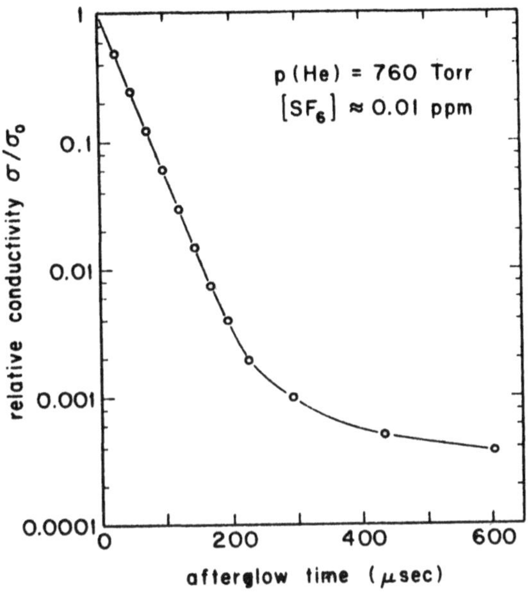

Figure 5. Decay of the conductivity in a He-SF_6 plasma.

Ion-Ion Recombination

As described in the previous section, it is possible to create a (+/-) ion plasma by attachment to SF_6. The conductivity of this ion-ion plasma was measurable. The question then arises, if the measured decay of the conductivity can be analyzed to obtain rate coefficients for ion-ion recombination. If we assume that only a single ion species of either polarity is present in the plasma, the ion density should decay in the form of a recombination law, i.e.,:

$$1/n_i = 1/n_{io} + \gamma t \tag{9}$$

where γ denotes the recombination coefficient. In order to relate the measured conductivity to the ion density (see

Eq. 1), we need to know the ionic mobilities in helium. For the type of ions likely to be present, reduced mobilities of about 15 $cm^2/Vsec$ seemed reasonable, and were used in the analysis.

An example of an ion decay, measured in the late afterglow of a $He-SF_6$ plasma, is shown in Fig. 6 (in the form $1/n_i$ vs t). Clearly, the functional form expected for recombination is well obeyed. In Fig. 7 measured recombination coefficients are shown for different helium pressures (the symbols distinguish different runs). The results show an increase of γ with pressure, as expected, but the reproducibility of the data is much poorer than the precision of the points. No

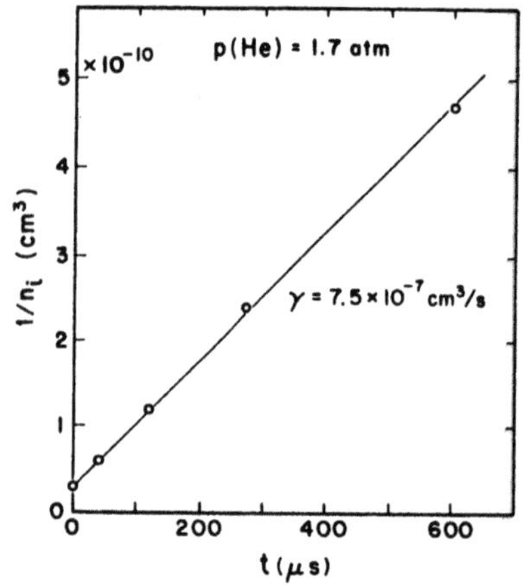

Figure 6. Decay of the ion density in the late $He-SF_6$ afterglow.

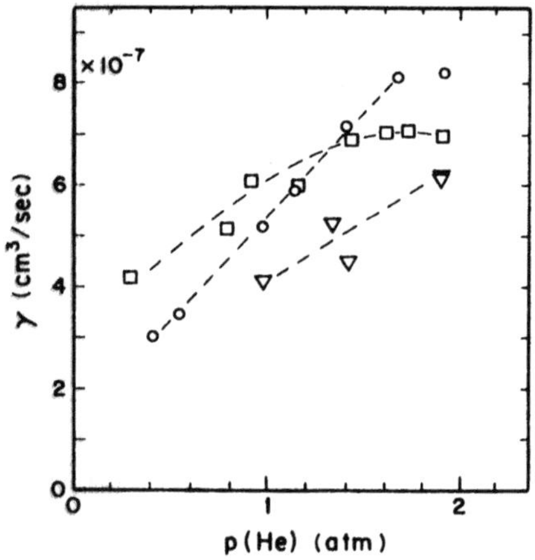

Figure 7. Measured ion-ion recombination coefficients as a function of He pressure.

mass analysis of the ions was performed, but it is likely that the ion composition varied in different runs and that γ depended on the ionic species present at the time of the measurement. Obviously, such measurements should be carried out with simultaneous mass analysis of the ions.

The magnitude of the recombination coefficient is about what one would expect [1].

Measurements of Electron-Ion Recombination Without Mass Analysis of the Ions

As stated earlier, recombination measurements without mass analysis are probably not very meaningful, but some data were taken in helium with trace additions of water vapor. Under those conditions, one can be reasonably sure that the ions will be of the form $H_3O^+(H_2O)_n$, for which the recombination rates are known from earlier work [12].

In a recombination-controlled afterglow, the electron density decays as

$$1/n_e = 1/n_{eo} + \alpha t \qquad (10)$$

where α is the effective recombination coefficient. The experimental data exhibited this behavior over n_e ranges of at least 16 and often more. The recombination rates ranged from about 3×10^{-6} to 7×10^{-6}, depending on H_2O concentration (larger clusters present at higher $[H_2O]$ have larger α).

The absolute values are in good agreement with those obtained in low-pressure afterglows [12], which is not unexpected, since third-body effects in helium make only a small contribution to the total recombination rate. These measurements were intended to show that the electron densities measured by the rf-method are of the correct absolute magnitude.

Decay of the Plasma Conductivity Over a Wide Range

For measurements of recombination coefficients it is not necessary to follow the plasma decay for more than a factor of eight or sixteen perhaps, but in some cases it was measured over more than a factor of 10^5. An example of such data (helium at 1 at. with some SF_6 and H_2O impurities left from earlier experiments) is shown in Fig. 8. The initial electron density was $\sim 2.3 \times 10^{10}$ cm^{-3}.

One can distinguish three different regimes: The initial, fast decay is most likely due to recombination of electrons. The subsequent, exponential decay arises predominantly from attachment of electrons to impurities (SF_6 probably). The final, slower decay reflects the ionic conductivity, which decays by ion-ion recombination. A quantitative model of the plasma conductivity decay was made and the relevant coefficients were adjusted to fit the measured decay. An attachment frequency of 6.25×10^2, ion-ion rec. coefficient of 5×10^{-7}, and electron-ion rec. coefficient of 4×10^{-6} cm^3/sec yield the solid curve at late times and the dashed part at early times. A better fit at early times results if one uses a variable electron-ion rec. coefficient. (It was assumed that an ion with $\alpha = 2 \times 10^{-6}$ converts into one with $\alpha = 8 \times 10^{-6}$ with a reaction time constant of 100 μsec). This then yields the solid line also at early times.

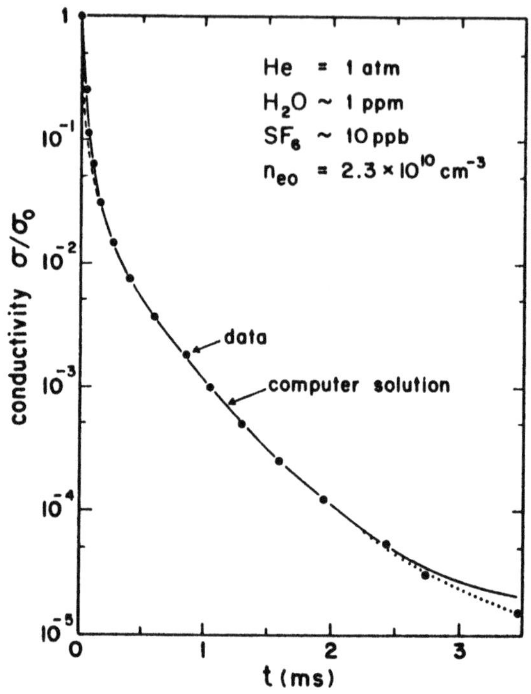

Figure 8. Decay of the electrical conductivity over a wide range.

Recombination of Mass Identified Ions with Electrons

After the mass spectrometer had been completed, an attempt was made to produce plasmas containing predominantly a single ion species, and to determine the electron ion recombination coefficient for this species. The conditions were reasonably well approached in mixtures of helium with small additions of argon. Ar_2^+ was the dominant ion at early afterglow times, but some impurity ions (N_4^+ or CO^+ CO, and N_2^+Ar or CO^+Ar) were present in the later afterglow. The effect of the impurity ions on the determination of $\alpha(Ar_2^+)$ was not large, but it was taken into account in the data analysis.

In the pressure range from 0.5 to 1 at. of helium,

$\alpha(Ar_2^+)$ was found to be 8×10^{-7} cm^3/sec independent of pressure, quite close to that inferred from low-pressure afterglow experiments ($\alpha = 9 \times 10^{-7}$ cm^3/sec) [7]. Even at atmospheric pressure the three-body contribution to the total recombination coefficient would be small compared to the two-body rate.

These measurements were intended to show that the experimental method would not give pressure-dependent results in cases where such should not be expected. The results indicate the absence of such "false" pressure dependences.

Some data were taken for recombination of Kr_2^+ ions with electrons in atmospheric pressure helium. The measured $\alpha(Kr_2^+) = 1 \times 10^{-6}$ cm^3/sec is somewhat smaller than that obtained in low-pressure measurements.

Decay of a Pure Helium Plasma

Few afterglow phenomena have been as difficult to understand and have raised as much controversy as the helium afterglow (see [13] for a review). The temptation to revive the subject was hard to resist, and we made some attempts to measure and model the electron density decay in the early afterglow. It became clear quickly that the decay of n_e in the early afterglow was not describable by a simple recombination law. Rather, n_e decreased much more rapidly initially than explainable by recombination.

It has recently been shown that metastable, negative He_2^- ions exist [14] with autodetachment lifetimes greater than 100 μsec. In principle, it seems possible to form He_2^- ions by attaching electrons to He_2^M or He_3^M molecular metastables, if the latter exist. Since molecular metastables are likely to be abundant in the early phase of high-pressure helium afterglows, one might speculate that attachment of electrons to these is responsible for the observed fast electron loss. The functional form of the electron-decay curves was well reproduced in model calculations of attachment to helium metastables, but a mass spectrometric search for He_2^- gave negative results. The existence of "loosely bound" electrons in helium afterglows has been postulated earlier [15] on the basis of experimental observations. Perhaps the subject deserves some new attention.

Summary

It was the purpose of this work to develop a relatively simple, high-pressure afterglow technique with a range of capabilities similar to that of the more common low-pressure methods. The experimental methods developed have been tested and have been shown to give results that are reasonable and consistent with earlier work.

It is clear that the initial measurements should be followed by more systematic studies. For instance, studies of ion-ion recombination should be carried out for known ion species, and electron-ion recombination should be studied in molecular ambient gases. We are planning to carry out such measurements in the future.

Acknowledgment

This work has been supported, in part, by the U. S. Army Research Office.

References

[1] Flannery, M. R. Atomic Processes and Applications 1976, Burke, P. G.; Moiseiwitch, B. L.; Eds., North Holland, Amsterdam, page 408.
[2] Morgan, W.; Recent Advances in Electron-Ion and Ion-Ion Recombination, 1985, this volume.
[3] Warman, J. M.; Sennhauser, E. S.; Armstrong, D. A; 1979, J. Chem. Phys. 70, 995.
[4] Kuo, C. -Y.; Keto, J. W.; 1983, J. Chem. Phys. 78, 1851.
[5] Littlewood, I. M.; Cornell, M. C.; Nygaard, K. J.; 1984 J. Chem. Phys. 81, 1264.
[6] Maier, H. N.; Fessenden, R. W.; 1973, J. Chem. Phys. 62, 4790.
[7] see also review articles by Wiegland, W. J.; 1982, and Biondi, M. A.; 1982, both in Applied Atomic Collision Physics, Vol. 3, Academic Press.
[8] Lee, F. W.; Collins, C. B.; Waller, R. A.; 1976, J. Chem. Phys. 65, 1605.
[9] Hayashi, M.; Ushiroda, S.; 1983, J. Chem. Phys. 78, 2621.
[10] Seguin, H. J.; Tulip, J.; McKen, D. C.; 1974, IEEE J. Quantum Electronics, QE-10, 311.
[11] Gusinow, M. A.; Gerber, R. A.; Gerardo, J. B.; 1970, Phys. Rev. Lett. 25, 1248.
[12] Leu, M. T.; Biondi, M. A., Johnsen, R.; 1973, Phys. Rev. A 7, 292.

[13] Deloche, R.; Monchicourt, P.; Cheret, M.; Lambert, F.; 1974, Phys. Rev. A 13, 1140.
[14] Bae, Y. K.; Coggiola, M. J.; Peterson, J. R.; 1984, Phys. Rev. Lett. 52, 747.
[15] Kaplafka, J. P.; Merkelo, H.; Goldstein, L.; 1968 Phys. Rev. Lett. 21, 970.

ELECTRON-ION RECOMBINATION IN HIGH-PRESSURE GASES

Wm. Lowell Morgan
Lawrence Livermore National Laboratory
and
Department of Applied Science
University of California at Davis-Livermore
Livermore, CA 94550

This paper reviews recent experimental and theoretical research on recombination of electrons and ions in molecular gases at high pressure.

Introduction

Recombination of electrons and ions in low-pressure gases has been reviewed extensively in the past (see, for example Ref. 1). Theoretical treatments of the effects of neutral gas pressure on recombination [2,3] preceded experimental observation [4] of such effects. The process being discussed here is neutral stabilized collisional radiative recombination, in which the recombining electron loses energy in collisions (elastic or inelastic) with atoms or molecules. The state of ionic recombination theory through 1975 has been ably reviewed by Flannery [5]. (Note that Bates [6] has very recently reviewed the theory of ion-ion recombination, a process possessing much of the same physics being discussed here). Since 1979 there has been a flurry of both experimental and theoretical research on electron recombination in high-pressure molecular gases where inelastic electron-molecular collisions serve to enhance recombination rates. This research is the subject of this review.

Recombination at Moderate Pressures

Warman, Sennhauser, and Armstrong [7,8] have measured electron recombination rate coefficients as functions of pressure in a variety of molecular gases. The recombination rates were obtained by microwave absorption measurement of the decay of conductivity after pulse ionization. Their measured values of α ($cm^3\ s^{-1}$) the total recombination rate coefficient, for CO_2, H_2O, and NH_3 are shown in Fig. 1.

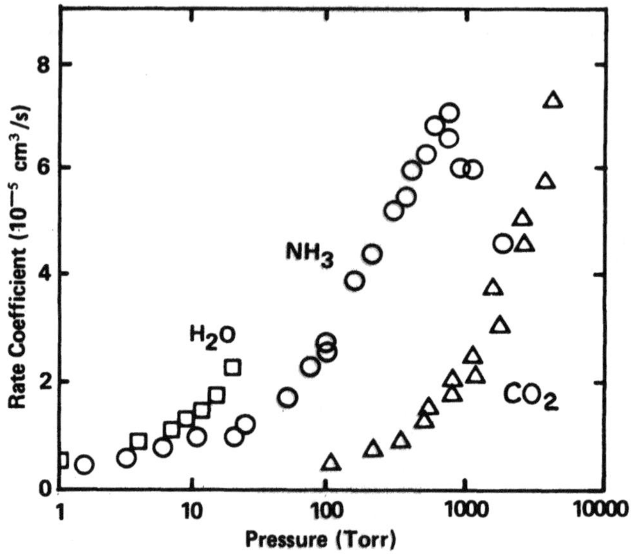

Figure 1. Measured recombination rate coefficients for electrons in molecular gases [7,8].

At pressures below that of the maximum rate coefficient α increases linearly with pressure. This is the region that is most difficult to describe theoretically. At high pressures the rate of recombination is simply limited by the rate at which electrons diffuse toward ions. This process, studied variously by Langevin, Harper, and Debye in the older literature and discussed recently from a new perspective by Bates [9,10], is described by the simple relation,

$$\alpha_D = 4\pi e\mu/\varepsilon \qquad (1)$$

where μ is the electron mobility and ε is the dielectric constant.

Theory of recombination below the diffusion limit. Warman, et al. [7] and Phelps [11] have proposed a model that is useful in estimating α and in illuminating the physics of the recombination process. Assuming that $\alpha = \alpha_2 + \alpha_3 N$ where α_2 is a two-body dissociative recombination rate and α_3 is a three-body rate (N is the gas density) they proposed writing

$$\alpha_3 = C r_T^3 (\nu_u/N) \qquad (2)$$

This expression is based on similar formulae derived by Massey and Burhop [12] ($C = 4\pi/3$) in analogy to Thomson ion-ion recombination theory (see Refs. 5 and 6) and by Pitaevskii [13] ($C = 9\pi$) from diffusion theory for recombination in a monatomic gas. Here $r_T = 2e^2/3kT$ is the Thomson radius. The experimentally measurable parameter ν_u/N is the energy exchange collision frequency [14] defined by

$$\nu_u/N = \frac{e(\mu N)\,(E/N)^2}{\varepsilon_K - kT} \qquad (3)$$

where ε_K is the characteristic energy, which is related to the mobility and the diffusion coefficient by $\varepsilon_K = eD/\mu$. The rate of energy exchange by electrons is expressed in terms of this collision frequency by [14]

$$-\frac{d\bar{u}}{dt} = (\nu_u/N)\,(\varepsilon_K - kT)\,N$$

Experimental values of D and μ as functions of E/N are available for many gases [15].

Fig. 2 [16] shows the predictions of eqn. (2) with the Massey and Burhop and the Pitaevskii constants. Some theoretical and measured values of α_3 are also plotted in Fig. 2. Additional theoretical values for the rare gases based on the Bates and Khan theory [2] are given in Ref. [22] . We see that the simple model expressed by eqn. (3) yields the correct qualitative trend in α_3 as a function of transport coefficients. Considering the simplicity of the model, the quantitative estimates of α_3 are reasonable. There are two major deficiencies of this model. First, ν_u/N is sometimes a steep function of ε_K so that α_3 is sensitive to the choice of ε_K. It is not even clear what the best method of choosing ε_K is. Second, D and μ are defined for constant electric fields and steady state conditions, whereas the recombination process below the diffusion limit involves strong fields and steep gradients.

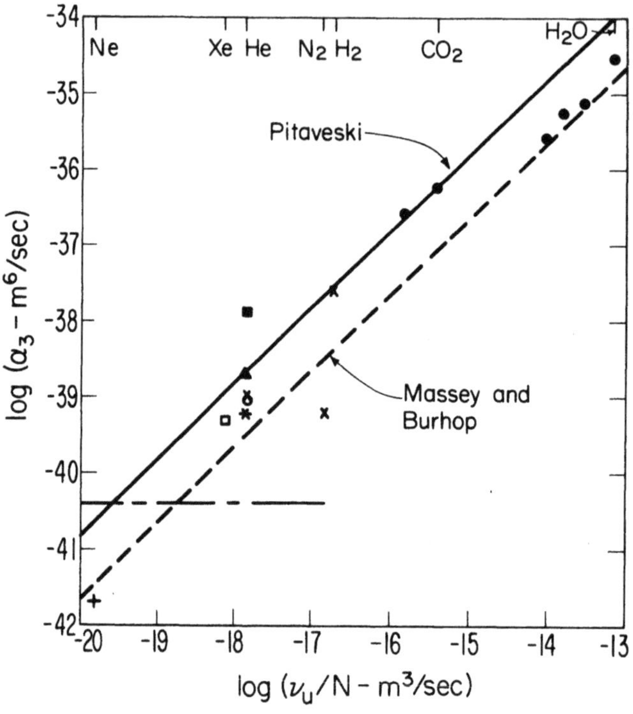

Figure 2. Three-body recombination rate coefficients versus energy exchange collision frequency. The references to the points plotted are: X [2,3], ● [7], ▲ [17], ■ [18], + [19], O [20], * [21]. Reproduced with permission [16].

Computer simulations of recombination. Computer simulation, which has been very useful in the study of ion-ion recombination (see Ref. [6] and the references contained therein), has been applied to the modeling of electron-ion recombination by Morgan and Bardsley [23] and by Morgan [24,25]. Two types of simulation have been used. First are Monte Carlo (MC) simulations [23,24] that compute the classical trajectories of electrons approaching ions from afar until they either recombine or escape. This is just the two-body central force problem plus elastic and inelastic collisions between electrons and molecules. The second kind of simulation is molecular dynamics (MD) [25], in which 50 mutually interacting electron-ion pairs are followed as a function of time. As in MC elastic and inelastic collisions between electrons and neutral molecules are included. The simultaneous calculation of the motions of 100 charged particles allows us to study the effects of collective interactions and external fields on the recombination process. Details of the simulation techniques are given in references [23-25].

Recombination in CO_2, CH_4, NH_3 and H_2O has been simulated by Monte Carlo. Results for electrons in CO_2 are shown in Fig. 3. Two-body dissociative recombination, a curve-crossing process,

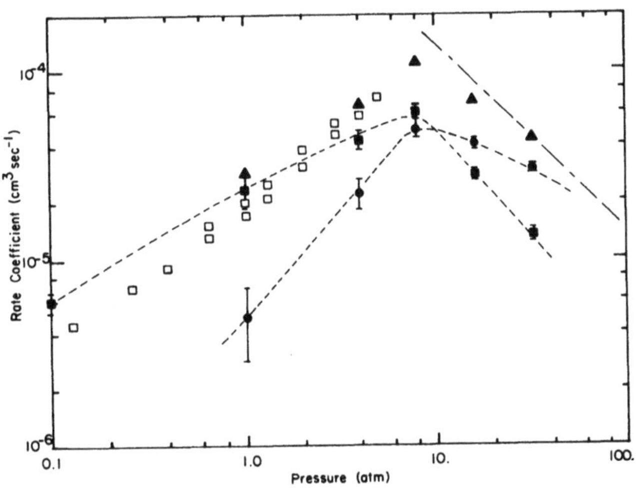

Figure 3. Recombination rate coefficients in CO_2. Points ■ and ● are the two-body and three-body contributions to the total (▲) calculated coefficient. The points □ are the measured coefficients from Ref. [8]. The line -.-.- is the diffusion limited rate, α_D.

is simulated by putting an absorbing sphere around the positive ion. The radius of the sphere is determined by matching the computed rate coefficient at very low pressure to a measured dissociative recombination rate coefficient. Three-body recombination is said to have occurred when the total energy of an electron in the neighborhood of an ion falls below -12 kT. In Fig. 3 we see that the computed total rate coefficient agrees well with the measurements of Armstrong et al [8]. The two-body recombination is seen to be enhanced by pressure effects, as has been postulated by Bates [26]. Collisions reduce the speed of an electron approaching an ion, making it more likely to pass within the absorbing sphere and dissociatively recombine. Three-body recombination dominates at higher pressures and eventually $\alpha \to \alpha_D$ as shown in Fig. 3. Similar results, in reasonable agreement with experiment [7,8], are obtained for recombination in NH_3 [23] and H_2O [24].

The computer simulations demonstrate the importance of energy exchange collisions in the recombination process,

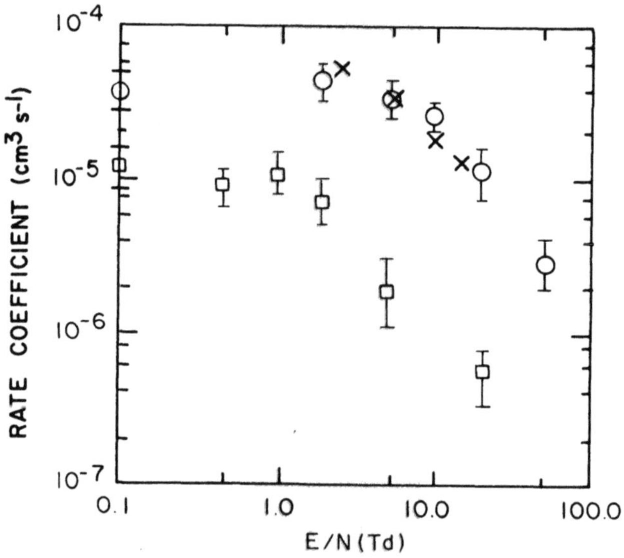

Figure 4. Computed total rate coefficient for recombination of electrons and ions in a discharge as a function of E/N for CO_2 (O) and CH_4 (□). The experimental results of Rev. [27] are denoted by the symbol X.

as was discussed above in the development of $\alpha \propto (\nu_u/N)$ (Eqn. 2). The MC calculations are, subject to the availability of good quality cross-sectional data, quantitatively more accurate than Eqn. (3). There are exceptions, however, as we will see below in the discussion of recombination in CH_4.

Littlewood, et al. [27,28] have measured recombination rates in $CO_2/N_2/He$ mixtures using a pulsed e-beam sustained discharge [29]. Knowledge of the e-beam ionization rate and the electron drift velocity, along with measurement of discharge current, yields the recombination rate. The measurements in CO_2 from Ref. [27] are plotted along with MD results [25] in Fig. 4, where we see excellent agreement.

Recombination at High Pressure: CH_4

Recently Nakamura, Shinsaka, and Hatano [30] (NSH) have measured the electron mobility and recombination rate coefficient in gaseous, liquid, and solid methane. Pulse radiolysis techniques [31] were used in the measurements. We see the rate coefficients measured by NSH plotted in Fig. 5 along with the MC calculations of Morgan [25]. For all but the very highest densities in the gas phase α is much less than α_D in both the measurements and the calculations. Except near the critical point, the calculations and the measurements disagree spectacularly. The calculations show α to peak at about 2 x 10^{21} cm^{-3}. As pressure goes to zero the calculated α goes to the measured two-body rate [4] of 3.9 x 10^{-6} cm^3/sec. In the process of modeling an e-beam controlled discharge switch, Kline [32] has estimated the recombination rate coefficient to be about 2 x 10^{-6} cm^3/sec for E/N values of a few Td. This estimate agrees with the MD calculation of α for CH_4 in an electric field as shown in Fig. 4.

At this point the reason for the great quantitative and qualitative difference between the measured and the calculated recombination rates is unknown. There are some possible theoretical causes. Examination of the static structure factor, S(k = 0), for CH_4 [23] shows that structure effects (significant pair correlations set in at densities of several times 10^{21} cm^{-3}. In addition, as Braglia and Dallacasa [34] have discussed, uncertainties in cross-sectional data and high density effects conspire

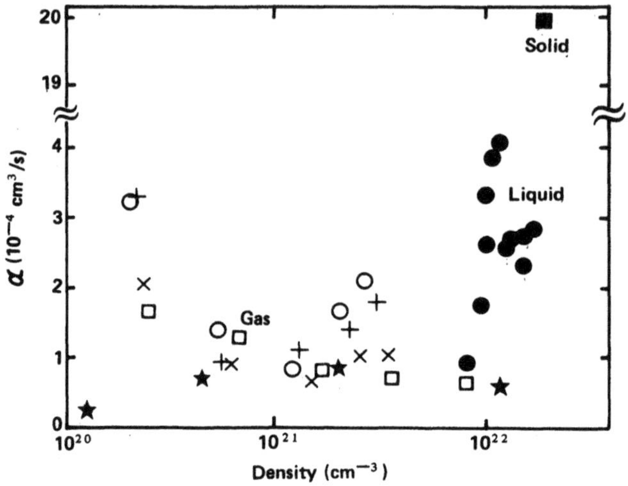

Figure 5. Recombination rate coefficients for electrons in methane. The points denoted by the symbol ★ are computed [25]. All the other points are measurements of Nakamura, et al. [30]. The gas phase measurements were performed at four temperatures ranging from 193 K to 295 K.

to make electron drift velocity quantitatively and qualitatively incorrect at densities as low as several times 10^{20} cm^{-3}. As the recent literature [35] on the subject demonstrates, the magnitude and energy dependence of the elastic cross-section at low energy (especially near the Ramsauer minimum) is still very uncertain, despite the numerous measurements of it dating back to 1925! There are enough theoretical uncertainties at these densities for us to be unable to state whether the difficulty is either experimental or theoretical. As a final note, unlike the situation with gas phase methane, recent MD simulations modeling recombination in liquid CH_4 [36] are in excellent agreement with the NSH measurements.

Summary

We see from the examples of CO_2, NH_3, and H_2O that the recombination process is reasonably well understood and modelable at low to moderate pressures. α_2 is enhanced by inelastic collisions with the neutral molecules. α_2 and α_3 are competitive with α_3 becoming dominant as pressure increases. At very high pressure, as we see with CH_4, structure effects and probably clustering (which also affect transport coefficients) make quantitative modeling difficult.

Acknowledgements

I wish to thank Prof. Y. Hatano and Dr. A. V. Phelps for helpful discussions. This work was performed in part under the auspices of the U.S. Department of Energy at the Lawrence Livermore National Laboratory under Contract No. W-7405-Eng-48.

References

[1] Mitchell, J.B.A.; McGowan, J. Wm. In "Physics of Ion-Ion and Electron-Ion Collisions"; Brouillard F.; McGowan, J. Wm., Eds; Plenum: New York, 1982. McGowan, J. Wm; Mitchell, J.B.A. In "Electron-Molecule Interactions and Their Applications"; Christophorou, L. G., Ed.; Academic Press: Orlando, 1984; Volume 2, Chapter 2.
[2] Bates, D. R.; Khare, S. P. Proc. R. Soc. (London) 1965, 85, 231.
[3] Bates, D. R.; Malaviya, V.; Young, N. A. Proc. R. Soc. (London) 1971, 320, 437.
[4] Maier, H. N.; Fessenden, R. W. J. Chem. Phys. 1975, 62, 4790.
[5] Flannery, M. R. In "Atomic Processes and Applications"; Burke, P. G.; Moiseiwitsch, B. L., Eds; North-Holland: Amsterdam, 1976; Chapter 12.
[6] Bates, D. R. In "Advances in Atomic and Molecular Physics"; Bates, D. R.; Bederson, B., Eds; Academic Press: Orlando, 1985; Volume 20.
[7] Warman, J. M.; Sennhauser, E. S.; Armstrong, D. A. J. Chem. Phys. 1979, 70, 995.
[8] Armstrong, D. A.; Sennhauser, E. S.; Warman, J. M.; Sowada, U. Chem. Phys. Lett. 1982, 86, 281.
[9] Bates, D. R. J. Phys. B 1975, 8, 2722.
[10] Bates, D. R. J. Phys. B 1983, 16, L295.
[11] Phelps, A. V. unpublished 1979; private communication.

[12] Massey, H.S.W.; Gilbody, H. B. "Electronic and Ionic Impact Phenomena, 2nd Edition"; Oxford: London, 1974; V. 4, p. 2149.
[13] Pitaevskii, L. P. Sov. Phys. JETP, 1962, 15, 5. Lifshitz, E. M.; Pitaevskii, L. P. "Physical Kinetics"; Pergamon Press: Oxford, 1981; Section 24.
[14] Frost, L. S.; Phelps, A. V. Phys. Rev. 1962, 127, 1621.
[15] Beaty, E. C.; Dutton, J.,; Pitchford, L. C. "A Bibliography of Election Swarm Data"; University of Colorado, Joint Institute for Laboratory Astrophysics Report No. 20, 1979.
[16] Phelps, A. V. In "Electrical Breakdown and Discharges in Gases"; Kunhardt, E. E.; Luessen, L. H., Eds.; Plenum: New York, 1981; Part A.
[17] Gousset, G.; Sayer, B.; Berlande, J. Phys. Rev. A 1977, 16, 1070.
[18] Johnson, A. W.; Gerardo, J. B. Phys. Rev. A 1972, 5, 1410.
[19] Borodin, V. M. Opt. Spectrosc. 1975, 38, 266.
[20] Collins, C. B. Phys. Rev. 1969, 177, 254.
[21] Deloche, R. Centre d'Etudes Nucléaires de Saclay Report CEA-R-3450 1968.
[22] Whitten, B. L.; Downes, L. W.; Wells, W. E. J. Appl. Phys. 1981, 52, 1255.
[23] Morgan, W. L.; Bardsley, J. N. Chem. Phys. Lett. 1983, 96, 93.
[24] Morgan, W. L. J. Chem. Phys. 1984, 80, 4564.
[25] Morgan, W. L. Phys. Rev. A 1984, 30, 979.
[26] Bates, D. R. J. Phys. B 1981, 14, 3525.
[27] Littlewood, I. M.; Cornell, M. C.; Clark, B. K.; Nygaard, K. J. J. Phys. D 1983, 16, 2113.
[28] Littlewood, I. M.; Cornell, M. C.; Nygaard, K. J. J. Chem. Phys. 1984, 81, 1264.
[29] Douglas-Hamilton, D. H. J. Chem. Phys. 1973, 58, 4820.
[30] Nakamura, Y.; Shinsaka, K.; Hatano, Y. J. Chem. Phys. 1983, 78, 5820.
[31] Wada, T.; Shinsaka, K.; Namba, H.; Hatano, Y. Can. J. Chem. 1977, 55, 2144.
[32] Kline, L. E. IEEE Trans. Plasma Sci. 1982, 10, 224.
[33] Christophorou, L. G.; McCorkle, D. L. Chem. Phys. Lett. 1976, 42, 533.
[34] Braglia, G. L.; Dallacasa, V. Phys. Rev. A 1982, 26, 902.
[35] Ferch, J.; Granitza, B.; Raith, W. J. Phys. B 1985, 18, L445.
[36] Morgan, W. L. J. Chem. Phys. 1986, to be published.

PART II

KINETIC THEORY

PRINCIPLES OF THE MEASUREMENT OF ELECTRON DRIFT VELOCITIES

H. Tagashira

Department of Electrical Engineering,
Hokkaido University, Sapporo,
Japan, 060

Introduction

It has been pointed out by Tagashira et al.[1] that the center-of-mass velocity W_r of an isolated swarm of electrons may assume a value which is different from the mean velocity W_v of the electrons when electron impact ionization or attachment exists, even if the value of E/N is kept the same. The significance of this point is that until then the value of the center-of-mass drift velocity W_r had been considered and treated to be the same as that of the mean velocity W_v both in experiment and theory. Since an estimation for argon [1] or for nitrogen [2] showed that W_r could be more than 30 % larger than W_v at E/N as high as 588 Td, it would be now important to specify what drift velocity is measured or calculated in an investigation, and in fact this point is clearly stated [3] in a few of the recent publications.

The purpose of this article is to treat some problems related to the principle of determination of the electron drift velocities in gases. First, the article treats the nature of the drift velocity which appears in the usual diffusion equation of electrons in hydrodynamic regime. Then, an exact formula for the mean-arrival-time drift velocity is shown. Finally, experiments for electron drift velocity measurements are studied by a simulation to find what drift velocity is determined in the experiments.

Constant Drift Velocity in Diffusion Equation

In a hydrodynamic regime, the diffusion equation in one dimension is usually written as below.

$$\frac{\partial n}{\partial t} = R_i n - W \frac{\partial n}{\partial z} + D_L \frac{\partial^2 n}{\partial z^2} + \text{higher order terms} \qquad (1)$$

Here, $n=n(z,t)$ is the one-dimensional electron number density, R_i is the electron ionization frequency, D_L is the longitudinal diffusion coefficient. The higher order terms represent the terms of the electron density differentiated with respect to space coordinate three times or more. W is the drift velocity to be studied. Use of a one-dimensional model does not lose generality of the problem. In a hydrodynamic regime, R_i, W, D_L and the coefficients for the higher order derivatives are all assumed constant. The direction of the electric field is assumed to lie along the z axis.

First, both sides of Eq.(1) are integrated with z. Then, the following equation is obtained.

$$\frac{d\,n(t)}{dt} = R_i\, n(t) \tag{2}$$

where

$$n(t) = \int_{-\infty}^{\infty} n(z,t)dx$$

A boundary-free case is considered here for clarity and simplicity. Eq.(2) shows that R_i is the relative rate of increase of the number of electrons, and when electron impact ionization is the cause of the increase, R_i represents the number of ionizations per unit time per electron, and thus the ionization frequency.

The hydrodynamic regime assumes that R_i and the other coefficients are constant. Therefore, it assumes that the number of electrons in an isolated swarm increases at a constant relative rate, i.e. exponentially.

Both sides of Eq.(1) are next multiplied by z and then integrated with z, to obtain,

$$\frac{d}{dt}\, n(t)\bar{z}(t) = R_i\, n(t)\bar{z}(t) + W\, n(t) \tag{3}$$

where

$$\bar{z}(t) = \frac{1}{n(t)} \int_{-\infty}^{\infty} z\, n(z,t)dz$$

is the center of mass of electrons at time t.

A simple manipulation using Eq.(3) gives

$$W = \frac{d}{dt}\,\bar{z}(t) = \frac{\bar{z}(t_2) - \bar{z}(t_1)}{t_2 - t_1}\ . \tag{4}$$

Eq.(4) proves that the drift velocity W which appears as the coefficient of the first-order derivative of the electron number density in the electron diffusion equation represents the velocity of the center of mass of electrons in an isolated swarm. Therefore,

$$W = W_r$$

It is noted that if one wishes to measure W_r as directly as possible, the position of the center of mass $\bar{z}(t)$ should be observed at least at two times t_2 and t_1, rather than the mean arrival time $\bar{t}(z)$ of electrons in a swarm at position z, as has been the convention (See next section.)
Another point that has to be mentioned here is that the drift velocity which appears in the well-known Gaussian shaped development of an isolated electron swarm is W_r and not W_v.

An Exact Expression for the Mean-arrival-time Drift Velocity

In many experimental studies, the drift velocity is determined by measuring the mean arrival time of electrons in an isolated swarm at two positions in the field direction and by calculating the velocity as,

$$W_m = \frac{dz}{d\bar{t}(z)} = \frac{z_2 - z_1}{\bar{t}(z_2) - \bar{t}(z_1)}$$

Here, z is the position in the field direction and $\bar{t}(z_j)$ is the mean arrival time at position z_j.
It has been known that

$$W_m = (W_r^2 - 4R_iD_L)^{1/2}$$

if the higher order terms in (1) are neglected. An effort was made to extend the above expression to the next higher order, i.e. to incorporate the effect of D_3 coefficient, which appears in (1) as $-D_3\partial^3n/\partial z^3$, but the result was a prohibitively lengthy expression.
Recently, however, an exact expression for W_m is obtained [4] in which higher order terms may be incorporated in principle as much as one wishes, though in a series form, as,

$$W_m = W_r - 2\alpha_T D_L + 3\alpha_T^2 D_3 - 4\alpha_T^4 D_4 + \cdots + (-1)^{n-1} n\, \alpha_T^n D_n + \cdots \quad (5)$$

where α_T is the Townsend first ionization coefficient and D_n is the n-th order coefficient in (1). If electron attachment is present, α_T is replaced by the effective ionization coefficient $\bar{\alpha}$ (=α-η), where α is the ionization coefficient and η is the electron attachment coefficient.

Equation (5) shows that if ionization or attachment is present, the mean-arrival-time drift velocity W_m generally assumes a different value from the center-of-mass drift velocity W_r. If ionization or attachment does not exist, W_m=W_r (=W_v); a happy situation !

Simulation Study of Drift Velocity Measurements

In the previous sections, it has been shown that at least three kinds of electron drift velocities may be considered theoretically. They are the center-of-mass drift velocity W_r, the average drift velocity W_v and the mean-arrival-time drift velocity W_m.

In this section, we try to find by a computer simulation which or what drift velocity is measured in some previous experimental studies.

In the simulation, an isolated group of electrons is represented by the formula below [5].

$$n(z,t) = \frac{N_0 \exp(R_i t)}{(4\pi D_L t)^{\frac{1}{2}} W_r t}\left[z \exp\{-\frac{(z-W_r t)^2}{4D_L t}\} + (z-2d) \times \exp\{-\frac{(z-W_r t)^2 + 4d(d-z)}{4D_L t}\}\right] \qquad (6)$$

Here, N_0 is the number of electrons generated at the cathode at time t=0, d is the gap length and R_i is the electron ionization frequency. This formula statisfies the cathode- and the anode boundary conditions to a good approximation [5]. A parallel -plane electrode system is assumed.

The external current I(t) due to the motion of the electrons may be represented to a good approximation by

$$I(t) = \frac{eW_v}{d} N_e(t)$$

where e is the electronic charge, W_v is the average velocity of electrons and

$$N_e(t) = \int_0^d n(z,t)dz$$

is the number of electrons in the gap.

In 1959, Frommhold [6] measured the electron drift velocity by following the procedure as described below using the notations defined above.

First, he integrated the gap current I(t) with respect to time. Namely, a capacitor was charged by I(t), therefore the terminal voltage was given by

$$v(t) = \frac{1}{C} \int_0^t I(\xi) d\xi$$

where C is the capacity of the capacitor. Note that for t large enough, v(t) tends to a constant v_1, say, since all the electrons disappear into the anode. Frommhold plotted ln v(t) against t; ln v(t) first increased linearly with t with a constant slope, then the slope gradually decreased and finally ln v(t) reached a constant value ln v_1. Note here that ln v(t) is a monotonical function. Frommhold deduced the drift velocity W_{FD} from

$$W_{FD} = \frac{d}{T_{FD}}$$

where T_{FD} is the time which corresponds to the intersection between the horizontal linear line, the value on the vertical axis of which = ln v_1, and the linear line which is the extension of the initial linear increase of ln v(t).

In 1960, Franke [7] measured the drift velocity by observing the light emission from the gap. The light intensity $N_{ex}(t)$ from a group of electrons traversing a gap may be represented by

$$N_{ex}(t) = \exp(-\frac{t}{\tau})[R_{ex} \int_0^t N_e(\xi)\exp(\frac{\xi}{\tau})d\xi + A] \quad (7)$$

where τ is the life time of the excited state and R_{ex} is the electron excitation frequency. Franke found that a plot of ln $N_{ex}(t)$ against t gave first an ascending linear line, then the slope decreased and assumed negative values relatively abruptly, and finally ln $N_{ex}(t)$ gave a descending linear line. Franke deduced the drift velocity from

$$W_{FK} = \frac{d}{T_{FK}}$$

where T_{FK} is the time corresponding to the intersection between the extrapolation of the ascending linear line and that of the descending linear line.

In 1965, Schlumbohm [8] measured the electron drift velocity. There he deduced the drift velocity from

$$W_{SC} = \frac{d}{T_{SC}}$$

where T_{SC} is the time corresponding to the peak of the gap current I(t).

These experiments are simulated in the present work. The gas is assumed to be methane. This is not too popular a gas, but the only common gas which these three investigators chose. For theoretical drift velocities, the center-of-mass drift velocity W_r, the average velocity W_v, the mean-arrival-time drift velocity W_m and the average velocity V_d for the steady state Townsend condition [1] are chosen. The simulation is performed for a gas pressure of 1 Torr at 0°C, and at gap lengths of 3.0 cm, 0.8 cm and 4.7 cm respectively for W_{FD}, W_{FK} and W_{SC}, as these were typical values for the respective studies. The results are shown in the table.

Table. Comparison between theoretical and simulated drift velocities in cm μs^{-1}. Methane, E/P_0=200 V cm^{-1} $Torr^{-1}$[9]

W_r	W_m	W_v	V_d	W_{FD}	W_{FK}	W_{SC}
45.0	36.0	34.6	31.2	40.1	40.1	38.0

In deduction of W_{FD}, W_{FK} and W_{SC} by the present simulation, the theoretical value of W_r, as in the table, and R_i (=1.50 x 10^8 s^{-1}) and D_L(=1.34 x 10^6 cm^2 s^{-1}), both at 1 Torr, are put in (6) for calculation of the number of electrons in the gap $N_e(t)$. τ in Eq.(7) is taken to be 7 ns.

What the results in the table signify is simple, yet remarkable. None of the simulated, and therefore experimentally determined, drift velocities agree with the theoretical ones. This result strongly suggests that attention should be paid to the nature of the drift velocity when an experimental investigation into the electron drift velocity is performed.

Conclusions

In the present article, problems encountered when an exact determination of the electron drift velocity is considered are discussed.

It has been shown that the drift velocity which appears as the coefficient of the first order concentration gradient of the electron density in the electron diffusion equation is the center-of-mass drift velocity. An exact expression for the mean-arrival-time drift velocity is introduced. A computer simulation has been performed to show that attention should be paid to identify the nature of the drift velocity measured in experimental studies.

Acknowledgments

The author wishes to thank Mr. Y. Ohmori of the author's department for much helpful discussion.

References

[1] Tagashira, H.; Sakai, Y.; Sakamoto, S. J. Phys. D: Appl. Phys. 1977 10, 1051.
[2] Taniguchi, T.; Tagashira, H. J. Phys. D: Appl. Phys. 1978 11, 1757.
[3] Yousfi, M.; Segur, P.; Vassiliadis, T. J. Phys. D: Appl. Phys. 1985 18, 359.
[4] Tagashira, H.; To be submitted to J. Phys. D.
[5] Huxley, L.G.H.; Crompton, R.W. "The Diffusion and Drift of Electrons in Gases"; Wiley-Interscience: New York, 1974; Chapter 5.
[6] Frommhold, L. Z. Phys. 1959 156, 144.
[7] Franke, W. Z. Phys. 1960 158, 96.
[8] Schlumbohm, H. Z. Phys. 1965 182, 306.
[9] Ohmori, Y.; Tagashira, H. Joint Convention Records, Hokkaido Branches, Electrical and Information Engineering Institutes, Japan 1985 No.201.

EFFECT OF BOUNDARIES AND FIELD INHOMOGENEITIES ON SWARMS

Kailash Kumar

Department of Theoretical Physics,
Research School of Physical Sciences,
The Australian National University,
G.P.O.Box 4, Canberra, A.C.T.2601.

Introduction

The effect of boundaries and finite enclosures on swarms has been the subject of a number of studies. We recall the work on electrons from a steady source incident on a plane-absorbing boundary [1], on diffusion cooling of electrons in finite enclosures [2], on attachment cooling [3] and on pressure dependence of mobility of ions (or runaway ions)[4]. The study of retarding potential difference analysis of the ion distribution function [5], and the considerations presented in the preceding talk by Tagashira [6], are also related to such effects.

From these studies it appears that the boundaries (or space dependent fields) have rather large effects on the distribution function. Perhaps more significantly, attempts to use a transport description in terms of the continuity equation for the number density are no longer straightforward. Thus for instance Lowke, Parker and Hall [1] interpreted their results in terms of space-dependent transport coefficients. In other cases one needs transport coefficients that depend on the electric field and the neutral density in a way which is different from that obtained from the kinetic theory of systems in free space, in the hydrodynamic limit.

In the following, an approach is described where starting from the Boltzmann equation, by appropriate specialisation, one may treat problems of this kind and see their connection to each other and to the free space problem in a systematic way.

It is, at least, a framework which may be useful in intuitive discussions - both theoretical and experimental. Whether it can be implemented with the currently available technical mathematical resources for some real systems remains to be seen.

Representation of the Boundary Effects

Our intuition concerning boundaries derives mainly from the experience with second order differential equations in oscillation problems and in classical hydrodynamics. In such theories one makes models for the expected behaviour of physical variables like the density and the current. A part of the model is contained in the differential operator and the rest in the boundary conditions. One interprets the boundary conditions as models of the physical processes acting at the boundary. Two rather different aims are served by prescribing the boundary conditions: one is to represent the physical process and the other to select the class of functions in which the solutions of the equation are to be sought. The distinction is perhaps unimportant in continuum theory.

In kinetic theory, modelling of physical phenomena is of a different type, and one should perhaps pay attention to associated questions in discussing the relation between kinetic theory and hydrodynamics in presence of boundaries.

If we take the view that in kinetic theory the physical processes should be represented in terms of forces acting on the particles, then we must represent the effects of boundaries in terms of such forces - that is, as an operator in the Boltzmann equation. The class of functions in which the solution is to be sought is given by fairly mild requirements, that the distribution function should be a probability density, that it should be integrable in velocity so that one may speak of a number density, and that it should decrease sufficiently rapidly with increasing $\underline{r}$ and $\underline{c}$ to be physically interesting. In such a view there are no conditions to be imposed on the values that the function may take at the physical boundaries.

It is useful to recall that boundaries may be simulated by suitable configuration of fields. Electric shutters used in experiments are based on this principle. In particular, we may think of the Tyndall-Powell shutters, where an electric field is created between two planes of wire gauze. Field applied in one direction closes the shutter and applied in the opposite direction opens it. The action is that of a repelling or absorbing boundary. Neither the repulsion nor the absorption is perfect. In the Boltzmann equation this will be represented by a term of the form $\underline{a}_1(\underline{r}).\underline{\partial}_c f$, where $\underline{a}_1(\underline{r})$ is a space-dependent field due to the shutter. It is clear in this case that there is no need for any constraints on the distribution function at the planes where the gauze is.

The retarding potential difference measurements [5] are another example where inhomogeneous electric fields play a similar role. By recognising this we relate boundary effects to another class of problems which may be more tractable.

The analogy actually goes much further in that the absorbing characteristics of metal surfaces may also be thought of as being caused by the operation of a potential difference - perfect absorption is then seen as a limiting case where the potential difference becomes very large.

Alternatively, one may represent the boundary effects by an operator constructed in analogy with the collision operator but containing a space-dependence corresponding to the location of boundaries. Again, a linear operator will be sufficient for describing many real situations.

In further discussion we shall suppose that the boundary effects may be represented by a linear operator with some such velocity and space dependence. Before taking up this discussion it is appropriate to first recapitulate the main points of the theory in free space, without boundaries.

Swarms in Free Space

The Boltzmann equation for a swarm without boundaries is

$$(\partial_t + \underline{c}.\nabla + \underline{a}.\partial_{\underline{c}} + J)f_0 = S \ , \qquad (1)$$

where $\underline{a}$ is the acceleration due to the field independent of $\underline{r}$, $\underline{c}$ and t , J is the collision operator and S a source term. Its solution may be written in the form

$$f_0 = G_0 S \qquad (2)$$

where the subscripts denote the case without boundaries and the Green's function G_0 satisfies the equation

$$(\partial_t + \underline{c}.\nabla + \underline{a}.\partial_{\underline{c}} + J)G_0 = \delta(\underline{r}-\underline{r}')\ \delta(\underline{c}-\underline{c}')\ \delta(t-t'). \qquad (3)$$

If the collision operator J has suitable properties the Green's function may be constructed in the form

$$G_0(\underline{r},\underline{c},t;\underline{r}'\underline{c}'t')$$
$$= \Theta(t-t') \sum_\alpha \int d\underline{k}\ \Psi_{\alpha,\underline{k}}(\underline{r},\underline{c})\ e^{(t-t')\omega(\alpha,\underline{k})}\ \Phi_{\alpha,\underline{k}}(\underline{r}',\underline{c}') \quad (4)$$

where the functions Ψ and Φ are solutions of the adjoint pair of eigenvalue equations

$$(\underline{c}.\nabla + \underline{a}.\partial_{\underline{c}} + J + \omega(\alpha,\underline{k}))\Psi_{\alpha,\underline{k}} = 0 \quad (5a)$$

$$(-\underline{c}.\nabla - \underline{a}.\partial_{\underline{c}} + \tilde{J} + \omega(\alpha,\underline{k}))\Phi_{\alpha,\underline{k}} = 0\,. \quad (5b)$$

Since $\underline{a}$ and J are independent of $\underline{r}$ the space and velocity dependence of these functions factorises in the form

$$\Psi_{\alpha,\underline{k}} \sim e^{i\underline{k}.\underline{r}}\psi_{\alpha,\underline{k}}(\underline{c})\ ;\ \Phi_{\alpha,\underline{k}} \sim e^{-i\underline{k}.\underline{r}}\phi_{\alpha,\underline{k}}(\underline{c})\,. \quad (6)$$

For each $\underline{k}$ we have an infinite number of discrete eigenvalues $\omega(\alpha,\underline{k})$ and a biorthogonality relation holds between Ψ and Φ functions

$$\int \phi_{\alpha,\underline{k}}(\underline{c})\ \psi_{\alpha',\underline{k}}(\underline{c})\ d\underline{c} = \delta_{\alpha\alpha'}\,. \quad (7)$$

Transport coefficients $\underline{\omega}_0^{(n)}$ are identified from the power series expansion of the lowest eigenvalue

$$\omega_{0,\underline{k}} \xrightarrow{\underline{k}\to -i\nabla} \omega_0(\nabla) = \omega_0^{(0)} - \underline{\omega}_0^{(1)}.\nabla + \underline{\omega}_0^{(2)}:\nabla\nabla + \ldots\,. \quad (8)$$

They may be expressed as integrals involving functions furnished by similar power series expansions of the corresponding eigenfunctions

$$\psi_{0,\underline{k}} \xrightarrow{\underline{k}\to -i\nabla} \psi_0(\underline{c},\nabla) = \sum_n \psi_0^{(n)}(\underline{c}) \odot (-\nabla)^n\,, \quad (9a)$$

$$\phi_{0,\underline{k}} \longrightarrow \phi_0(\underline{c},\nabla) = \sum_n \phi_0^{(n)}(\underline{c}) \odot (-\nabla)^n\,. \quad (9b)$$

Thus, for instance, the drift velocity is given by

$$\underline{\omega}_0^{(1)} = \int \phi_0^{(0)}(\underline{c})\ \underline{c}\ \psi_0^{(0)}(\underline{c})\ d\underline{c}\,. \quad (10)$$

This is the formula which after suitable specialisation in two-term theory of electron swarms becomes

$$nW = -\frac{8\pi eE}{3m^2N}\int_0^\infty \frac{\varepsilon}{Q}\frac{\partial f^{(0)}}{\partial\varepsilon}\,d\varepsilon \tag{11}$$

Here the function $f^{(0)}$ comes from a part of the eigenfunction $\Psi_0^{(0)}(\underline{c})$. It is not the distribution function itself, as is sometimes thought.

This identification of the transport coefficients is based upon the observation that for an initial value problem with $S = \delta(t)f(\underline{r},\underline{c})$, if at sometime the distribution function has the form

$$f_0(\underline{r},\underline{c},t) = \Psi_0(\underline{c},\nabla)\; n(\underline{r},t)\;, \tag{12a}$$

$$n(\underline{r},t) = e^{t\omega_0(\nabla)}\; n(\underline{r},0)\;, \tag{12b}$$

$$(\partial_t - \omega_0(\nabla))\; n(\underline{r},t) = 0\;, \tag{12c}$$

then it continues to have this form at all subsequent times by virtue of equations (2) and (3). The continuity equation (12c) for the number density is then simply interpreted in terms of the expansion (8) of the lowest eigenvalue. This is the description of the hydrodynamic approximation to kinetic theory.

One can not change the above interpretation of formulas like (11) without reconsidering the whole question of the relationship between a kinetic theory description and that obtained by considering a phenomenological continuity equation for the density alone.

The kinetic theory solution (2) and (3) actually involves all the eigenvalues $\omega(\alpha,\underline{k})$. The criteria for the validity of the hydrodynamic approximation are therefore to be sought in the properties of these eigenvalues and eigenfunctions.

Figure 1 shows two possible behaviours of the two lowest eigenvalues as functions of $\underline{k}$. This is very schematic, since $\underline{k}$ is actually three dimensional and eigenvalues can be complex. In the case (a) the eigenvalues are well separated for all values of $\underline{k}$ and a time scale τ_c can be defined by this separation. One can then say that for times $t \gg \tau_c$ only the lowest eigenvalue matters and the hydrodynamic regime holds. In this case there are no restrictions on the values of $\underline{k}$.

If, however, the situation is like that in case (b) the hydrodynamic approximation can occur only if $|\underline{k}| << |\underline{k}_0|$ and the time scale is determined by the separation around $|\underline{k}| \sim 0$.

This is in keeping with the view that hydrodynamic regime holds for long times and small gradients. Here we have a way of exploring these limitations. Perhaps experiments would reveal whether one or more eigenvalues are needed for fitting certain data.

While perturbation theory is always suspect in such circumstances, it may nevertheless be possible to get some idea about the behaviour of these eigenvalues by constructing the power series with existing methods.

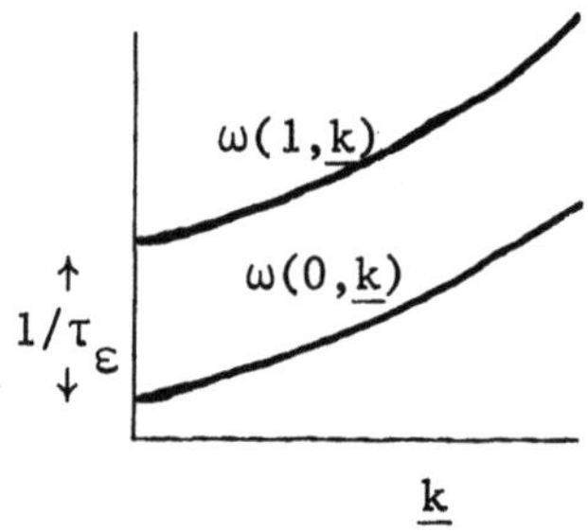

Figure 1(a)

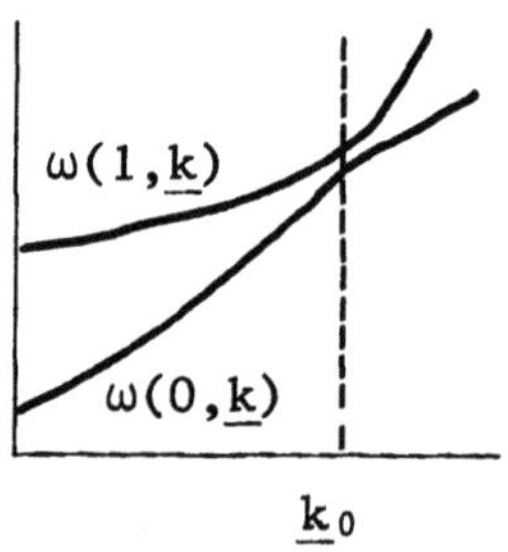

Figure 1(b)

Effects on Boundaries and Field Inhomogeneities

As discussed before these effects may be represented by a linear operator B in the Boltzmann equation

$$(\partial_t + \underline{c}.\nabla + \underline{a}.\partial_{\underline{c}} + J + B)f = S \ . \tag{13}$$

For any f the function Bf is non-zero only in the part of space where the boundary or the field inhomogeneity is located. We may include a multiplicative parameter to vary the strength of the associated effects.

As before, the solution may be expressed by help of the Green's function, which now satisfies the equation

$$(\partial_t + \underline{c}.\nabla + \underline{a}.\partial_{\underline{c}} + J + B)G = \delta(\underline{r}-\underline{r}')\ \delta(\underline{c}-\underline{c}')\ \delta(t-t').\tag{14}$$

The relation to the free-space problem is given by

$$\begin{aligned} f &= GS \\ &= f_0 + G_0Bf_0 + G_0B\,G_0B\,f_0 + \ldots \end{aligned} \quad (15)$$

The first term f_0 is the free-space distribution function given by equation (2). In the second term the operator B effectively truncates f_0 to the boundary region, possibly changing the velocity distribution and G_0 then propagates the resulting pulse as if there were no boundaries. The process is iterated in subsequent terms.

Consider a plane boundary and the initial value problem in which at time $t = 0$ the distribution function is confined to a narrow region a distance d from the plane. The solution f remains substantially close to f_0 until by free propagation f_0 builds up sufficient amplitude in the region of B. If the hydrodynamic approximation holds for the boundary-free case then

$$f(t) \cong f_0(t) \quad \text{for} \quad \tau_D > t \gg \tau_c \; . \quad (16)$$

where $\tau_D \cong d/|\underline{\omega}_0^{(1)}|$ is the drift time.

For such values of t the system will be well described by the free-space hydrodynamic approximation, and the measured transport coefficients will be those given by the theory of the previous section. This is more or less independent of what B may be, since only the smallness of f_0 in the region of B was required. The result is in keeping with what has been assumed all along for the time-of-flight experiments. It could be further refined by working out the restrictions on $\underline{k}$ and t mentioned in the last section.

It is best to imagine that the system is observed as in a photographic process, since a further step is required to obtain currents in an external circuit.

We expect that the expansion (15) will not be good for $t \cong \tau_D$ and we have to seek some other expression for G.

Let us suppose that G may still be constructed in the form

$$\begin{aligned} &G(\underline{r},\underline{c},t;\underline{r}',\underline{c}',t) \\ &= \Theta(t-t') \sum_\alpha \int d\underline{k}\; \Psi^B_{\alpha,\underline{k}}(\underline{r},\underline{c})\; e^{(t-t')\omega^B(\alpha,\underline{k})}\; \Phi^B_{\alpha,\underline{k}}(\underline{r}',\underline{c}') \, , \end{aligned} \quad (17)$$

with the new eigenvalues and eigenfunctions still indexed by α and $\underline{k}$.

The new equations are

$$(\underline{c}.\nabla + \underline{a}.\partial_{\underline{c}} + J + B + \omega^B(\alpha,\underline{k}))\ \Psi^B_{\alpha,\underline{k}}(\underline{r},\underline{c}) = 0\ , \quad (18a)$$

$$(-\underline{c}.\nabla - \underline{a}.\partial_{\underline{c}} + \tilde{J} + \tilde{B} + \omega^B(\alpha,\underline{k}))\ \Phi^B_{\alpha,\underline{k}}(\underline{r},\underline{c}) = 0\ . \quad (18b)$$

The eigenfunctions are no longer factorable. We have for instance,

$$\Psi^B_{\alpha,\underline{k}}(\underline{r},\underline{c}) \sim e^{i\underline{k}.\underline{r}}\ \Psi_{\alpha,\underline{k}}(\underline{c}) + \hat{\Psi}_{\alpha,\underline{k}}(\underline{r},\underline{c})\ , \quad (19)$$

where $\hat{\Psi}$ is the non-factorable part due to B. If B is small then $\hat{\Psi}$ is a small correction localised near the region of B. It will in general extend beyond this region and may have oscillations. These functions determine how far the boundary effects extend. When B is large the $\hat{\Psi}$ component in (19) will dominate, and even more importantly, changes may occur in the nature of the spectrum.

For the initial value problem with $S = \delta(t) f(\underline{r},\underline{c})$

$$f = \sum_\alpha \int d\underline{k}\ \Psi^B_{\alpha,\underline{k}}(\underline{r},\underline{c})\ e^{t\omega^B(\alpha,\underline{k})}\ f_{\alpha,\underline{k}}\ , \quad (20a)$$

$$f_{\alpha,\underline{k}} = \int f(\underline{r},\underline{c})\ \Phi^B_{\alpha,\underline{k}}(\underline{r},\underline{c})\ d\underline{r}\ d\underline{c}\ . \quad (20b)$$

Whether the long-time limits may be described by the lowest eigenvalues and the limitations on the times and gradients will now be determined by the new quantities $\omega^B(\alpha,\underline{k})$.

If it is still possible to obtain a differential equation for the number density from $\omega^B(0,\underline{k})$ the coefficients occurring in it will depend on the operator B, and will therefore not have the simple dependence on ratio E/n_0 familiar from the free space theory.

For a steady source of the form $S = S(\underline{r},\underline{c})$

$$f = \sum_\alpha \int d\underline{k}\ \Psi^B_{\alpha,\underline{k}}(\underline{r},\underline{c})\ \frac{e^{t\omega^B(\alpha,\underline{k})} - 1}{\omega^B(\alpha,\underline{k})}\ S_{\alpha,\underline{k}}\ , \quad (21a)$$

$$S_{\alpha,\underline{k}} = \int \Phi^{B}_{\alpha,\underline{k}}(\underline{r},\underline{c})\ S(\underline{r},\underline{c})\ d\underline{r}\,d\underline{c} \quad . \tag{21b}$$

To model the steady-state behaviour of the corresponding number density

$$n(\underline{r}) = \lim_{t\to\infty} \int f(\underline{r},\underline{c},t)\ d\underline{c} \quad , \tag{22}$$

according to hydrodynamical ideas one would like to use an equation of the form

$$(\omega(\nabla) + \overline{B})\ n(\underline{r}) = \overline{S}(\underline{r}) \quad . \tag{23}$$

Even if a single eigenvalue in (21a) should give a good approximation for $n(\underline{r})$ it is not evident how to obtain $\omega(\nabla)$ from it. The effects of B are distributed over all three terms in (23). One will have to reconsider how to interpret such a model on the basis of kinetic theory. That is to say the transport coefficients obtained in the boundary-free case cannot be directly used in such a model of steady state $n(\underline{r})$, at least not without some consideration of $\overline{B}$ and $\overline{S}$ terms also.

It may be more profitable therefore to deal directly with the solution (21) and seek simplifications of some other kind.

Summary and Conclusion

The problem of transmission of a swarm through a localised field inhomogeneity is of importance in many swarm experiments. It is also connected to boundary effects. In both cases it is necessary to look at the solutions provided by kinetic theory in order to assess the models that may be made according to the ideas of hydrodynamics. These problems may be approached by expressing the solutions in terms of Green's functions and the associated eigenvalue problems. This approach was illustrated by making some assumptions about the eigenvalues and eigenfunction.

References

[1] Lowke, J.J.; Parker, J.H., Hall, C.A. Phys.Rev. 1977, A15, 1237;
Robson, R.E. Aust. J. Phys. 1981, 34, 223;
Braglia, G.L. Phys. Rev. 1982, A25, 1214;
Chantry, P.J. Phys. Rev. 1982, A25, 1209.

[2] Parker, J.H. Phys. Rev. 1965, A139, 1792;
Leemon, H.I.; Kumar, K. Aust. J. Phys. 1975, 28, 25;
Robson, R.E. Phys. Rev. 1976, A13, 1536.

[3] Skullerud, H.R. Aust. J. Phys. 1983, 36, 845;
Crompton, R.W.; Hegerberg, R. Aust. J. Phys. 1983, 36, 831.

[4] Mason, E.A.; Waldman, M. In "Electron and Ion Swarms"; Christophorou, L.G., Ed.; Pergamon Press: New York, 1981; p147.;
Morruzi, J.L.; Kondo, Y. Jap. J. App. Phys. 1980, 19, 1411.

[5] Skullerud, H.R.; Holmstrom, S. "Retarding potential difference analysis of ion swarm velocity distributions: A computer simulation study", in these proceedings.

[6] Tagashira, H. "Principle of measurement of electron drift velocities", in these proceedings.

PART III

ABSTRACTS OF CONTRIBUTED SWARM PAPERS

PROGRAM OF THE JILA ATOMIC COLLISIONS CROSS-SECTION DATA CENTER

Jean W. Gallagher

Joint Institute for Laboratory Astrophysics
Boulder, CO

Since publication in 1983 of an evaluated compilation of electron swarm data in electromagnetic gases, the JILA Data Center has completed two reviews on data describing heavy particle collisions; i.e., one on proton impact ionization in gaseous targets and another on charge transfer of hydrogen ions and atoms with metal vapors. The Data Center is currently involved in several projects concerning electron impact excitation. To complement this work,a data base for both bibliographic and numerical data has been designed, and entry of cross-sections, collision strengths, reaction rates, and swarm parameters into the data base has begun. Planned projects include an evaluated compilation of data describing electron collisions with small molecules.

COLLISIONALLY INDUCED DISSOCIATION OF IONS IN A STRONG DRIFT FIELD AND ITS USE AS AN ANALYSIS TOOL

Fred L. Eisele

Georgia Institute of Technology

Drift fields of 100-1000 Td. were used to collisionally dissociate ions (which were directly sampled from the lower atmosphere) to aid in their chemical identification. The ions being studied make up only one part in 10^{17} of the gas in which they are found, and their mass spectrometric analysis is accomplished near the high sensitivity limit of the present apparatus. Since the chemical identity of the sampled ions can often not be determined from their mass alone, additional identification methods are needed. They must not, however, cause ionization in the far more abundant neutral gas or reduce the number of ions being detected. Under the appropriate conditions collisional dissociation of the ions being studied can accomplish both of the above goals, and also make available a continuously variable range of collision energies. At low field strengths ligand detachment from the parent ion can be observed, while fragmentation of the more strongly bound core ion is accomplished at fields near 1000 Td. An example of ligand removal is the gradual conversion of an atmospheric ion family believed to be $NH_4^+ \cdot (H_2O)_n$ (mass 36, 54, 73...) into a single mass peak at mass 18 at 300 Td. in a nitrogen buffer gas. At 800 Td. an ion believed to be NO_3^- is seen to be largely dissociated into a fragment at mass 46 (probably NO_2^-) thus helping to confirm the identification of the parent ion as NO_3^-.

Acknowledgments

Supported by NSF Grant No. ATM-8210902 and ATM-8414198.

RETARDING POTENTIAL DIFFERENCE ANALYSIS OF ION SWARM VELOCITY DISTRIBUTIONS: A COMPUTER SIMULATION STUDY

H.R. Skullerud and S. Holmstrom

Department of Applied Physics
Norwegian Institute of Technology
N7034 Trondheim, Norway

Several RPD measurements of ion swarm velocity distributions have been reported the last years, but there is a nearly total lack of related theoretical works. To improve our understanding of these measurements, we have therefore performed a computer simulation study of Li^+ ions moving in and sampled from a He gas. Some typical results are shown below.

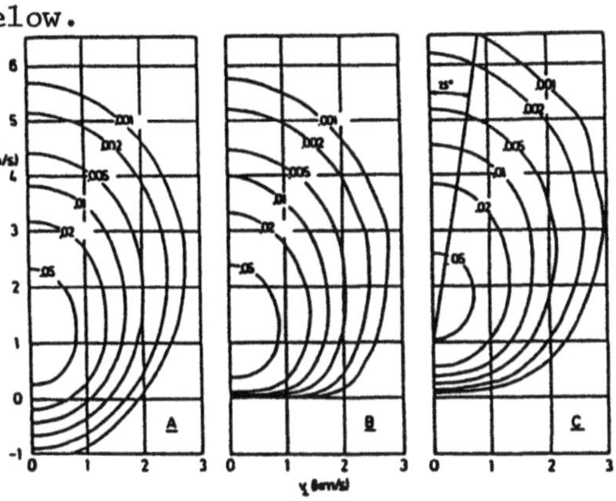

Fig. 1. Free-space distribution (A), distribution close to a cathode (B), and distribution of ions hitting the cathode (C) (25 Td, 300K).

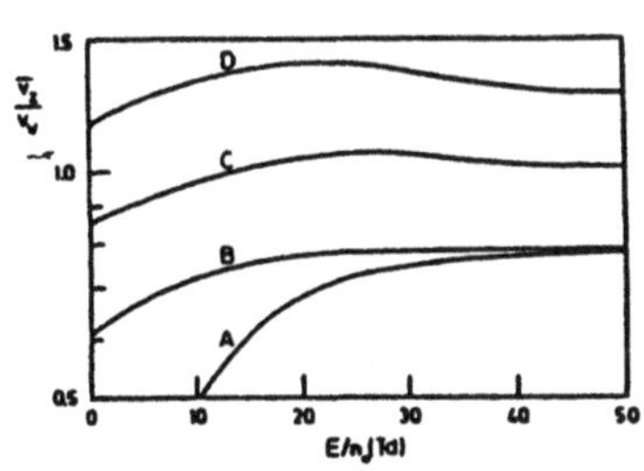

Fig. 2. Corresponding mean velocities (A,B,C) and mean velocity for ions sampled through a 7.5° cone (C), relative to a velocity $v_w = (v_{dr}^2 + 3kT/m)^{1/2}$.

The mean velocity and the velocity distribution of sampled ions are clearly very different from the corresponding free-space quantities.

TRANSVERSE DIFFUSION OF LITHIUM IONS IN HELIUM

H.R. Skullerud, T. Eide and Thorarinn Stefansson

Department of Applied Physics
Norwegian Institute of Technology
N7034 Trondheim, Norway

The ratio D_T/μ between transverse diffusion coefficient and mobility has been measured for Li^+ ions in He, at 295 K and with $5 < E/n_o < 200$ Td.

The basis for the experiment is the asymptotically valid relation $\langle x^2 \rangle \cong 2\,(D_T/\mu)\, z/E + x_o^2$. Care was taken to ensure proper correction for end-effects and for non-Gaussian density profiles.

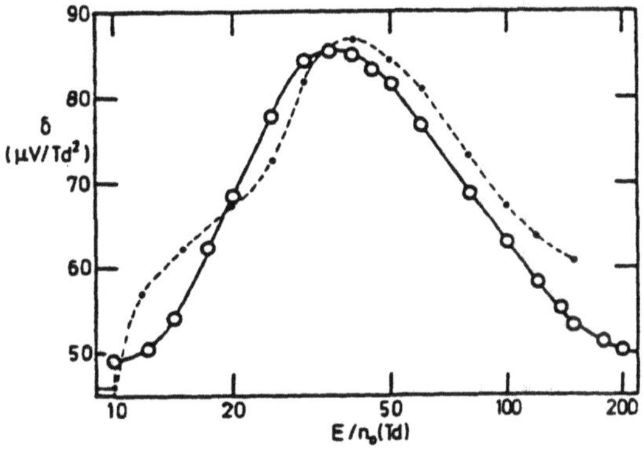

The figure shows the quantity $\delta = (D_T/\mu - kT/e)\,(E/n_o)^2$ as a function of E/n_o. The full line indicates experimental values, while the dotted line represents theoretical values calculated by Viehland (1983)[1] from a mobility-derived potential.

The discrepancy between the curves above the maximum suggests that either our D_T/μ - data or the mobility data used by Viehland are seriously in error at high E/n_o - values.

References

[1] Viehland, L.A. (1983) Chem. Phys. 78 279-94

ELECTRON ATTACHMENT TO F_2

D.L. McCorkle and L.G. Christophorou

Department of Physics and Astronomy
The University of Tennessee
Knoxville, Tennessee 37996

There has been considerable recent experimental and theoretical interest in the electron attachment properties of F_2 principally because of its use in rare gas-halide lasers. We undertook a study of electron attachment to F_2 in order (i) to provide accurate data on the magnitude and energy dependence of the electron attachment rate constant k_a ($<\varepsilon>$) as a function of the mean electron energy $<\varepsilon>$, (ii) to resolve the existing differences [1] between the swarm and between the swarm and beam data below ≅ 0.2 eV, (iii) to investigate the effect of temperature T on k_a ($<\varepsilon>$) and (iv) to relate the experimental results to the results of recent theoretical calculations on the negative ion states of F_2. We measured the k_a ($<\varepsilon>$) for F_2 in mixtures with N_2 at T ≅ 233, 263, 298 and 373K and in the energy range ≅ 0.04 to 0.8 eV. The $<\varepsilon>$ dependence of k_a is consistent with that we calculated from the beam cross-section [2] using our electron energy distributions for N_2 and confirms the existence of a large attachment process at thermal energies. The $<\varepsilon>$ dependence of k_a we measured is also consistent with other swarm data [1] for $<\varepsilon>$ >0.2 eV but it indicates that the published swarm data [3] below 0.2 eV are in error. We have experienced enormous difficulty in determining accurately the magnitude of k_a principally because the measured F_2 pressure was a function of the passivation, but varied from one passivation technique to another by, often, more than a factor of 2 (e.g. at 0.1 eV the k_a values range between 0.8 and 1.9 x $10^{-8}cm^3s^{-1}$). The measured k_a ($<\varepsilon>$) increase only slightly with T which is consistent with the dissociative attachment to F_2 being exoergic and the potential energy curve of $F_2^-(^2\Sigma_u^+)$ crossing that of $F_2(^1\Sigma_g^+)$ close to the equilibrium separation. The theory failed to predict the "zero" energy process, whose existence is indicated by both the beam [2] and the present swarm data.

Acknowledgments

Research sponsored by the Department of Energy under contract DE-AS05-76EV04703.

References

[1] Christophorou, L.G., McCorkle, D.L.;Christodoulides, A.A. in Electron-Molecule Interactions and Their Applications, edited by Christophorou, L.G. (Academic Press, New York, 1984), Vol. 1, Chap. 6, pp. 477-617.

[2] Chantry, P.J. in Applied Atomic Collision Physics, edited by McDaniel, E.W. and Nigham, W.L. (Academic Press, New York, 1982), Vol. 3, Chap. 2, pp. 35-70.

[3] Sides, G.D., Tiernan, T.O.; Hanrahan, R.J. Chem. Phys. 65, 1966 (1976).

THERMAL ELECTRON ATTACHMENT TO VAN DER WAALS MOLECULES ($O_2 \cdot N_2$)

Minoru Toriumi and Yoshihiko Hatano

Department of Chemistry
Tokyo Institute of Technology
Meguro-ku, Tokyo 152, Japan

This is an extension of our work reported in the previous swarm meeting. [1] Electron attachment to van der Waals (vdw) molecules ($O_2 \cdot N_2$) has been investigated at temperatures from 77 K to 373 K and N_2 densities from 2.0×10^{18} molecules/cm^3 to 3.5×10^{19} molecules/cm^3 using a microwave conductivity technique combined with pulse radiolysis. [2] The initial electron attachment rate constants to O_2 and ($O_2 \cdot N_2$) at 300 K are obtained as 3×10^{-11} cm^3/s and 3×10^{-8} cm^3/s, respectively. The resonance energy for the electron attachment to ($O_2 \cdot N_2$) is reduced to 20 ± 1 meV, while the resonance width is broadened to 800 ± 100 μeV. The reason why the electron attachment rate constant for ($O_2 \cdot N_2$) becomes about 1×10^3 times larger than that for O_2 is (1) the increase of electron density at the resonance energy for attachment, (2) the increase of attachment cross-section with decreasing the resonance energy, and (3) that by vdw interaction partial wave electrons with lower angular momentum can attach to vdw molecules.

References

[1] Toriumi, M., Suzuki, E.; Hatano, Y. Proc. 3rd Int. Swarm Seminar, Aug. 3-5, 1983, Innsbruck, Austria (1983), p. 24.

[2] Toriumi, M.;Hatano, Y. J. Chem. Phys., 82, 254 (1985).

TIME-RESOLVED MEASUREMENTS OF THERMALIZATION PROCESSES OF SUBEXCITATION ELECTRONS IN RARE GASES

Etsuhito Suzuki, Yoshinori Hirako and Yoshihiko Hatano

Department of Chemistry
Tokyo Institute of Technology
Meguro-ku, Tokyo 152, Japan

Thermalization processes of subexcitation electrons in Ar, Kr and Xe have been measured using a microwave conductivity technique combined with pulse radiolysis, which is almost the same as that used for electron attachment experiments.[1]

A microwave-cavity resonance-frequency shift has been measured as a function of time after nsec pulsed X-ray irradiation upon rare gases, from which the time-dependent changes of the electron temperature, T_e, have been obtained assuming the Maxwellian electron-energy distribution. The excess electron temperature, T_e-T_g, where T_g is the gas temperature, decays exponentially in the region of 10^3 K>T_e>T_g from which the decay constant, $\tau_{th}\cdot P$, where P is the gas pressure, are almost constant at pressures of 300 Torr>P>30 Torr, and have been compared with recent theories. [2]

References

[1] Hatano, Y.; Shimamori, H. Electron and Ion Swarms, ed. by Christophorou, L.G., Pergamon (1981), p. 103.

[2] Mozumder, A., J. Chem. Phys. 72, 1657 (1980), ibid. 72, 6289 (1980);(b) Shizgal, G. J. Phys. Chem. 88, 4854 (1984); (c) Koura, K., J. Chem. Phys. 81, 4180 (1984); Koura, K., J. Chem. Phys. (to be published).

ELECTRON SWARM STUDY IN NITROGEN-RARE-GAS MIXTURES AND VIBRATIONAL EXCITATION IN NITROGEN

Yoshiharu Nakamura

Department of Electrical Engineering
Faculty of Science and Technology, Keio University
3-14-1 Hiyosi-cho, Yokohama 223, Japan

The drift velocities and the longitudinal diffusion coefficients of electrons were measured in nitrogen-helium, nitrogen-neon, and nitrogen-argon mixtures over the range of E/N from 0.7 to 70 Td by using a pulse-operated double shutter drift tube. The objectives of these measurements were (1) to discuss the effects of the molecular additives on electron swarm parameters in rare gases, and (2) to investigate the new possibilities of the electron swarm study in gas mixtures.

The stress was especially placed on the second objective.

The results of the present study were summarized as follows:

(1) A multiplying factor 1.5 ± 0.1 to the existing vibrational cross sections of nitrogen molecule (Schultz et al 1964 and 1973) was found to give the most reasonable consistency among the measured and calculated drift velocities in these nitrogen-rare-gas mixtures. The existing momentum transfer cross-section of nitrogen molecule (Hayashi 1979) was slightly modified accordingly.

(2) The present study showed that the uniqueness of the collision cross-section of a gas can be tested more effectively in the mixtures with other gas than in the gas by itself. The electron drift velocity in the mixture will be the most convenient swarm parameter to test the uniqueness, and argon gas is a very efficient buffer gas to amplify the effect of low-lying inelastic cross-sections of an atom or a molecule on the electron swarm parameters.

EXCITED-STATE QUENCHING PHENOMENA IN WEAKLY IONIZED NITROGEN

M.T. Elford,* A. Ernest† and S.C. Haydon†

*Research School of Physical Sciences
A.N.U., Canberra

†Department of Physics
U.N.E., Armidale, N.S.W.

The reduction in impurity level by the application of prolonged outgassing procedures causes some remarkable changes in the ionization behaviour of molecular nitrogen. Intensive studies of this behaviour have revealed the overriding importance of neutral excited metastable particles created by direct electron impact excitation or cascade processes. Whilst the dominant influence of the $N_2(A^3\Sigma_u^+)$ metastable particle has already been well established, evidence has accumulated that a second metastable particle also contributes significantly when the purity of the gas samples is sufficiently high.

This paper is concerned with the nature of the quenching processes responsible for the very high sensitivity of the pre-breakdown ionization currents to trace amounts of impurities in the gas samples. Whilst the energies of the metastable particles responsible for the phenomena are not large enough to create Penning ionization of impurities in the gaseous volume, they are sufficiently energetic to create secondary electrons at boundary surfaces, the actual production depending on the state of the surface. Careful preparation of stable surface boundary conditions has therefore been essential before attempting to determine the precise influence of trace impurities. The procedures developed for this purpose are described, and the results of doping the N_2 gas samples at parts per million levels are given. The dramatic effects of very low levels of CO are demonstrated, in contrast to the relatively small influence of corresponding levels of H_2 gas.

ELECTRON DRIFT VELOCITY AND ATTACHMENT AND IONIZATION COEFFICIENTS IN CH_4, CF_4, C_2F_6, C_3F_8 AND n-C_4F_{10}

S.R. Hunter, J.G. Carter and L.G. Christophorou

Atomic, Molecular and High Voltage Physics Group
Health and Safety Research Division
Oak Ridge National Laboratory
Oak Ridge, Tennessee 37831

The perfluoroalkane series of molecules n-C_NF_{2N+2} (N = 1 to 4) have received increasing attention in recent ears, due to a large extent to their very desirable electron ransport and rate coefficient characteristics for use in a ariety of applications including high voltage gaseous dielectrics, lasma etching, radiation counters and diffuse discharge witches.[1] Despite the diverse nature of the applications o which these molecules have been used, very few accurate ransport and rate coefficient measurements have been made in hese molecular gases, particularly at low E/N values.

In this paper, we present measurements of the attachment η_T/N) and ionization (α_T/N) coefficients for the perfluoroalkanes ver the E/N range $2 \times 10^{-16} \leq E/N \leq 10^{-15}$ V cm^2 using modified Pulsed Townsend (PT) technique. The electron drift elocity w has also been measured for these molecules and CH_4 ver the E/N range $1 \times 10^{-19} \leq E/N \leq x\ 10^{-15}$ V cm^2 using the technique. The present measurements of w and η_T/N are e first to be obtained for C_2F_6, C_3F_8 and n-C_4F_{10} at low N values. We have observed that η_T/N in C_3F_8 and n-C_4F_{10} are ependent on gas pressure over a wide E/N range and have also bserved that w is pressure dependent in C_3F_8 and n-C_4F_{10} at igh E/N, even after allowing for diffusion corrections to he electron swarm transit time. This is the first observation f a pressure dependence in w in C_3F_8 and n-C_4F_{10} and is hought to be due to changes in the electron energy distribution unction with gas pressure due to the increase in η_T/N with ressure as was outlined in a recent paper on the influence f electron attachment on the transport coefficients of an lectronegative gas.[2]

Acknowledgments

esearch sponsored in part by the Office of Health and nvironmental Research, U.S. Department of Energy, under ontract DE-AC05-84OR21400 and in part by the Office of Naval esearch under contract 42 01 24 60 2 with Martin Marietta nergy Systems, Inc.

References

[1] Christophorou, L.G. and Hunter, S.R. in Electron-Molecule Interactions and Their Applications, Vol. 2, Chapt. 5, Academic Press, New York, 1984.

[2] Blevin, H.A., Fletcher, J. and Hunter, S.R. Phys. Rev. A, April 1985.

COMPARISON OF CALCULATED AND EXPERIMENTAL THERMAL ATTACHMENT RATE CONSTANTS FOR SF_6 IN THE TEMPERATURE RANGE 200-600K

O.J. Orient and A. Chutjian

Jet Propulsion Laboratory
California Institute of Technology
Pasadena, CA 91109

Electron attachment cross-sections are calculated for the process $e + SF_6 \rightarrow SF_6^-$ in the energy range 1-200 millielectron volts. A local electron-scattering approximation [1,2] is used in which diatomic-like potential energy curves near the equilibrium SF_6 ground state are constructed from recent spectroscopic data. Good agreement is found with experimental attachment cross-sections [3,4] at a temperature of 300K. The same calculation, with appropriate adjustment of the thermal populations, is used to calculate attachment rate constants in the range 50-600K. An additional cross-section contribution from the channel $e + SF_6 \rightarrow SF_5^- + F$, obtained from experimental data of others, is then added to the SF_6^-/SF_6 channel cross-section and comparisons made with measured rate constants [5-7] in the range 200-600K. An estimate of the photodetachment energy of SF_6^- (X^2A_{1g}) is also given.

Acknowledgments

This work was jointly sponsored by the AF Office of Scientific Research, the Department of Energy, and the NSF, and was carried out at the Jet Propulsion Laboratory, California Institute of Technology, through agreement with NASA.

References

[1] O'Malley, T.F., Phys. Rev. 150, 14 (1966); 155, 59 (1967).

[2] Chutjian, A., J. Phys. Chem. 86, 3518 (1982).

[3] Chutjian, A.; Alajajian, S.H., Phys. Rev. A 31, 2885 (1985).

[4] Zollars, B.G.,Higgs, C., Lu, F., Walter, C.W., Gray, L.G., Smith, K.A., Dunning, F.B.; Stebbings, R.F., "Ionization in Rydberg Atom-SF_6 Collisions at High n," Phys. Rev. A, in press.

[5] Spence, D.; Schulz, G.J., J. Chem. Phys. 58, 1800 (1973).

References Continued

[6] Smith, D., Adams, N.G.; Alge, E., J. Phys. B 17, 461 (1984).

[7] Petrovic, Z. Lj.; Crompton, R.W., J. Phys. B 18, 000 (1985).

MEASUREMENTS OF ELECTRON AND ION TRANSPORT DATA IN SF_6 AND SF_6/N_2 MIXTURES

T. Aschwanden

Swiss Federal Institute of Technology, ETH
High Voltage Engineering Group
ETH-Zentrum, Physikstrasse 3,
CH-8092 Zurich, Switzerland

The drift velocity v_e and the longitudinal diffusion coefficient ND_L for electrons and the mobility of positive (K_p) and negative ions (K_n) have been determined in SF_6 and SF_6-mixtures containing N_2. A time-resolved swarm method (TRT) was used to cover a wide range of E/N: 10 to 900 Td.

In pure SF_6 the observed E/N-dependence of the positive ion mobility with a marked mobility peak around 350 Td can be explained by the transition from the induced dipole interaction to hard sphere collision interaction at high E/N. The reduced negative ion mobility K_{on} is dependent on the gas density N used in the experiment. In the low field region this can be attributed to the formation of unstable cluster ions, whereas in the high field regime (>300 Td) an ion conversion (presumably to F^-) is observed, which increases the apparent mobility with increasing N.

In SF_6/N_2 mixtures Blanc's law is strictly obeyed for the ions in the low field region (at constant total density) and for the electron mobility K_{eo} at higher E/N. The values of v_e and ND_L in the mixture at the limiting field $(E/N)_{lim}$, where the $k_a = k_i$, have a maximum for mixtures containing 20 to 40% SF_6. As a function of the SF_6-content the characteristics of the ratio D_L/K_e at $(E/N)_{lim}$ exhibits a similar synergism as it was observed for the values of $(E/N)_{lim}$ in the mixture.

COMMON PARAMETERIZATIONS OF SWARM AND BREAKDOWN DATA FOR BINARY ELECTRONEGATIVE-NONELECTRONEGATIVE GAS MIXTURES

R.J. Van Brunt and M.C. Siddagangappa

National Bureau of Standards
Gaithersburg, MD 20899

It has been shown [1] that a simultaneous parameterization of swarm and breakdown data appears possible for mixtures of N_2 with various electronegative gases such as SF_6 and CCl_2F_2. In particular it was found that experimental data could be fit with the forms

(1) $\bar{\varepsilon}_m = \bar{\varepsilon}_{N_2} + (\bar{\varepsilon}_A - \bar{\varepsilon}_{N_2})g(F)$

for mean electron energies,

(2) $\eta_m = \eta_A g(F)$

for attachment coefficients, and

(3) $(E/N)_{cm} = (E/N)_{cN_2} + [(E/N)_{cA} - (E/N)_{cN_2}]g(F)$

for critical field-to-gas density ratios ($\alpha = \eta$), where $g(F)$ is a common parameterized function of gas mixture ratio F only which satisfies the conditions $g(F) > 0$ for $1 > F > 0$ and $g(F) = 1$ and 0 for $F = 1$ and 0 respectively. It is shown here that such common parameterizations involving separation of the variables F and E/N are not in general possible. The conditions for which these parameterizations are possible will be considered, and the physical significance of the fitting parameters discussed.

References

[1] Siddagangappa, M.C., Lakshminarasimha, C.S., Naidu, M.A. J. Phys. D, 16, 1595 (1983).

BOLTZMANN EQUATION ANALYSIS OF THE SYNERGISM OF UNIFORM FIELD BREAKDOWN VOLTAGE FOR SF_6-CCL_2F_2 AND SF_6-SO_2 MIXTURES

Makoto Hayashi

Nagoya Institute of Technology
Gokiso-cho, Showa-ku, Nagoya

The electron collision cross-section set for SF_6 is given in my paper. [1] There are many uncertainties about the values of cross-sections for SF_6, as discussed in the paper. Recently, we determined the cross-section sets for CCl_2F_2 and SO_2 by Boltzmann equation analysis, using the available cross-section and also electron swarm data.

Calculations were carried out for the effective ionization coefficients $(\alpha - \eta)/N$ for the mixtures of SF_6-CCl_2F_2 and SF_6-SO_2 as a function of gas concentrations. The results are shown in the figures. There is a minimum for the values of $(\alpha - \eta)/N$.

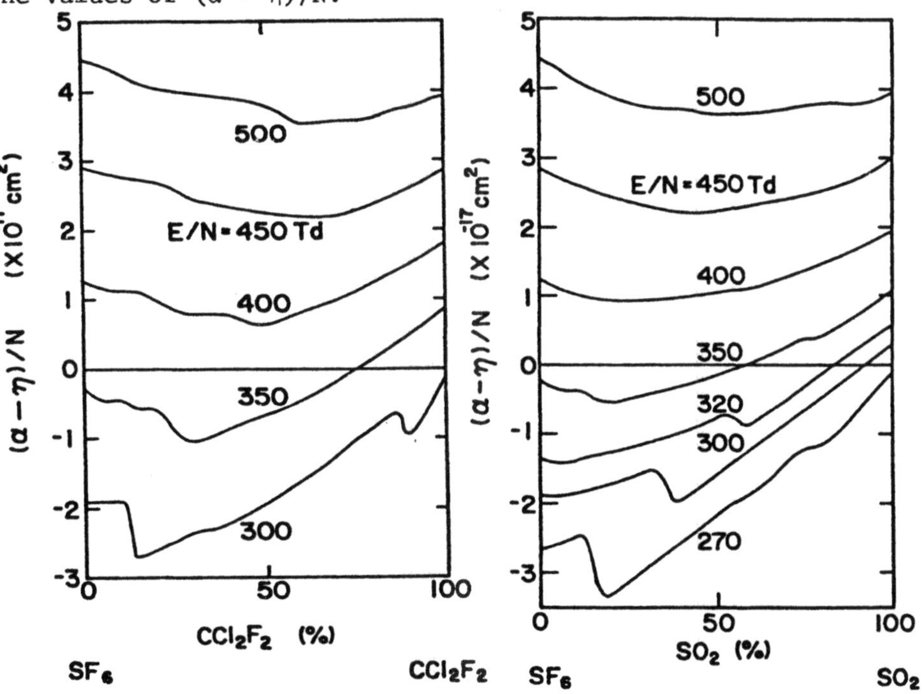

References

[1] Hayashi, M., and Nimura, T. J. Phys. D17, 2215-2223 (1984).

EFFECT OF GAS IMPURITY FOR THE ELECTRON DRIFT VELOCITIES IN INERT GASES BY BOLTZMANN EQUATION ANALYSIS

Makoto Hayashi

Nagoya Institute of Technology
Gokiso-cho, Showa-ku, Nagoya

There is a possibility that the molecules CF_4 have a very large vibrational excitation cross-section.

Around 1955, many scientists discussed the electron drift velocities for argon and many papers were published. It was found that the drift velocities for argon were very sensitive to the concentration of the impurity molecules.

Calculations are made by Boltzmann equation analysis to the effect of impurity molecules in He, Ne and Ar. We found that the impurity concentration of CF_4 of smaller than 1 ppm affect the values of drift velocities in Ar. This is due to the small momentum transfer cross-sections for Ar of around 0.3 eV and large vibrational excitation cross-sections of impurity molecules.

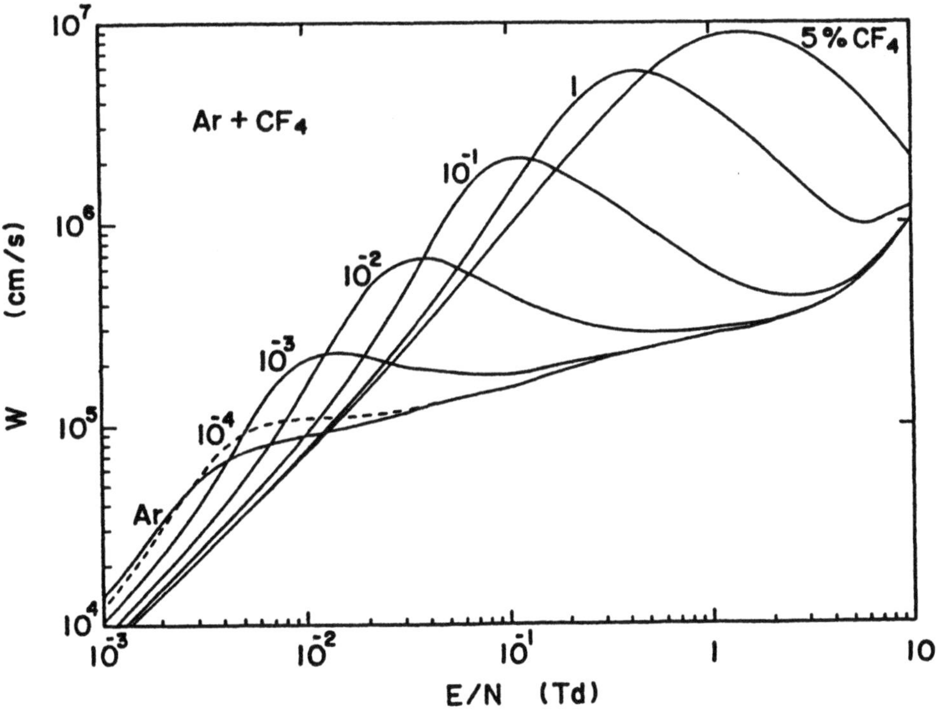

SPATIAL VARIATIONS IN THE ENERGY DISTRIBUTION FUNCTION FOR STEADY STATE TOWNSEND DISCHARGES

H.A. Blevin, J. Fletcher, L.J. Kelly and A.B. Wedding

School of Physical Sciences
Flinders University of South Australia
Bedford Park 5042

Within a steady-state nitrogen discharge the relative excitation rates for two electronic states; the 3371 A band of the second positive system and the 3914 A band of the first negative system have been measured.

The excitation thresholds of these two states, 11.05 eV and 18.75 eV respectively, are significantly different as to expect that their relative intensities may be sensitive to small changes in the mean energy of the electrons.

The ratio of these excitation rates shows a spatial dependence which can be explained in terms of an electron concentration gradient expansion of the energy distribution function.

EXCITATION AND IONIZATION RATES IN HYDROGEN AT VERY HIGH E/n

G.N. Hays, L.C. Pitchford*, J.B. Gerardo,
J.T. Verdeyen**, and Y.M. Li*

Sandia National Laboratories
Albuquerque, New Mexico

We will report time-dependent electron density and excited state density measurements at very high E/n with a temporal resolution of 1 nsec. We create electron avalanching in gases of interest using a fast-rising (3-6 nsec) microwave (S-band) pulse, while simultaneously monitoring the electron-density with a low-power, S-band probe signal. This latter signal is in an interferometer configuration having sufficient sensitivity to monitor densities as low as $\cong 10^6$ cm^{-3}. The avalanche ionization takes place in a quartz cell contained in the S-band waveguide and having no electrodes or externally applied fields, except for a longitudinally applied magnetic field tuned to electron-cyclotron resonance at the high-power pulse frequency. The equivalent effective field under these conditions reduces to a DC field at one-half the peak value.

We have examined the approach to local field equilibrium of the growth of ionization in pure N_2 (37th Gaseous Electronics Conference). These results have been extended to include studies in H_2. We show both experimental and computational results of the time-dependent and local field values of the ionization and certain excitation rate coefficients.

Acknowledgments

This work was performed at Sandia National Laboratories, Albuquerque, New Mexico, supported by the U.S. Department of Energy under contract number DE-AC04-76DP00789.

*GTE Laboratories, Inc., Waltham, Massachusetts 02254.

**University of Illinois, Champaign, Illinois 61820.

EVOLUTION EQUATION AND TRANSPORT COEFFICIENTS OF SWARMS IN SHORT TIME DEVELOPMENT OF INITIAL RELAXATION PROCESSES

Keiichi Kondo

Department of Electrical Engineering,
Anan College of Technology
265, Aoki, Minibayashi, Anan, Tokushima, 774, Japan

An evolution equation and transport coefficient expressing the short time development of swarms were derived from the space-time dependent Boltzmann equation by introducing the "projection operator" which acts on the velocity distribution function. The evolution equation of the density $n(\vec{r},t)$ injected at the time t=0 with the initial distribution $f(\vec{r},\vec{v},t=0) = f0(\vec{v})*n(\vec{r},t=0)$ can be generally written as follows: $\partial_t n(r,t) =$

$$\sum_{k=1} \int_0^t d\tau\, \Omega^{(k)}(t-\tau)\odot(-\nabla_r)^k n(r,\tau) + \sum_{k=1} \omega_0^{(k)}(t)\odot(-\nabla_r)^k n(r,t=0)$$

The second part of this equation expresses the memory effects of the initial velocity distribution. If $f0(\vec{v})$ is chosen to be the same as the final steady state $\varphi_0(v)$, $\omega_0^{(k)}(t)$ becomes zero and no effects due to the memory of initial distribution exist at all. Time-dependent transport coefficients $\omega^{(k)}(t)$ of swarms starting with delta function profile $f(\vec{r},\vec{v},t=0) = f0(\vec{v})*\delta(\vec{r})$ can be obtained as follows:

$$\omega^{(k)}(t) = \int_0^t \Omega^{(k)}(\tau)\,d\tau + \omega_0^{(k)}(t)$$

Assuming the BGK model of collision operator, time-dependent drift velocity W(t) and diffusion coefficients D(t) can be easily obtained as follows:

$$W(t) = (\langle v\rangle - \langle v_0\rangle)e^{-\nu t} + \langle v\rangle$$

$$D(t) =$$

$$(\langle v^2\rangle - \langle v\rangle^2)\int_0^t e^{-\nu\tau}d\tau + \left\{\langle v_0^2\rangle - \langle v^2\rangle + \langle v\rangle(\langle v\rangle - \langle v_0\rangle)\right\} t e^{-\nu t}$$

where ν is the collision frequency of particles, $\langle v \rangle = \int v \varphi_0(v) dv$. After sufficient time has elapsed, equation (1) gives the ordinary continuity equation with constant coefficients. In the duration approaching to their hydrodynamic regime, it should be noted that equation (1) can be transformed to the form of the continuity equation of time-dependent coefficients.

TRANSIENT HOT ELECTRON MOBILITY IN ETHENE AND CYCLOPROPANE

Bernie Shizgal

Department of Chemistry
University of British Columbia
Vancouver, British Columbia V6T 1 Y6

The transient behaviour of an initial nonequilibrium distribution of electrons in either ethene or cyclopropane is studied with solutions of the Boltzmann equation. Recently, Gee and Freeman [1] extracted momentum transfer cross-sections for these moderators from experimental measurements of the temperature dependence of the electron mobility. Since inelastic processes probably contribute in the temperature range of the experiments, the cross-sections determined in this way are effective momentum transfer cross-sections. Their results indicate pronounced Ramsauer-Townsend minima at 0.08 eV for ethene and 0.07 eV for cyclopropane.

The present paper examines the transient electron mobility for these moderators with the cross-sections reported by Gee and Freeman [1]. The method of solution of the Boltzmann equation employed in the present work follows closely that used in a previous paper [2]. The time dependence of the electron mobility is extremely sensitive to the details of the momentum transfer cross-section. In particular, a negative transient mobility is observed owing to the Ramsauer-Townsend minima in the cross-sections. The present calculations demonstrate, as was shown previously [2], that the transient mobility is more sensitive to the details in the cross-sections than is the temperature dependence of the thermal mobility. The results suggest that transient mobility measurements for these moderators, such as those reported recently by Warman et al [3] for Xenon, might yield definitive information with regard to the Ramsauer-Townsend minima in the e-ethene and e-cyclopropane (effective) momentum transfer cross-sections.

References

[1] Gee, N. and Freeman, G.R., J. Chem. Phys. 81, (1984).

[2] McMahon, D.R.A. and Shizgal, B., Phys. Rev. A 31, 1894 (1985).

[3] Warman, J.M., Sowada, U.; De Haas, M.P., Phys. Rev. A 31, 1974 (1985).

NONEQUILIBRIUM BEHAVIOR OF ELECTRONS IN STRONG ELECTRIC FIELDS

B.M. Jelenkovic* and A.V. Phelps†

Joint Institute for Laboratory Astrophysics
University of Colorado and National Bureau of Standards
Boulder, Colorado 80309

Experimental tests are made of theoretical predications of the behavior of electrons under highly nonequilibrium conditions. We have chosen to monitor the spatial dependence of the electrons by observing the radiation emitted by molecules and atoms excited by the electrons. Excited states having excitation cross-sections peaking at different energies are used to bring out the behavior of different parts of the electron energy distribution. Planar electrodes and the use of low currents (<1 μA) ensure a uniform electric field. Close fitting electrodes inside a quartz cylinder enable one to obtain a radially uniform, high voltage discharge at low pressures (left-hand side of the Pashen minimum). We present data obtained in D_2 at E/N from 400 to 10,000 Td and applied voltages up to several keV. Results have been obtained for molecular singlet and triplet transitions and for atomic Balmer lines. Data taken at discharge currents of 1 μA and less are independent of current, since space charge effects should be small. We make comparisons between our experimental data, the fluid model of Muller,[1] the numerical solutions of Pitchford et al.[2] and our simple electron beam model.

Acknowledgements

Supported in part by Lawrence Livermore National Laboratories.

References

[1] Muller, K.G., Z. Physik 169, 432 (1962).
[2] Pitchford, L.C., Segur, P. and Yousfi, M. (private communication, 1985).

---

*On leave from Institute of Physics, Belgrade, Yugoslavia.
†Staff Member, Quantum Physics Division, National Bureau of Standards.

STEADY-STATE, SPATIALLY DEPENDENT ELECTRON TRANSPORT AND RATE COEFFICIENTS

L.C. Pitchford*, T.J. Moratz**, P. Segur† and M. Yousfi†

*GTE Laboratories, Inc.
Waltham, Massachusetts

**University of Pittsburgh
Pittsburgh, Pennsylvania

†Université Paul Sabatier
Toulouse, France

Space dependent Boltzmann calculations have been performed to study the relaxation of the electron transport and rate coefficients in the presence of boundaries and for very high values of E/n. The parameters chosen for these calculations are such that comparisons can be made with the recent experimental results of Jelenkovic and Phelps.[1]

The electron velocity distribution function is calculated using an iterative method to solve the integral form of the Boltzmann equation. This method does not rely on the commonly used Legendre expansion of the angular dependence of the velocity, and can easily incorporate the conditions at the physical boundaries.[2]

The simple model of electron multiplication in a gap under very high field conditions proposed by Muller [3] yields qualitative agreement with the calculated multiplication. Comparisons are shown with the experimental "apparent" rate coefficients from Jelenkovic and Phelps over a range of E/n up to 10000 Td. There are some differences between these calculations and the experimental data. In an attempt to understand the sources of the discrepancies, calculations are shown for different reflection properties of the cathode and anode and for slight variations in the cross-sections. Another possible source of the differences which cannot be tested with the present numerical scheme is the assumption of isotropic electron scattering.

References

[1] Jelenkovic, B. and Phelps, A.V. this meeting.
[2] Segur, P. et al, J. Comp. Phys. 50,116 (1983).
[3] Muller, K.G., Z. Physik 169, 432 (1962).

TRANSPORT COEFFICIENTS OF ELECTRON SWARMS IN GASES IN THE PRESENCE OF NONCONSERVATIVE COLLISIONS

K.F. Ness* and R.E. Robson

Physics Department
James Cook University
Townsville, Australia

*Chemistry Department
University British Columbia
Vancouver, Canada

The moment method of solution of Boltzmann's equation for electron swarms, as developed originally by Lin, Robson and Mason, has been extended to include various nonconservative processes (attachment, ionization), and the validity of the two-term approximation for calculation of transport and rate coefficients has been investigated. A second-order density gradient expansion is required. Improvements to the original LRM computer code have been made, enabling more accurate calculations for cases where cross-sections vary rapidly with energy. Velocity distribution functions are also found from the moment method, showing good agreement with finite element methods. Transport coefficients, as identified by various coefficients of ∇n in the equation of continuity, are argued to be universal constants, independent of the particular experimental arrangement used to extract them.

CONVENIENT EXPRESSIONS OF DIFFUSION COEFFICIENTS FOR FREE ELECTRONS IN GASES IN PRESENCE OF RAMSAUER EFFECTS

G. Cavalleri

Istituto di Matematica dell'Universita Cattolica di Milano
Sede di Brescia, Italy

G. Mingari Scarpello

CISE S.p.A.
Milano, Italy

The lateral and longitudinal diffusion coefficients for free electrons in weakly ionized gases (or in intrinsic semiconductors) under a steady-state and uniform electric field $\vec{E}$ and with elastic collision frequency ν much larger than the inelastic one are obtained by the correlation function of the electron velocities. The mean free path method is used, since it allows expansions without divergences in the small number $W^2/\langle v^2\rangle$ where W is the electron drift velocity and $\langle v^2\rangle$ the mean square value of the electron speed v. The obtained explicit expressions are particularly convenient when there are Ramsauer effects which can produce up to 40% errors in the results obtained by the usual Boltzmann - Legendre expansion. Moreover, the obtained expressions are immediately generalized so as to give noise spectral densities including the "convective noise" found by Gurevich, which corresponds to the difference between the generalized longitudinal and lateral diffusion coefficients.

THE APPROACH TO EOUILIBRIUM OF ELECTRON SWARMS IN NON-PLANAR GEOMETRIES

M.J. Kushner

Spectra Technology, Inc.
(Formerly Mathematical Sciences Northwest, Inc.)
2755 Northup Way
Bellevue, Washington 98004

The application to the modeling of gas discharges of electron transport coefficients obtained from swarm measurements and characterized by the parameter E/N (electric field/number density) requires one to invoke the local field approximation (LFA); that is, the electron swarm is in equilibrium with the local value of the electric field. For conditions where the electron mean free path, λ, is large compared to the characteristic distance over which the electric field changes, or the electron collision frequency is small compared to the time rate of change of the electric field, equilibrium cannot typically be obtained and the LPA is no longer valid. This condition often occurs in non-planar geometries. In this paper, electron properties (eg. electron distribution function, average electron energy) are examined in non-planar geometries with results from a time and spatially dependent Monte-Carlo plasma simulation code. Secondary electrons are explicitly included in the simulation. The quantity $\delta = \lambda \cdot (\nabla E/E)$ is discussed as a scaling parameter for when the LFA analysis is valid; small values of δ correspond to conditions where the LFA may be used. The form of the differential cross-section (ie. distribution of scattering angles) is found to be an important consideration. Dominantly isotropic scattering increases the residence time of electrons in a region having a particular electric field, thereby effectively reducing δ; dominantly forward scattering effectively increases δ.

MAGNETIC CONTROL OF LOW PRESSURE DIFFUSE DISCHARGES

J.R. Cooper, K. Schoenbach and G. Schaefer

Department of Electrical and Engineering
Texas Tech University
Lubbock, Texas 79409

The application of crossed E and B fields in a low pressure gas discharge shifts the electron energy distribution function f(ε) to lower mean energies. Monte Carlo calculations demonstrate that the dominant effect is a strong reduction in the number of high energy electrons. The low energy part of the distribution can be influenced by the addition of an attacher which peaks at energies below the mean energy of the distribution. These changes in the electron energy distribution affect significantly the electron mobility and the ionization and attachment coefficients. The resulting variations in the discharge characteristics were observed experimentally. They are in reasonable agreement with the results generated by the computer calculations for gas mixtures containing He, SF_6 and N_2.

Acknowledgments

This research was supported by GTE Laboratories and Texas Tech University.

PART IV

RELATIONS BETWEEN SINGLE AND MULTICOLLISION PHENOMENA

RELATIONS BETWEEN ELECTRON-MOLECULE SCATTERING AND SWARM EXPERIMENTS AND ANALYSIS

A. V. Phelps*

Joint Institute for Laboratory Astrophysics,
National Bureau of Standards and University of Colorado,
Boulder, Colorado 80309-0440

The purposes of this paper are to review the basic relations between electron-molecule collision cross-sections and electron transport and reaction coefficients and to summarize the current state of research in the area of the determination of cross-sections from swarm data.

Several thorough reviews of swarm experiments and their interpretation have been published [1-4]. In order to minimize the reference list, we will refer to these reviews rather than to the original publications whenever possible.

Basic Relations

We have chosen to illustrate the connection between cross-sections and swarm coefficients using results obtained for H_2 because of the variety of distinguishable inelastic collision processes [5,6], because of the availability of precision transport data [2], because of the availability [7,8] of experimental and theoretical cross-sections, and because of the current interest as represented by the following paper in this volume and by recent theory [9].

When the swarm coefficients from measurements [2,3] or from calculations of the electron drift velocity w and the ratio of the transverse diffusion coefficient D_T to the mobility $\mu = w/E$ are combined as in Fig. 1, they have been shown [5] to give collision frequencies per molecule which effectively separate elastic momentum transfer and energy exchange collisions. Here E is the electric field. The elastic momentum transfer collision frequency per molecule is

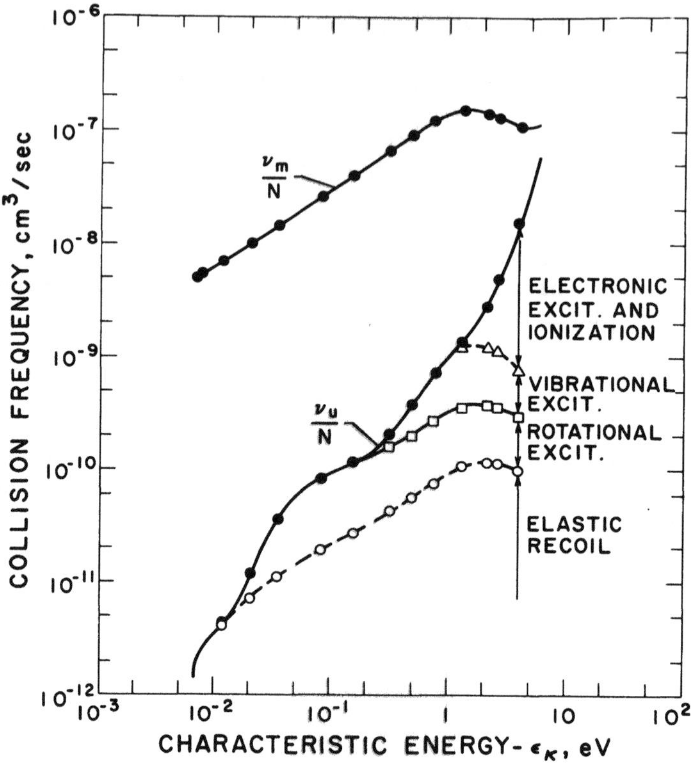

Figure 1. Momentum transfer and energy exchange collision frequencies per molecule for electrons in H_2. The solid curves are averages through experiment and the points are the results of calculations as described in the text. Reproduced with permission [6].

defined by

$$\frac{\nu_m}{n} = \frac{e}{m\mu n} = \frac{e}{mw}\left(\frac{E}{n}\right) , \qquad (1)$$

where n is the gas density. The energy exchange collision frequency per molecule is defined by

$$\frac{\nu_u}{n} = \frac{ew(E/n)}{\dfrac{eD_T}{\mu} - kT} . \qquad (2)$$

The upper solid curve shows the momentum transfer collision frequencies for electrons in H_2 determined from measurements, while the nearby points are calculated as discussed below. The collision frequencies are plotted as a function of the characteristic energy $\varepsilon_K = eD_T/\mu$. This quantity is the only readily determined experimental measure of the energy of the electrons in the swarm. In the special case of a Maxwellian distribution of electron energies the characteristic energy equals the electron temperature in eV. Techniques for experimentally determining the mean electron energy have been used only in very special cases. The lower solid curve shows the energy exchange collision frequency per molecule from experiment. The structure in this curve is indicative of various inelastic processes and will be considered in detail below.

A second type of swarm coefficient data is illustrated in Fig. 2, where measured [3] excitation and ionization coefficients for electrons in H_2 are shown as points. Unfortunately, the experimental data shown for far uv production and

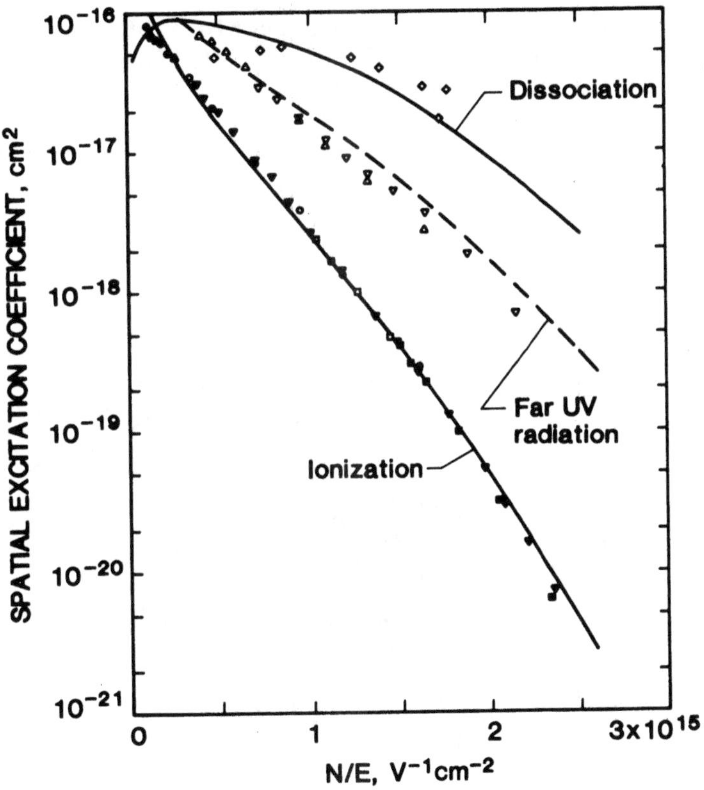

Figure 2. Ionization and excitation coefficients for electrons in H_2. The points are experimental data and the curves are the results of calculations described in Ref. 8.

dissociative excitation processes for H_2 are relatively poorly defined and/or badly scattered.

A recently published [8] set of electron-H_2 collision cross-sections consistent with swarm data [1-5] in H_2 at low and moderate E/n is shown in Fig. 3. In order to simplify the figure we have summed the excitation cross-sections for the triplet states and for the singlet states and have shown only the more important rotational excitation processes for H_2 at 77 K. Also note that the effective momentum transfer cross-section shown is the sum of the elastic momentum transfer cross-section and the total inelastic and ionization cross-sections. As pointed out in Ref. 8, the cross-sections shown in Fig. 3 are in unusually good agreement with theoretical predictions and/or electron beam experiments. An exception is the vibrational excitation cross-section near threshold, as will be discussed in the paper by Morrison and Crompton at this meeting [10].

The connection between the cross-sections of Fig. 3 and the transport and reaction coefficients of Figs. 1 and 2 is through calculations of the measured coefficients using an

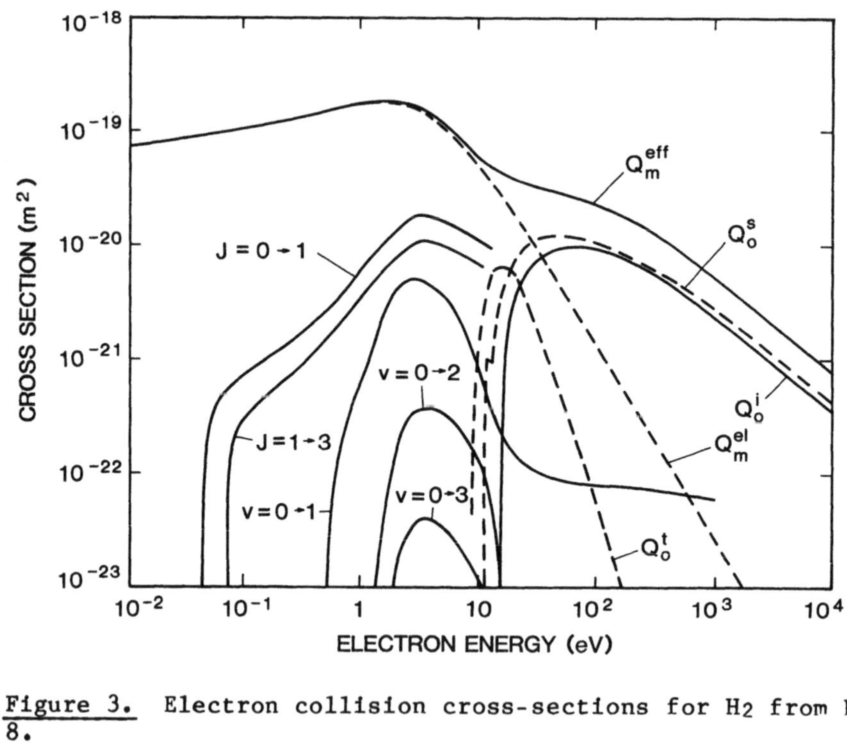

Figure 3. Electron collision cross-sections for H_2 from Ref. 8.

assumed cross-section set and either numerical solutions of the Boltzmann equation or Monte Carlo simulation of the experiment [1,2,4,5,8]. The procedure utilizing the Boltzmann equation is to calculate the electron energy distribution consistent with the assumed cross-sections and applied electric field. Then the individual transport or reaction coefficients are calculated using weighted averages of the cross-sections appropriate to a particular measurement. Note the parallels between this procedure and that for the calculation of electron scattering by a molecule, i.e., the input cross-section corresponds to the molecular potential, the energy distribution to the wave function, and the resultant transport and reaction coefficients to the energy levels and scattering cross-sections. The Boltzmann equation has been discussed in great detail at this meeting and in the literature [1-5]. The results of numerical solutions of the Boltzmann equation and of subsequent calculations of the coefficients are shown by the solid points of Fig. 1 and the solid curves of Fig. 2. These calculations made use of the two-term spherical harmonic expansion technique,which has been shown [8,11] to be satisfactory for electrons in H_2 at the E/n values used.

We now wish to comment on the physical significance of the structure in the energy exchange collision frequency data shown by the lower set of data shown in Fig. 1. We will do this by adding energy loss processes to the Boltzmann calculation, one at a time. Thus, the open circles of Fig. 1 are obtained when using only the elastic momentum collision cross-section as shown at the top of Fig. 3. Comparison of the solid curve from experiment with these points shows that for 77 K and ε_K less than 0.015 eV the electron energy loss is dominated by elastic recoil of the molecule.

When rotational excitation cross-sections are added,the calculations yield the energy exchange collision frequencies shown by the solid circles at $\varepsilon_k < 0.2$ and open squares at higher ε_k. We see that rotational energy losses dominate for ε_K between 0.025 and 0.3 eV, so that swarm experiments are a sensitive measure of rotational excitation cross-sections for energies up to at least the vibrational excitation threshold near 0.5 eV. Crompton and coworkers [2] have obtained accurate rotational excitation cross-sections from precision swarm data obtained with H_2 at 77 K.

Energy losses caused by vibrational excitation are seen to dominate for ε_K between 0.5 and 2 eV. This is the energy region of greatest importance in the following paper [10]. One should keep in mind that in spite of the limited ε_K range over which vibrational excitation dominates, the high accuracy of the experimental data of Crompton and coworkers [2] for H_2 and H_2-rare gas mixtures allows an accurate determination of the vibrational excitation cross-section near threshold.

Finally, the energy losses by electronic excitation and ionization become dominant for ε_K greater than 2 eV. Although not shown in Fig. 1, the momentum transfer collision frequencies calculated using the various incomplete cross-section sets agree with those shown to within the size of the points, and therefore demonstrate the effective separation of the momentum transfer and energy loss collisions for the case of H_2. Furthermore, our previous work [5] has shown that at least for ε_K less than 1 eV the momentum transfer and energy loss frequencies are directly proportional to their respective cross-sections.

An important corollary to the regions of dominance for the various inelastic collision processes is that in each of these regions there are basically two unknowns, the momentum transfer and the inelastic cross-section, and at least two measurements, the electron drift velocity and the characteristic energy. Where there is more than one inelastic cross-section of a given type, e.g. the vibrational excitation cross-sections of Fig. 3 for H_2, one usually uses theory or electron beam results to determine the relative cross-sections. In order to determine a cross-section for energies outside the range of dominance, it is usually necessary to use theory to extrapolate the cross-sections for competing processes [2].

The actual determination of cross-sections from the swarm coefficients is usually accomplished by assuming a set of energy-dependent, input cross-sections based on any available data, solving the Boltzmann equation and transport integrals using this set, comparing the calculated transport and reaction coefficients with experiment, and adjusting the input cross-sections to improve the fit. Although this iterative procedure involves considerable effort on the part of the scientist, automated procedures have been used only for cases in which elastic collisions dominate [2].

An often discussed [1-5] aspect of swarm experiments and analysis is the relatively low energy resolution resulting from the broad electron energy distributions produced by the combined effects of the collisions and the electric field. While this is certainly true, experience has shown that the effective energy resolution is much better than the half width of the energy distribution for the threshold region of an inelastic process, and is rather poor for a cross-section which decreases with increasing electron energy. The energy resolution problem has led to the beam-swarm technique [5] in favorable cases such as electron attachment. Here the energy dependence for electron attachment is determined by electron beam techniques, and the magnitude determined by swarm measurements and analysis using small concentrations of the attaching gas in a well characterized buffer gas such as N_2 or Ar. This procedure takes advantage of the high accuracy with which swarm measurements can be made and the high energy resolution of the beam technique.

In this discussion I have not reviewed the improvements in the determination of cross-sections resulting from the inclusion in the analysis of the effects of (a) electron density gradients introduced in the measurement of transport coefficients; (b) attachment and ionization; (c) the various types of electron drift velocities and diffusion coefficients obtained from different experiments; or d) the use of the multiterm spherical harmonic approximation in solutions of the Boltzmann equation. These effects are discussed in Refs. 2, 4, and 12 and are particularly important for gases with large inelastic excitation cross-sections and at high E/n.

Momentum Transfer Cross-Sections

A second aspect of the connection between swarm coefficients and cross-sections which I have been asked to review is that of the theoretical cross-sections needed for the more accurate swarm calculations now being made. Emphasis is on the distinction between the two definitions of the momentum transfer cross-section for inelastic processes which have been used in the literature. This discussion is an extension of earlier work [11-13] on the effects of anisotropic scattering, and is based on that of Appendix A of Ref. 12.

In order to simplify the discussion we consider in Eq. (1) only the steady-state, spatially independent Boltzmann equation, in which the left-hand side represents the flow in velocity space and the right-hand side represents the effects of instantaneous elastic and inelastic collisions:

$$\vec{a}\cdot\nabla_v F(v,\theta) = -n\sum_k vF(v,\theta)2\pi \int_0^\pi I^k(v,\theta)\sin\theta\, d\theta$$

$$+ n\sum_k v'2\pi \int_0^\pi F(v',\theta)\ I^k(v',\psi)\sin\theta\, d\theta \quad . \tag{3}$$

Here $F(v,\theta)$ is the velocity distribution, $\vec{a}$ is the electron acceleration due to the electric field, v and θ are the magnitude and direction of the velocity relative to the direction of acceleration of the electrons, $I^k(v,\psi)$ is the differential cross-section for the k'th collision process, v' is the velocity before the collision and ψ is the angle through which the electron is scattered. The first term represents electrons scattered out of an element of velocity space, while the second represents electrons entering the distribution after a collision.

Spherical Harmonic Expansion. The first definition of momentum transfer cross-section which we consider is that appropriate to analyses of swarm data using the two-term spherical harmonic or Lorentz approximation for solving the Boltzmann equation [1-5]. In this approximation $F(v,\theta)$ is expanded in spherical harmonics, i.e.,

$$F = \sum_j f_j(v) P_j(\cos\theta) \quad , \tag{4}$$

where $P_j(\cos\theta)$ is the zero-order Legendre polynomial of degree j. When Eq. (4) is substituted into Eq. (3) one obtains a differential equation for the velocity-dependent coefficient of each Legendre polynomial. The right-hand side of the equation for $f_1(v)$, i.e., the collision term, can be written as

$$C[f_1(v)] = -\sum_k \left[v f_1(v) Q_0^k(v) - v' f_1(v') Q_1^k(v')\right] \quad , \tag{5}$$

where

$$Q_1^k(v') = 2\pi \int_0^\pi I^k(v',\phi) P_1(\cos\phi) \sin\phi \, d\phi \quad . \tag{6}$$

For elastic scattering, where v = v', Eq. (6) reduces to

$$C[f_1(v)]\big|_0 = v f_1(v) Q_m^0(v) \quad , \tag{7}$$

where

$$Q_m^0(v) = 2\pi \int_0^\pi I^0(v,\phi)[1-\cos\phi]\sin\phi \, d\phi \quad . \tag{8}$$

Equation (8) defines what is often called the momentum transfer cross-section and is occasionally called the diffusion or transport cross-section. However, in gases such as N_2 or CO near the peak in the vibrational cross-section or methane at energies near the Ramsauer minimum, or for any gas near and above the maximum in the ionization cross-section, the contribution of the inelastic cross-sections to Eq. (5) may be comparable with the elastic momentum transfer cross-section. In this case one cannot define a unique momentum transfer cross-section. Two limits are of interest. In one limit, e.g., when the mean electron energy is much less than the excitation energy, the value of $f_1(v')$ is much smaller than $f_1(v)$ and the contribution of the entering electrons can be neglected. In such cases the effective momentum transfer cross-section $Q_m^e(v)$ is equal to the sum of the elastic momentum transfer cross-section and the "total" or spherically symmetric component of the cross-sections for the inelastic processes [14]. In the limit where the fractional change in v on collision is small, as at very high E/n, the inelastic contribution is calculated using Eq. (8) with $I^0(v,\theta)$ replaced by $I^k(v,\theta)$. In this limit the momentum transfer cross-section for an allowed excitation process decreases rapidly with energy, i.e., like that for elastic scattering. For N_2 the contribution of inelastic collisions [12] to $\Sigma Q_m^k(v)$ relative to that for elastic collisions is small above 100 eV, while for H_2 the inelastic contribution [8] is several times the elastic contribution. These limiting cross-sections are shown in Fig. 4 for Na as calculated from the

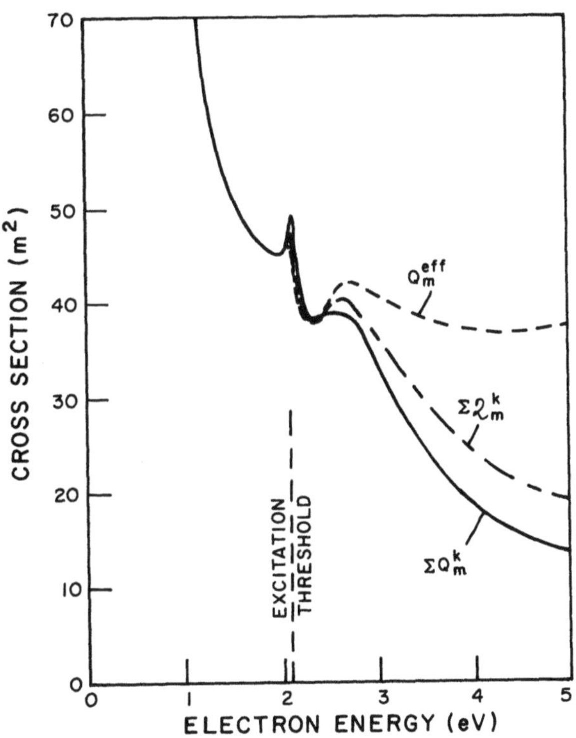

Figure 4. Electron collision cross-sections for Na. Starting from the top the cross-section curves are the upper limit to Q_m^e, $\Sigma \mathcal{Q}_m$, and the lower limit to Q_m^e, as discussed in the text.

theory of Moores and Norcross [15]. Note that in the multi-term spherical harmonic solutions of the Boltzmann equation which consider anisotropic scattering [11], the effects of $Q_1^k(v')$ in Eq. (5) are included properly.

Velocity moment. The second definition of momentum transfer cross-section occurs when the Boltzmann equation is solved using the velocity moment method common to fluid models of electron motion in highly nonequilibrium situations [16], e.g., the cathode fall of an electric discharge or very short pulse discharges. The first moment or momentum equation is obtained by multiplying Eq. (1) by the electron momentum and integrating over all velocities. This procedure yields Eq. (9) for the contribution of the k'th collision process

$$m \int v \cos\theta \, C[F(\vec{v})] d^3v \Big|_k$$

$$= -2\pi m \int_0^{\pi} \int_{v_k}^{\infty} v^3 \mathcal{Q}_m^k(v) F(v,\theta) dv \cos\theta \sin\theta d\theta$$

$$= -m \int_{v_k}^{\infty} v^3 \mathcal{Q}_m^k(v) f_1(v) dv \quad , \tag{9}$$

where

$$\mathcal{Q}_m^k(v) = 2\pi \int_0^{\pi} I^k(v,\phi)\left[1 - \left(\frac{v^2 - v_k^2}{v^2}\right)^{1/2} \cos\phi\right] \sin\phi d\phi \quad . \tag{10}$$

Here m is the electron mass and v_k is the velocity corresponding to the excitation energy. For elastic collisions Eqs. (8) and (10) yield the same result. For inelastic collisions the square root factor is very important near the excitation threshold as is seen in Fig. 4, where we show the $Q_m^e(v,)$ $\Sigma Q_m^k(v)$, and $\Sigma \mathcal{Q}_m^k(v)$ for Na. At high electron energies, where the Born approximation is valid, we find that $\Sigma \mathcal{Q}_m^k(v)$ decreases somewhat less rapidly than $\Sigma Q_m^k(\varepsilon)$, i.e., as the inelastic cross-section divided by the electron energy. In most fluid models Eq. (9) has been approximated by the product of an average of $Q_m^0(v)$ and the mean momentum.

The lesson to be learned from this discussion is that when inelastic collision processes make a significant contribution to the collisional relaxation of the anisotropic portion of the electron energy distribution, an accurate modél requires that the anisotropic portion of the differential cross-section for each important collision process be included. For calculations utilizing the spherical harmonic approach or an equivalent angular representation, this is most readily accomplished by analyzing the differential cross-section into spherical harmonic components as in Refs. 12 and 17. In view of the multitude of inelastic cross-sections required to describe the energy loss processes, it should not surprise one that the use of a single momentum transfer cross-section is in principle unsatisfactory in situations where momentum loss due to inelastic processes is important. Although several investigators [11-13] have considered the effects of anisotropic inelastic scattering on solutions of the Boltzmann equation and on transport coefficients,it is not yet possible to give general rules as to when anisotropic scattering needs to be included. One can say that when the sum of the inelastic cross-sections is small compared to the elastic momentum transfer cross-section,only the total or spherically symmetric component of the inelastic cross-sections is needed.

Status of Swarm Cross-Section Determinations

In this section we will cite some of the successes in the determination of electron collision cross-sections from swarm data and some of the current problem areas. Table 1 lists some of the cross-sections determined from swarm data which have since been verified using theory or electron beam experiments. The object of this discussion is to try to convince some of the skeptics that in favorable cases swarm experiment and analysis are capable of yielding precision cross-section information. Of course the first requirement for success is an accurate and precise determination of the swarm coefficients. Crompton and coworkers [2] have demon-

Table 1. Some Verified Determinations of Cross-Sections from Swarm Experiment

1. Precision values of $Q_m^0(v)$ for He confirmed by theory and beam experiment.
2. Rotational excitation of H_2 in agreement with theory.
3. Vibrational excitation of N_2 near threshold confirmed by beam experiment.
4. Free-free emission from Ar in agreement with theory.
5. Electronic excitation of H_2 in agreement with theory or experiment.

strated that measurements of swarm coefficients can be made to an accuracy of better than 1% in many gases and gas mixtures of interest.

The most successful of the swarm cross-section determinations is that by Crompton et al. [2] of $Q_m^0(v)$ for He at energies below about 4 eV. Theory and, more recently, beam experiment have converged on low energy cross-sections which are in very good agreement with the values determined from swarm experiments. In this case the analysis is aided by the fact that the energy-dependent elastic momentum transfer cross-section is the only unknown.

Our second example is that of the determination [2] of the rotational excitation cross-sections for para-H_2 at 77 K. In this case the unknowns are the energy-dependent elastic momentum transfer cross-section and the energy dependent cross-section for rotational excitation from J = 0 to J = 2. The rotational excitation cross-section which results from the iterative fitting of calculated and measured drift velocities and characteristic energies is in very good agreement with theory [18]. Using the results for the J = 0 to J = 2 transition, the analysis was then successfully extended to normal H_2 at 77 K so as to yield the cross-section for the J = 1 to J = 3 transition.

The very early determination [19] of the vibrational excitation cross-section for N_2 near threshold has recently been confirmed using electron beam techniques [20]. Figure 4 of Ref. 21 shows that the agreement of measured and calculated free-free emission coefficients for electrons in Ar is within the experimental uncertainty of ±15%. A very recent determination [7] of a consistent set of electronic excitation cross-sections for H_2 has resulted in remarkably good agreement with theory and beam experiments, although much of the detailed structure in the cross-sections could not be tested by swarm experiments because of their limited energy resolution.

Table 2. Cross-sections determined from swarm experiment which disagree with other determinations.

1.	Threshold vibrational excitation for H_2.
2.	Vibrational excitation of O_2 and NO via resonances.
3.	Metastable excitation of He.

Table 2 lists cross-sections which have been determined from analysis of swarm measurements and for which there are significant disagreements with cross-sections determined either from theory or from beam experients. The question of vibrational excitation of H_2 near the excitation threshold will be discussed by Morrison and Crompton [10]. The cross-sections for vibrational excitation of O_2 via resonances or temporary negative ion states derived from swarm data [22] are a factor of 2.5 times those derived from electron beam experiments [23]. Although the measurements of characteristic energy for NO are very old and probably less reliable than for O_2, it appears [24] that the resonant vibrational excitation cross-sections from swarm data for NO are significantly smaller than those from beam experiments. Finally, we note the disagreement between the metastable excitation cross-sections derived from multiple scattering experiments and those derived from swarm analyses [25].

Table 3 lists some of the cross-sections determined from analyses of swarm data for which there are little or no beam experiments or theory for comparison, or for which the comparisons are incomplete, e.g., momentum transfer vs. total scattering cross-sections. Particularly noteworthy is the large number of determinations of momentum transfer cross-sections for polar and organic gases [2]. Finally we mention

Table 3. Other cross-sections determined from swarm experiments.

1. $Q_m^e(v)$ for Ne, Ar, Kr, Xe, H_2, N_2, and O_2 [2,4,10].
2. Rotational excitation of N_2 and D_2 [2,12].
3. Vibrational excitation of N_2, CO, CO_2, and D_2 [2,4,8,12].
4. Electronic excitation of N_2 and O_2 [4,18,22].
5. Metastable excitation of Ne and Ar [25].
6. $Q_m^e(v)$, vibrational and elctronic excitation of SF_6, CH_4, etc. [26,27].
7. Rotational excitation of polar molecules [4].
8. Attachment cross-sections using the swarm-beam technique [4].

again the determination of attachment cross-sections for a large number of gases using a combination of swarm and beam techniques [4].

Summary

This paper has briefly reviewed the determination of electron collision cross-sections from swarm data and has summarized the verified successes, outstanding discrepancies, and numerous applications. In addition, we have called attention to the need for a more complete and more easily tabulated description of the angular dependence of theoretically calculated electron scattering cross-sections for gases in which the inelastic collision cross-sections are comparable with the momentum transfer cross-section.

References

*Staff Member, Quantum Physics Division, National Bureau of Standards and Lecturer, Department of Physics, University of Colorado.

[1] Gilardini, A.L. "Low-Electron Collisions in Gases"; Wiley: New York, 1972.

[2] Huxley, L.G.H.; Crompton, R.W. "The Diffusion and Drift of Electrons in Gases"; Wiley: New York, 1974.

[3] Dutton, J. J. Phys. Chem. Ref. Data 1975, 4, 577; Gallagher, J.W.; Beaty, E.C.; Dutton, J.; Pitchford, L.C. J. Phys. Chem. Ref. Data 1983, 12, 109.

[4] Hunter, S.R.; Christophorou, L.G. In "Electron-Molecule Interactions and Their Applications"; Christophorou, L.G., Ed.; Academic; Orlando, 1984; Vol. II, Chap. 3.

[5] Frost, L.S.; Phelps, A.V. Phys. Rev. 1962, 127, 1621; Engelhardt, A.G.; Phelps A.V. Phys. Rev. 1963, 131, 2115.

[6] Phelps, A.V. In "Proceedings of the International Seminar on Swarm Experiments in Atomic Collision Research"; Ogawa, I., Ed.; Tokyo, 1979; p. 23.

[7] Trajmar, S.; Register, D.F.; Chutjan, A. Phys. Rept. 1983, 97, 219.

[8] Buckman, S.J.; Phelps, A.V. J. Chem. Phys. 1985, 82, 4999.

[9] Holly, T.H.,; Chung, S.; Lin, C.C.; Lee, E.T.P. Phys. Rev. A 1982, 26, 1852: Mu-Tao, L.; Lucchese, R.R.; McKoy, V. Phys. Rev. A 1982, 26, 3240.

[10] Crompton, R.W.; Morrison, M.A. this volume.

[11] Reid, I.D. Aust. J. Phys. 1979, 32, 231: Haddad, G.N.; Lin, S.L.; Robson, R.E. Aust. J. Phys. 1981, 34, 243.
[12] Phelps, A.V.; Pitchford, L.C. Phys. Rev. A 1985, 31, 2932.
[13] Tzeng, Y.; Kunhardt, E.E. Bull. Am. Phys. Soc. 1983, 28, 180.
[14] Yoshida, S.; Phelps, A.V.; Pitchford, L.C. Phys. Rev. A 1983, 27, 2858.
[15] Moores, D.L.; Norcross, D.W. J. Phys. D 1972, 5, 1482: Norcross, D.W. (private communication).
[16] Segur, P.; Yousfi, M.; Boeuf, J.P.; Marode, E.; Davies, A.J.; Evans, J.G. In "Electrical Breakdown and Discharges in Gases"; Kunhardt, E.E.; Luessen, L.H., Eds.; Plenum: New York, 1983; p. 331.
[17] Chandra, N.; Temkin, A. Phys. Rev. A 1976, 14, 507: NASA Tech. Note TN D-8347, 1976 (unpublished).
[18] Crompton, R.W.; McIntosh, A.I. Aust. J. Phys. 1968, 21, 637: Henry, R.J.W.; Lane, N.F. Phys. Rev. 1969, 183, 221: Robertson, A.G. Aust. J. Phys. 1971, 24, 445.
[19] Engelhardt, A.G.; Phelps, A.V.; Risk, C.G. Phys. Rev. 1964, 135, A1566.
[20] Allan, A. (private communication).
[21] Yamabe, C.; Buckman, S.J.; Phelps, A.V. Phys. Rev. A 1983, 27, 1345.
[22] Lawton, S.A.; Phelps, A.V. J. Chem. Phys. 1978, 69, 1055.
[23] Linder, F.; Schmidt, H. Z. Naturforsch. 1971, A26, 1617.
[24] Phelps, A.V. (unpublished).
[25] Burrow, P.D. J. Phys. B 1983, 16, 613.
[26] Yoshizawa, T.; Saki, Y.; Tagashira, H.; Sakamoto, S. J. Phys. D 1979, 12, 1839: Kline, L.E.; Davies, D.K.; Chen, C.L.; Chantry, P.J. J. Appl. Phys. 1979, 50, 6789: Dincer, M.S.; Govinda Raju, G.R. J. Appl. Phys. 1983, 54, 6311: Novak, J.P.; Frechette, M. J. Appl. Phys. 1984, 55, 107: Hayashi, M.; Nimura, T. J. Phys. D 1984, 17, 2215: Yousfi, M.; Segur, P.; Vassiliadis, T. J. Phys. D 1985, 18, 359.
[27] Hayashi, M. this volume.

EXPERIMENTAL AND THEORETICAL INVESTIGATION OF NEAR-THRESHOLD e-H_2 COLLISIONS

Robert W. Crompton* and Michael A. Morrison†

*Research School of Physical Sciences, Australian National University, Canberra, A.C.T. 2600, AUSTRALIA

†Department of Physics and Astronomy, University of Oklahoma, Norman, Oklahoma 73019

Introduction

For several reasons threshold rotational and vibrational excitation of molecules by electron impact has received much less attention than resonance phenomena in these processes. More than five years ago, H. S. W. Massey noted that [1]

> ...there has been so much interest in the resonance contributions that one wonders sometimes whether it has been forgotten that we also need to know quite accurately what happens in direct scattering.

For non-polar diatomic molecules, the near-threshold cross sections lack structure, and are therefore less interesting to many experimentalists and theorists. Moreover, until recently, energies near threshold had been generally inaccessible to beam experiments - even those for vibrational excitation, where the thresholds often exceed 100 meV, while absolute determinations of cross-sections are difficult because they are small. Direct excitation has also been largely neglected by the theoretical community, and widely-used methods such as adiabatic-nuclei theory, which yield accurate cross-sections under some circumstances, have been shown to give grossly inaccurate results near threshold [2]. Consequently, experimental and theoretical data have been sparse and often conflicting. On the other hand, direct rotational and vibrational excitation often play a crucial role in electron transport, and hence in determining the characteristics of phenomena or devices that depend on electrical conductivity in gases [3]. Thus neglect of direct excitation, while understandable, is regrettable both from scientific and technological viewpoints.

This paper is a report of research carried out over the past five years jointly between groups at the Australian National University (swarm experiments) and at the University of Oklahoma (theoretical calculations). The aim of the program is threefold:

1. to carry out accurate calculations of cross-sections for elastic scattering, total momentum transfer, and rotational and vibrational excitation;

2. to use these data as benchmarks for testing approximate collision theories, and to compare them with experimental data;

3. to refine the swarm analysis by

   (a) using the new theoretical rotational cross-sections above the vibrational threshold, where the swarm analysis cannot yield an accurate result for the cross-section,

   (b) taking account of rotational inelasticity in vibrational excitation, again using new theoretical results,

   (c) applying, if necessary, more rigorous transport theory than has heretofore been implemented.

In the present paper we report on e-H_2 scattering. This research was intended as the first phase of an extensive investigation of several diatomic systems; but, as we shall show, there is a hiatus at present which is as intriguing as it is frustrating.

Background to the Analysis of Swarm Experiments

We first recall a few facts relating to the extraction of cross-sections from swarm experiments. To simplify the argument, we consider the unique case of electron transport in parahydrogen at 77K where virtually all the molecules are in the ground ro-vibrational state. Figure 1 illustrates the variation of the electron energy distribution function $f_o(\varepsilon)$ with E/N and the varying influence of the two inelastic processes. (For ease of illustration, the distribution functions have been arbitrarily normalized.) In the same figure are shown the cross-sections for momentum transfer and for the $j_o=0 \rightarrow j=2$ rotational and $v_o=0 \rightarrow v=1$ vibrational excitations.

At the lowest value of E/N [curve (a): E/N = 0.01 Td] very few electrons in the distribution have energies above the rotational threshold. Therefore at these low values of E/N the energy distribution of the electrons in the swarm, and consequently the transport coefficients, are determined primarily by the value of E/N and the elastic cross-section.

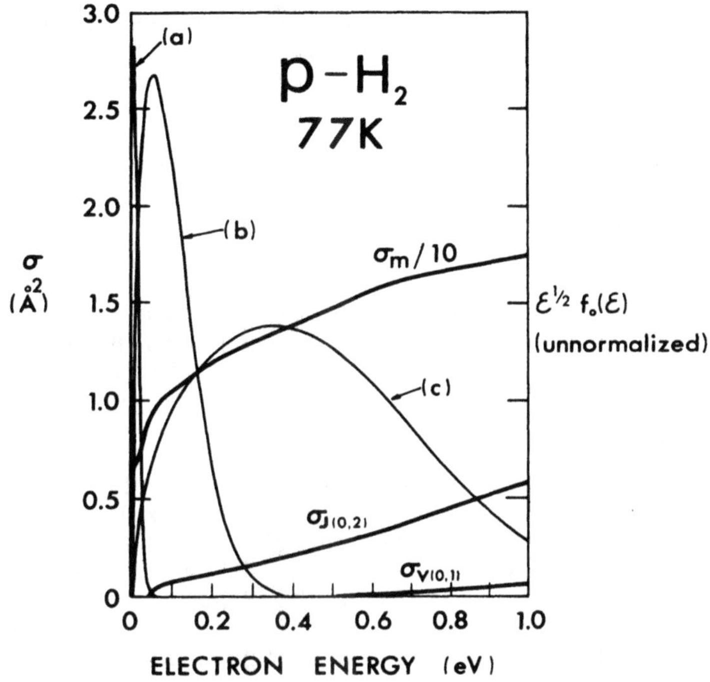

Figure 1. Electron energy distribution functions in parahydrogen at 77K, at three values of E/N, shown in relation to the cross-sections that influence them.

At an intermediate value of E/N [curve (b): E/N = 1.0 Td] very few electrons have energies above the vibrational threshold, but a large fraction can excite molecular rotation. Because about 45 meV of energy is lost in each rotational collision, compared to a small fraction of a millivolt in each elastic collision, collisions resulting in rotational excitation play a dominant role in determining $f_o(\varepsilon)$, even though the cross-section for this process is relatively small. Therefore the energy dependence of this cross-section (and of the momentum transfer cross-section [4,5]) can be probed by varying E/N within a range of values about 1.0 Td, because doing so causes the foot of the distribution function to sweep across the threshold.

At the highest value of E/N shown [curve (c): E/N = 7.0 Td] 40% of the electrons exceed the vibrational threshold, so vibrational as well as rotational excitation plays a role in determining $f_0(\varepsilon)$ and hence the motion of the swarm. As we shall show, the competition of these processes renders the interpretation of the results ambiguous if only data for drift velocities and diffusion coefficients are available.

The varying influence of rotational and vibrational excitation as E/N is changed can be further illustrated by determining how the total power fed into the swarm by the electric field is distributed among the collision processes [6]. At E/N ≈ 0.5 Td, for example, 20% of the power is dissipated in elastic collisions, and 80% in rotational excitation; virtually none of the power heats the gas through vibrational excitation. On the other hand, at E/N ≈ 6 Td, roughly equal fractions are absorbed in rotational and vibrational excitation, the remaining 20% being dissipated through elastic collisions.

As a final qualitative demonstration of the relative importance of rotational and vibrational excitation, we can omit either of these processes and observe the changes in the calculated drift velocities and diffusion coefficients. Doing so yields the results shown in Figure 2. (Data for the ratio D/μ of the transverse diffusion coefficient D to the mobility μ rather than for D are shown, since the required experimental data exist only for this ratio.) To obtain the data shown in this figure, the drift velocities v_{dr} and the D/μ ratios were first calculated with the final set of cross sections; these data agree to within a percent or two with the experimental data, since the cross-section set was obtained in the first instance by analyzing the experimental data. Then the rotational and vibrational cross-sections were set equal to zero in turn and the transport coefficients recalculated. The differences, expressed as percentages of the original values, are plotted as functions of E/N.

In the vicinity of 0.5 Td the transport coefficients are extremely sensitive to rotational excitation, so it is not surprising that this cross-section can be accurately determined. This sensitivity diminishes at higher E/N as increasing power is fed into vibrational excitation. The effect of restoring rotational excitation and omitting vibrational excitation is not as dramatic as that of removing rotational excitation, but its influence on the transport coefficients is very large compared to the accuracy with which they can be measured, conservatively estimated to be within 1 or 2%.

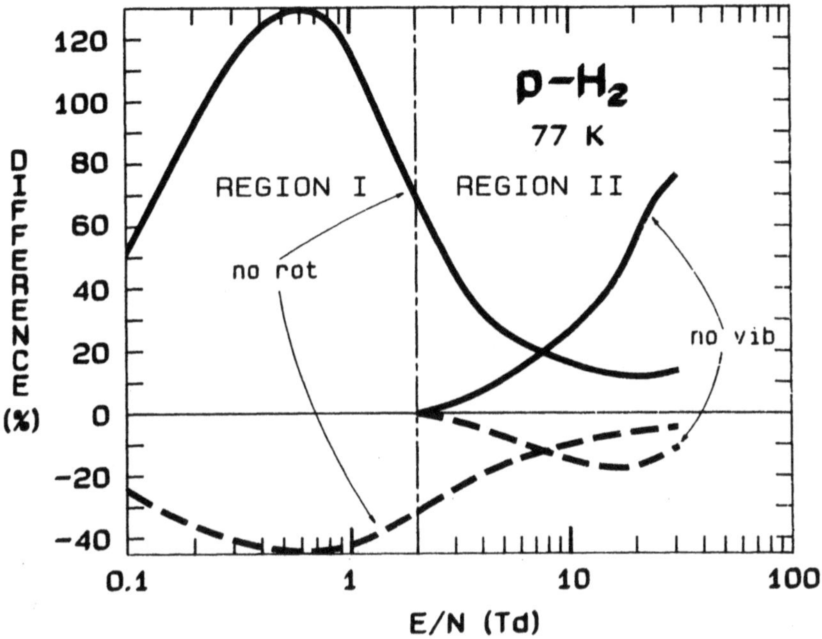

Figure 2. Changes in the transport coefficients resulting from the suppression of either rotational or vibrational excitation: D/μ (solid curve); v_{dr} (dashed curve).

Three other important points are evident in Figure 2:

1. The removal of the vibrational cross-section has no discernible effect on the calculated transport coefficients in the range of E/N from 0 to 2 Td, which we shall call region I. Hence, when initially adjusting the cross-section set to fit the measured transport data in this region, only changes in the momentum transfer and rotational excitation cross-sections influence the calculated transport coefficients. Consequently, cross sections determined from analysis of the transport data are nearly unique - and accurate - in the energy range that significantly influences the transport coefficients. This range extends from threshold to about 0.35 eV.

2. In the range of E/N above 2 Td, which we shall call region II, the changing importance of rotational and vibrational excitation is apparent. The effect of rotation diminishes, and it becomes impossible to determine uniquely the cross-section at energies above about 0.35 eV because of the competing influence of vibrational excitation.

3. Although the effect of rotation becomes smaller with increasing E/N, it never becomes negligible. Therefore, unlike the rotational cross-section, the vibrational cross-section can never be determined uniquely in the threshold region without knowing the rotational cross section reasonably accurately. But, as we have just shown, swarm experiments and their analysis cannot provide an accurate rotational cross-section at energies above the vibrational threshold because of the uniqueness problem. Input from another source is required - in the present program, from the theoretical calculations described below. This close interaction between the theoretical and swarm analysis is an underlying theme of our research.

Analysis of Transport Data for Parahydrogen

The simplicity of the analysis for parahydrogen makes it desirable to analyze transport data for para- rather than normal hydrogen. The D/μ data for $6 < E/N\ (Td) \leq 30$ Td have not, however, been measured, but their absence presents no real problem, because the data can be readily and accurately derived from those for normal hydrogen. As E/N increases to 6 Td, the experimental values of D/μ for para- and normal hydrogen converge to within 1% [7]. This convergence is confirmed by calculations of D/μ using the cross-sections appropriate to the two forms of hydrogen, and is primarily due to the waning influence of rotational excitation as E/N increases. Because the adjustments to the D/μ data for normal hydrogen that are required to obtain the data for parahydrogen are so small, they can be made using the differences between the calculated values for the two forms [8] with insignificant error introduced through inaccuracies in the cross-section set used for the calculations.

Having completed the data set for parahydrogen, we adopted the following strategy to analyze the data and so derive the cross-sections:

1. We first determined the rotational cross-section using

the data in region I from threshold to about 0.35 eV. Above this energy the loss of uniqueness leads to the rapid widening of the error bounds shown in Figure 3 [6]. The agreement between the swarm-derived and theoretical $j_o=0 \rightarrow j=2$ cross-section in this energy range is satisfactory. The maximum disagreement is about 9%; at the "high energy" end of this range the discrepancy is fairly constant.

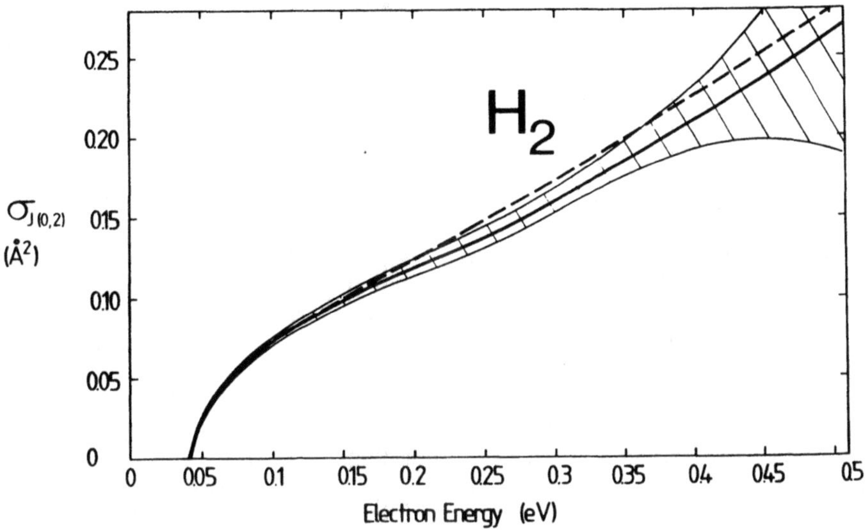

Figure 3. The swarm-derived cross-section for the $j_o=0 \rightarrow j=2$ rotational excitation. The shaded area shows the estimated uncertainty; above 0.3 eV the estimate is based on a number of assumptions (see Ref. [5]). Also shown, as the dashed line, is the theoretical cross-section.

2. We then used the theoretical cross-section to construct the rotational cross-section above this energy for subsequent use in the determination of the vibrational cross-section. We extrapolated the experimental rotational cross-section by smoothly grafting the experimental and theoretical results over the energy range 0.35 to 0.5 eV. (Alternatively, we could have "normalized" the theoretical cross-section by multiplying it by a factor of 0.9 to align it with the experimental result at higher energies, but doing so would require justification.) Because the energy loss in vibrational excitation is roughly 10 times that in rotational excitation, the effect on the vibrational cross-section

of this uncertainty in the rotational cross-section is not large - about 10% at 1 eV and 4% at 2 eV.

3. Finally, with the rotational cross-section determined we analyzed region II to determine the vibrational cross section. A series of tests was performed to show that

   (a) the transport data were not influenced by energy losses in dissociation in the range of E/N used for the analysis (i.e., up to E/N = 30 Td), and

   (b) the transport data were primarily influenced by the vibrational cross-section in the range 0.51 to 2 eV.

To determine vibrational cross-sections of maximum accuracy, the transport analysis requires input in addition to rotational excitation cross-sections: the branching ratios for rotationally elastic and inelastic vibrational excitation. But because the additional energy loss in a ro-vibrational collision is only 10 to 15% of that sustained in a pure vibrational collision, no great error is involved even if "rotational splitting" is ignored, and it is assumed that the energy loss in all collisions that result in vibrational excitation is the same as that lost in rotationally elastic collisions, 0.51 eV. Calculations showed that the maximum error in v_{dr} due to neglect of rotational splitting is 0.5%, and that in D/μ just over 1% [8]. These errors are well within experimental error; nevertheless rotational splitting has been taken into account to avoid compounding errors. Because slightly larger energy losses are associated with the simultaneous transitions, allowing for rotational splitting lowers somewhat the total vibrational cross-section with respect to the cross-section derived when it is neglected [9].

As will be shown later, the result of this analysis is disappointing, with the experimental cross-section lying well below the theoretical result. However, the important question is whether the disagreement lies outside the uncertainty of the theoretical and experimental results. We will return to this matter after a brief description of the theory used to calculate cross-sections, and a comparison of the resulting cross-sections with swarm and beam experimental data.

Theory and Comparison to Experimental Data

The objectives of the theoretical component of this program are twofold: first, to assess the accuracy of several approximate scattering theories for near-threshold ro-vibrational excitation [10,11], with special emphasis on the widely used adiabatic nuclei method [12]; second, to join with swarm and beam experimentalists in a concerted, long-term attempt to determine benchmark cross-sections for electronically-elastic electron-molecule collisions, beginning with the e-H_2 system. The results of both aspects of this program have proven, to say the least, startling, but only the second falls within the purview of this paper [13,14].

Theory

The interaction potential. We determined the static, exchange, and polarization components of the electron-H_2 interaction potential from near-Hartree-Fock electronic wave functions for the $X^1\Sigma_g^+$ state of H_2. These wave functions were calculated at 11 values of the internuclear separation R, ranging from 0.5 to 2.6 a_0, using a (5s2p/3s2p) extended basis set. The static term, for example, is just the average of the two-particle, bound-free Coulomb potentials over this electronic function.

Exchange effects are treated differently in our calculations of elastic and inelastic cross-sections. We include these effects exactly in our theory for elastic cross sections and via a model potential for inelastic cross sections. The particular model potential we use - the "tuned free-electron-gas exchange potential" (TFEGE) [15] - has been designed to accurately mimic exchange effects that are most important for inelastic scattering at energies below a few eV. This potential is defined so as to ensure that at each R it reproduces the exact static-exchange eigenphase sum in the Σ_u symmetry at a single energy, 0.54 eV. (This symmetry dominates the inelastic cross-sections at energies below about 5.0 eV.) It is important to emphasize that the procedure by which our exchange potential is determined in no way biases the results in favor of one or another experiment. Like the other constituents of our interaction potential, the exchange term is independent of any experimental data.

Polarization effects are incorporated via a local, energy-independent potential that takes account of adiabatic distortions of the target and, in an approximate way, of non-adiabatic effects, such as short-range bound-free correlation. The latter are represented by a non-penetrating approximation [16]. Unlike many approximate polarization potentials that have been used in earlier electron-molecule scattering calculations [10,11], our model contains no adjustable parameters.

Collision Theory. The interaction potential described above is inserted into the time-independent Schrödinger equation for the continuum states of the system; this equation must then be solved numerically. There are basically two ways to tackle this part of the theoretical calculation: the scattering theory used can be based either on performing straightforward eigenfunction expansions of the system wave function in a basis of nuclear and angular functions, which strategy leads to the laboratory-frame close-coupling (LFCC) method, or on implementing the Born-Oppenheimer approximation for the electron-molecule system, which leads to the adiabatic-nuclei (AN) theory and its variants [10,11]. The latter tack is preferable, since AN methods make more reasonable computational demands than do full LFCC calculations. Extensive investigation of AN methods, however, has shown them to yield grossly inaccurate ro-vibrational e-H_2 cross-sections at energies below a few eV [13,14]. Happily, the AN method produces very accurate elastic cross-sections, and, since it is comparatively easy to treat exchange exactly in an AN calculation, we use this method for elastic scattering. The numerical technique used for the evaluation of the body-frame/fixed-nuclei scattering matrices, which is the core of an AN calculation, is the iterative exchange procedure of Collins et al. [17].

The failure of standard AN techniques for ro-vibrational cross - sections reflects the importance of accurately incorporating in the scattering theory effects that derive from the dynamical interaction between the motion of the scattering electron and that of the nuclei. This "coupling" of the electron and nuclear motions is particularly vital to vibrational excitation. To ensure that it is treated exactly, we implement full laboratory-frame ro-vibrational close-coupling to determine the inelastic e-H_2 cross sections. In our implementation, exchange effects are included via the model potential described in the previous section. We imposed extremely stringent accuracy criteria on every aspect of the scattering calculations; extensive numerical tests and convergence studies indicate that our integrated and differential cross-sections are converged to 1% or better.

Comparison of Theoretical and Experimental Cross-Sections

We have performed scattering calculations at energies from threshold to roughly 10.0 eV, but it is important to keep in mind that our interaction potential is designed to be most accurate at energies below about 5.0 eV.

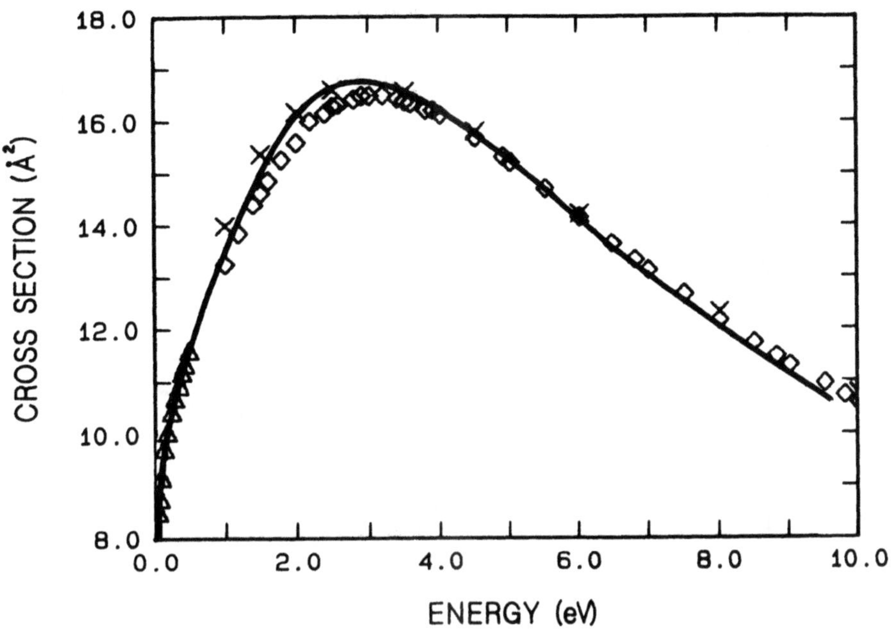

Figure 4. Total integrated cross-sections for e-H_2 scattering from theoretical (solid curve) and experimental studies. The experimental points are from the work of Ferch et al. [18] (triangles), Jones [20] (diamonds), and Dalba et al. [19] (crosses).

In Figure 4, the theoretical total integrated cross section is compared to a selection of experimental results obtained in beam experiments [18-20]. The excellent agreement between theory and experiment, which is apparent in this figure, is encouraging but somewhat peripheral to our immediate concerns, because σ_{tot} is dominated by elastic scattering at the energies we are considering.

Turning now to inelastic scattering, we show in Figure 5 differential cross-sections for the pure vibrational excitation $v_0=0 \rightarrow v=1$ at 1.5 and 3.5 eV. The beam experiments of Linder and Schmidt [21] were performed at a temperature of 76K, so to effect a comparison between their results and ours, we Boltzmann-averaged our ro-vibrational cross-sections $\left.\frac{d\sigma}{d\Omega}\right|_{j_0 0 \rightarrow j_0 1}$ over the distribution of initial rotational states j_0 at this temperature. An independent beam measurement for excitations from $v_0=0$ to $v=1$ and $v=2$ was carried out by Wong and Schulz [22]. The agreement between

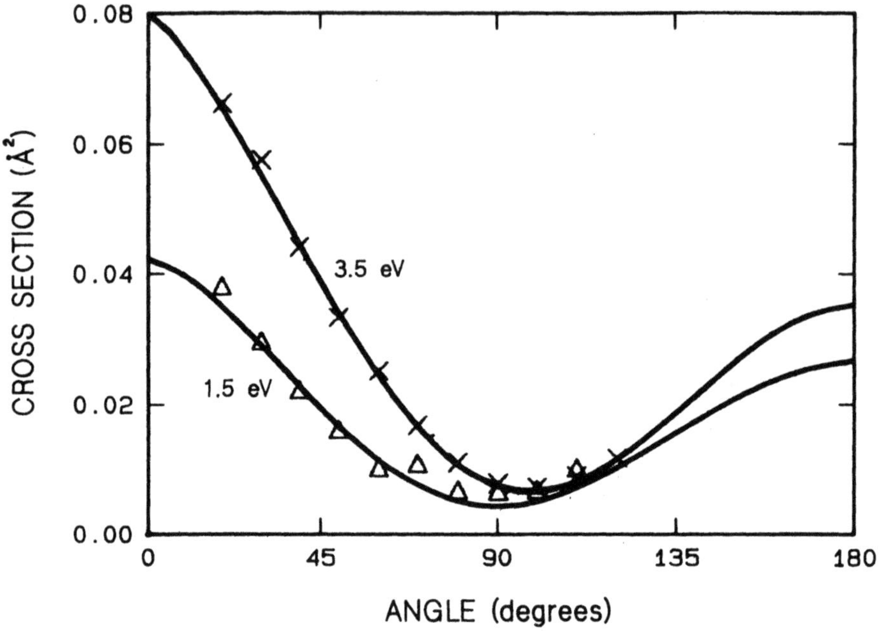

Figure 5. Theoretical (solid curves) differential cross sections for the pure vibrational excitation $v_0=0 \rightarrow v=1$ at 76 K for collision energies of 1.5 eV and 3.5 eV. Also shown are measured values of Linder and Schmidt [21].

theory and experiment for both excitations (not shown) is as good as that seen in Figure 5.

To begin the comparison of our theoretical results to swarm-derived cross-sections, we show in Figure 6 the total momentum transfer cross-section for e-H_2 scattering. The excellent agreement illustrated in this figure encourages us to press on to the inelastic results - where we find some surprises.

The situation for pure rotational excitation is quite satisfactory. As shown in Figure 3, theory and experiment for the 00 → 02 cross-section concur over the energy range where the swarm analysis yields unique values for these cross sections. Comparable agreement is obtained for the 01 → 03 excitation (not shown).

In Figure 7 theoretical vibrational excitation cross sections for $v_0= 0 \rightarrow v=1$ (summed over all final rotational states) over the entire energy range studied are compared to

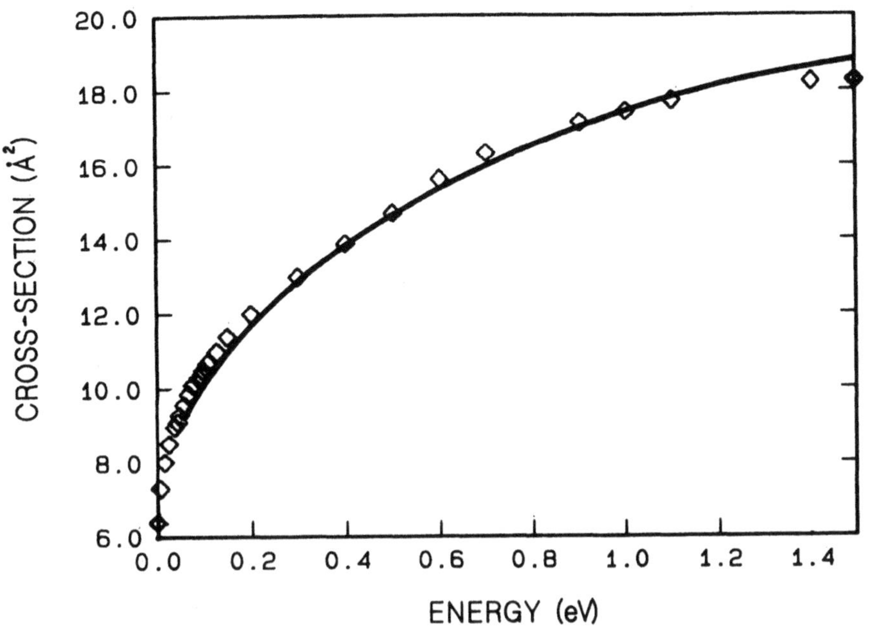

Figure 6. Theoretical (solid curve) and swarm-derived (diamonds) total momentum transfer cross-sections.

the beam results of Ehrhardt et al. [23] and to swarm-derived cross-sections. At first glance, all seems well: at energies very near threshold, where accurate beam measurements are particularly difficult, the theoretical cross-sections agree beautifully with the swarm-derived cross sections. As the energy increases, the theoretical cross section smoothly deviates from the swarm results, coming into agreement with the beam data. This behaviour is in accordance with the improved accuracy of the beam measurements with increasing energy.

On closer examination, however, a problem surfaces: the divergence of the theoretical and swarm cross-sections occurs at about 0.8 eV, an energy where the swarm analysis should still provide an accurate result. The swarm-derived cross sections should be accurate to 10-15% for the energies shown in Figure 7; but the disagreement between theory and experiment is as large as 50% at the highest energy.

Critique of the Theoretical Study

The superstructure of the theoretical calculation is summarized in Table 1. Playing devil's advocate to the

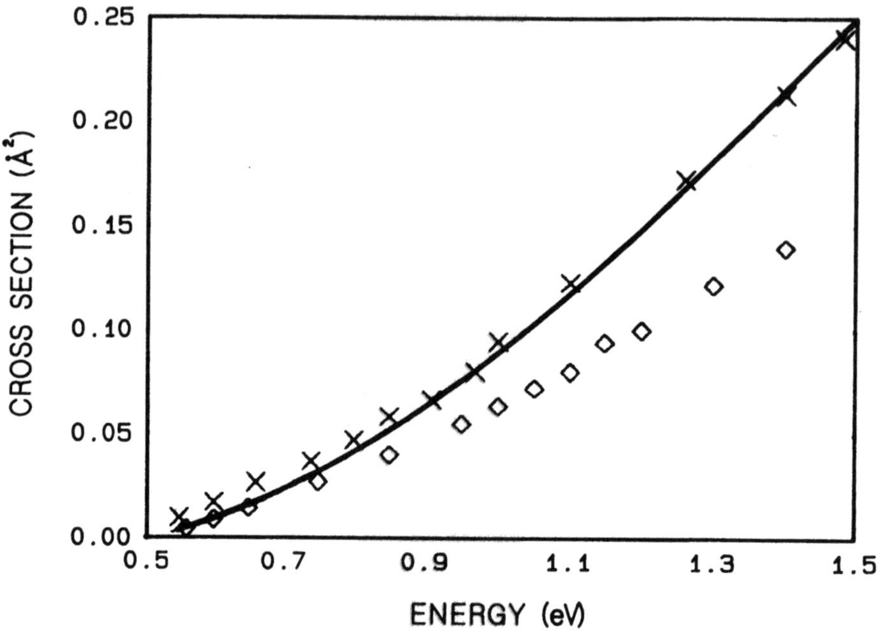

Figure 7. Cross-sections for the $v_0=0 \rightarrow v=1$ excitation of H_2 from theory (solid curve), beam experiments of Ehrhardt et al. [23] (crosses) and transport analysis (diamonds).

theoretical study, we can identify two possible sources of inaccuracy: the model exchange potential and the non-penetrating approximation used to represent non-adiabatic effects in the polarization potential. We now look briefly at each of these in turn.

Table 1. Overview of Theory

	Elastic	Inelastic
scattering	adiabatic-nuclei	LFCC
exchange	exact	TFEGE model

Extensive studies of the TFEGE potential [15] have shown it to be a very accurate approximation to exact exchange for a variety of molecules. All of these studies, however, were performed in the rigid-rotator approximation at the equilibrium geometry. This limitation raises a question: can the model potential accurately track the variations in the exchange interaction as the molecule vibrates? In search of an answer, we performed exact and model e-H_2 scattering calculations at several energies for internuclear geometries corresponding to the excursions the molecule makes in its lowest few vibrational levels. The most important electron-molecule symmetry for vibrational excitation at energies below a few eV is Σ_u, so in Figure 8 we show eigenphase sums in this symmetry from the two sets of calculations. (Recall that the TFEGE potential is defined by adjustment to an exact static-exchange eigenphase sum in this symmetry at 0.54 eV.) At these energies, the model potential accurately mimics the true exchange interaction as the molecule vibrates; additional studies (not shown) reveal that this level of accuracy is attained at scattering

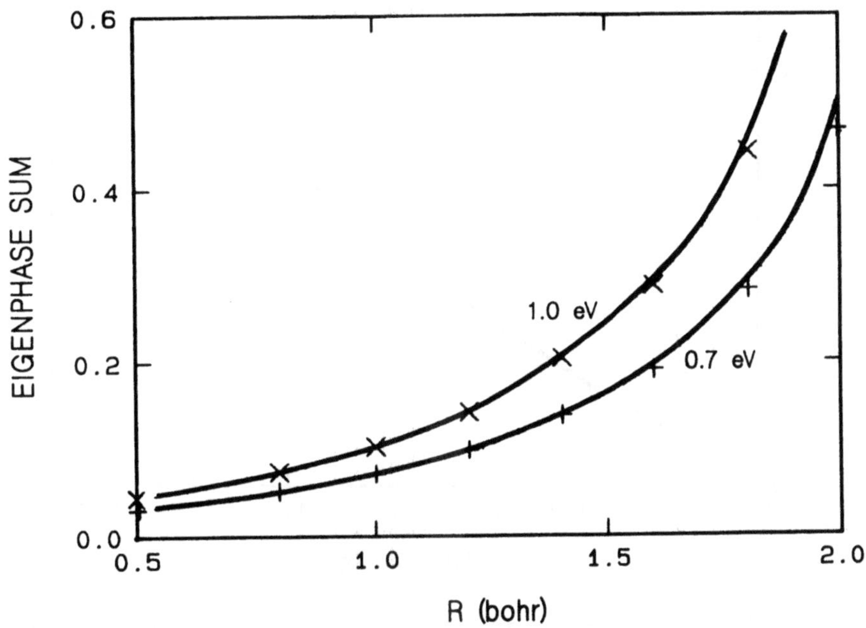

Figure 8. Eigenphase sums in the Σ_u e-H_2 symmetry at 0.7 and 1.0 eV. The results shown as solid curves were calculated using the TFEGE potential, while those shown as pluses and crosses were determined from an exact exchange calculation. Polarization is included in all results.

energies as high as 5.0 eV.

Calculating benchmark results to assess our polarization potential is much more difficult, as electron-molecule cross sections determined from a theory based on a more rigorous treatment of short-range bound-free correlation exist only at equilibrium. In Figure 9 we compare our partial cross sections in the Σ_u symmetry to those of the optical potential calculations of Schneider and Collins [24]. Although their theoretical formulation is more rigorous than the non-penetrating approximation, optical potential calculations, like any L^2 method, are subject to uncertainties of basis set convergence. Nonetheless, the agreement is excellent in the energy range under consideration. This comparison leaves unanswered the crucial question of whether our polarization potential is equally accurate at off-equilibrium geometries.

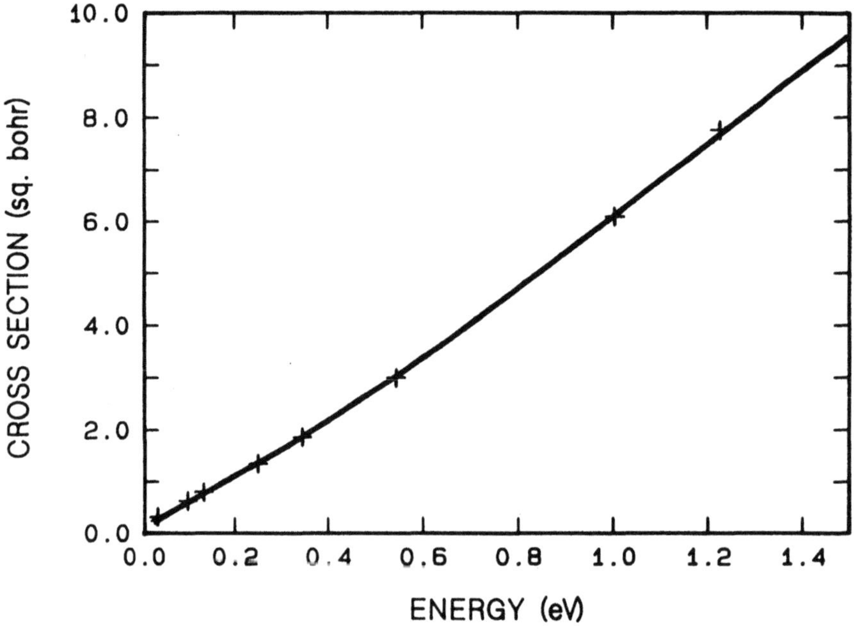

Figure 9. Partial e-H_2 cross-section in the Σ_u symmetry at equilibrium. The solid curve was calculated using a model polarization potential based on the non-penetrating approximation. The crosses are results from the optical potential calculations of Schneider and Collins [24].

Having argued that the theoretical results are unassailable, we shall now argue the same for swarm-derived cross sections!

Critique of the Swarm Results

Before examining the swarm experiments and analysis in some detail, we must ask: is the discrepancy between the theoretical and swarm-derived vibrational cross-sections significant? The most straightforward way to answer this question is to calculate the transport coefficients using the theoretical cross-sections and to compare the resulting data with those obtained by experiment. Such a comparison is shown in Figure 10. Also shown here is a similar comparison using the swarm-derived cross-sections. At the higher values of E/N, where the effect of vibrational excitation is dominant, the transport coefficients calculated using the theoretical cross-sections lie well outside the experimental uncertainties (about 1% in v_{dr} and 2% in D/μ). The difference is too large to be ignored.

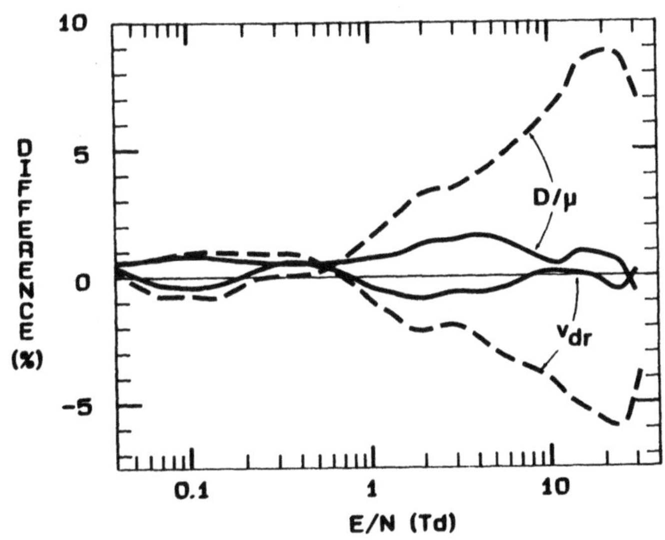

Figure 10. Percentage differences between the experimental results for v_{dr} and D/μ and those calculated using swarm-derived cross-sections (solid curves) and the theoretical cross-sections (dashed curves).

The comparisons shown in Figure 7 throw doubt on the swarm result at higher energies, where the reliability of the beam experiment improves. On the other hand, the swarm experiment and analysis can in principle yield definitive experimental data in this case - and a great deal of attention has been paid to getting such data. If the swarm results are incorrect, what might have gone wrong?

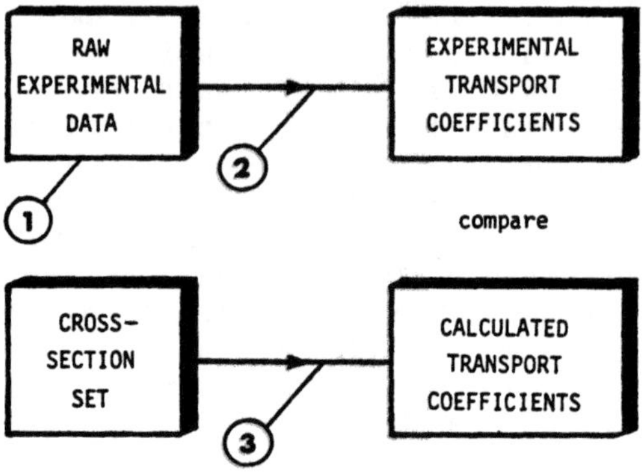

Figure 11. Block diagram of the steps in the analysis to determine cross-sections from data obtained in swarm experiments.

It is important to bear in mind the three-step process involved in determining the swarm-derived cross-sections - this process is summarized in Figure 11. In the swarm experiments used for this work the basic physical measurements are those of pressure, temperature, current and frequency. From these data the drift velocities and D/μ ratios are inferred. Starting from the computational side, the transport coefficients are calculated from an assumed set of cross-sections. A comparison is then made between calculated and measured values. The cross-section or cross sections are deemed to be "correct" when the two sets of transport coefficients agree at all values of E/N. Leaving aside for the moment the problem of uniqueness, which has already been discussed at length, there are essentially three stages at which errors can occur:

1. in the measurements themselves;
2. in inferring the transport coefficients from the raw experimental data;
3. in calculating the transport coefficients from the cross-sections.

An example of the subtleties in step 2 has been discussed in the chapter by H. Tagashira, which deals with the interpretation of drift velocity experiments. In the present case this step is essentially straightforward, since there is no ionization. Nevertheless, additional subtleties remain in the interpretation of the drift velocity experiments because of the effects of diffusion on the drifting electron groups; we believe, however, that these have been properly accounted for. The interpretation of the lateral diffusion experiments from which the D/μ ratios are derived is more subtle. We shall argue later, however, that the interpretation of these experiments is also valid to the claimed accuracy.

In recent years there has been much more debate about step 3, the calculation of the transport coefficients from the cross-sections. Several theoretical analyses and computer simulations [25] have been carried out to examine the validity of the transport theory that was universally used at the time of the original analysis of the parahydrogen transport data [6,9]. These studies proved that in certain circumstances transport coefficients calculated using that theory could be subject to errors much larger than those we are now seeking [25]. But this work also showed that the older transport theory, which is based on a two-term expansion of the electron velocity distribution function (the so-called "two-term approximation"), introduced very little error in the earlier analysis of the parahydrogen data. Nevertheless, in the calculations reported in this paper we used one of the new generation of transport codes, even though it was not strictly necessary to do so. These codes have been shown to calculate the transport coefficients to within 1% even when the older theory breaks down [25].

Rather than introduce the intermediate step of inferring transport coefficients from the experimental data, for subsequent comparison with data calculated from a set of cross-sections, it would be preferable to have a complete theoretical description or simulation that would bypass this step and produce the experimentally measured quantities directly from the cross-sections. Such calculations would show the influence of the assumed boundary conditions, a question discussed in the chapter by K. Kumar. So far such a procedure has not been developed. However, there is evidence to suggest that the existing procedure is valid to the claimed accuracy when the conditions are similar to those for the parahydrogen experiments. This evidence comes from a comparison of the transport coefficients inferred from drift and diffusion experiments in helium [5] with values calculated using Nesbet's *ab initio* theoretical momentum transfer cross-section [26]. Figure 12 shows the differences

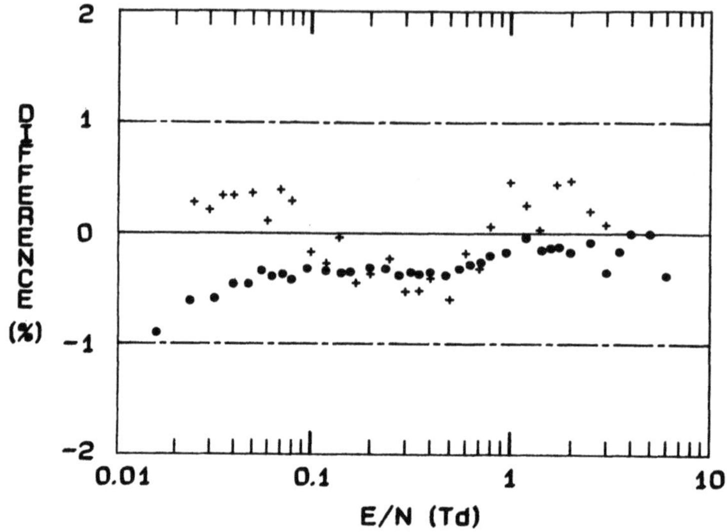

Figure 12. Percentage differences between the experimental results for v_{dr} and D/μ in helium and those calculated using Nesbet's theoretical momentum transfer cross-section [26]: for v_{dr} (dots); for D/μ (crosses).

between the experimental transport coefficients in helium and those calculated using the theoretical cross-section. The agreement confirms the validity of the currently used method of analysis for this case. In moving from this example, which involves elastic scattering only, to the case of hydrogen, the most suspect of the three steps is the third. But the possibility of an error creeping into this step has been effectively eliminated by the aforementioned theoretical analyses and computer simulations.

Despite the apparent soundness of the case just presented for the parahydrogen data, we felt the need of another approach. This approach and its results are described in the next section.

Measurements in Hydrogen-Helium Mixtures

The measurement and analysis of drift velocities in mixtures provides an alternative method of determining integral inelastic cross-sections [27,28]. For the present work we chose to use a mixture containing about 10% of H_2 in He as a buffer gas, and to adopt a difference technique.

Thus, drift velocities were first remeasured in pure He and then measured in the mixture using identical conditions of pressure, temperature, etc. The measured differences between the drift velocities in pure He and in the mixture were then compared with calculated values. There are four advantages of such a procedure:

1. The method is based on drift tube (time-of-flight) experiments, which are intrinsically the most accurate of all swarm experiments and are the most easily interpreted in terms of a transport coefficient.

2. In the comparison of calculated and experimental differences nearly all systematic experimental errors cancel, and even errors in σ_{mom} for He have only a second-order effect. Thus errors in pressure, temperature, and drift tube length are essentially eliminated. Errors in the frequency measurements used to determine the transit times are negligible. Hence, the only remaining significant experimental error is in the correction for diffusion.

3. By using a dilute mixture of H_2 in He we can isolate more effectively the effects of elastic and inelastic scattering in H_2. Since the momentum transfer cross sections for the two gases are comparable, momentum transfer is dominated by collisions with the dominant species, He, for which the cross-section is accurately known. Elastic scattering by H_2 thus plays only a minor role in both energy and momentum transfer and its effect, therefore, is of second order.

4. Two-term Boltzmann transport codes, with their advantages of computational speed and ease of implementation, can be used with negligible error [28].

Figure 13 shows the experimental differences between the drift velocities in pure He and in the mixture [8]. The corresponding differences calculated using the swarm-derived cross-sections for both H_2 and He lie just outside the experimental error bounds. Note that substitution of Nesbet's He momentum transfer cross-section [26] for the swarm-derived result would have little effect on the results shown in this figure.

Figure 13 also shows differences obtained when the theoretical e-H_2 cross-sections are substituted for the swarm-derived cross-sections. The convergence of the curves at low E/N clearly shows where vibrational excitation becomes

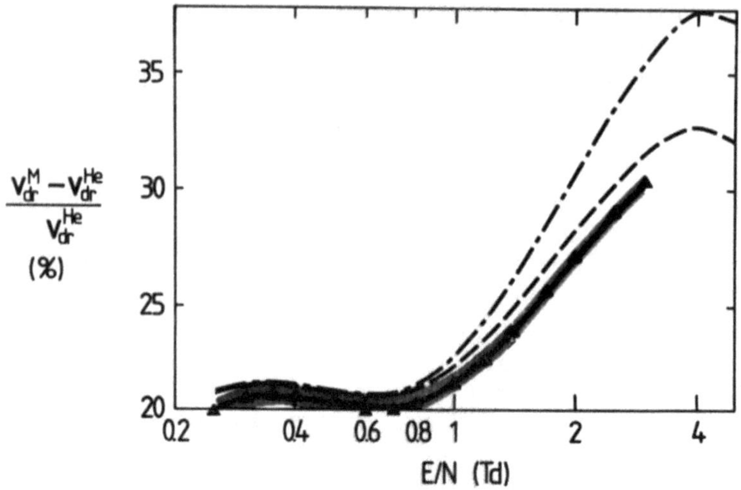

Figure 13. Percentage differences between drift velocities measured in He and in the H_2-He mixture (see text) plotted as a function of E/N. The shaded area shows the estimated uncertainty of ± 0.5%. Corresponding curves calculated using the swarm-derived cross-sections (dashed curve) and the theoretical cross-sections (dot-dashed curve) are shown for comparison.

unimportant. At higher E/N, however, where vibrational excitation dominates, the mixture results clearly favor the swarm-derived cross-section. A similar conclusion with respect to the discrepancy between the beam and swarm vibrational excitation cross-sections was reached by Haddad and Crompton from an analysis of their experimental transport data in H_2-Ar mixtures [28].

Where Do We Go From Here?

We have presented compelling cases for the validity of the theoretical and the swarm-derived vibrational excitation cross-sections, yet these cross-sections are clearly incompatible. This situation is quite unsatisfactory and must be resolved. At the time of writing, however, we remain baffled.

Acknowledgments

We are grateful to Z. Petrović, G. Haddad, and G. Braglia for contributions to various aspects of the swarm program and to T. Gibson and B. Saha for assistance with the theoretical calculations. In addition, we would like to acknowledge useful conversations with nearly everyone in the atomic physics community.

References

[1] Massey, H. S. W. in "Electron-Molecule Scattering"; Brown, S. C., Ed.; John Wiley & Sons: New York, 1979; Chapter 6.

[2] Morrison, M. A.; Saha, B. C.; Feldt, A. N. Phys. Rev. A 1984, 30, 2811.

[3] See, for example, Phelps, A. V. in "Electron-Molecule Scattering"; Brown, S. C., Ed.; John Wiley & Sons: New York, 1979; Chapter 3.

[4] Crompton, R. W. in "XVI International Conference on Phenomena in Ionized Gases - Invited Papers"; Bötticher, W.; Wenk, H.; Schulz-Gulde, E., Eds; University of Düsseldorf: Düsseldorf, 1984.

[5] Huxley, L. G. H.; Crompton, R. W. "The Diffusion and Drift of Electrons in Gases," John Wiley & Sons: New York, 1974; Chapter 13.

[6] Crompton, R. W.; Gibson, D. K.; McIntosh, A. I. Aust. J. Phys. 1969, 22, 715.

[7] See Reference [5], Chapter 14.

[8] Petrović, Z. Lj. Ph.D. Thesis; Australian National University: Canberra, 1985; Chapter 9.

[9] Crompton, R. W.; Gibson, D. K.; Robertson, A. G. Phys. Rev. A 1970, 2, 1386; see also Reference [6].

[10] Lane, N. F. Rev. Mod. Phys. 1980, 52, 29.

[11] Morrison, M. A. Aust. J. Phys. 1983, 36, 239.

[12] Temkin, A.; Sullivan, E. C. Phys. Rev. Lett. 1974, 33, 1057.

[13] See reference [2] and general discussion in [10] and [11].

[14] Morrison, M. A.; Gibson, T. L.; Austin, D. M. Phys. Rev. A 1984, 29, 2518.

[15] Morrison, M. A.; Collins, L. A. Phys. Rev. 1980, 23, 127.

[16] See Gibson, T. L.; Morrison, M. A. Phys. Rev. A 1984 29, 1497 and references therein.

[17] Collins, L. A.; Robb, W. D.; Morrison, M. A. Phys. Rev. A 1980, 21, 488.

[18] Ferch, J.; Raith, W.; Schröder, K. J. Phys. B. 1980, 13, 1481.

[19] Dalba, G.; Fornasini, P.; Lazzizzera, I.; Ranieri, G.; Zecca, A. J. Phys. B. 1980, 13, 2839.

[20] Jones, R. K. Phys. Rev. A. 1985, 31, 2898.

[21] Linder, F; Schmidt, H. Z. Naturforsch 1971, 26, 1603.

[22] Wong, S. F.; Schulz, G. J. Phys. Rev. Lett. 1974, 20, 1089.

[23] Ehrhardt, H.; Langhans, D. L.; Linder, F.; Taylor, H.S. Phys. Rev. 1968, 173, 222.

[24] Schneider, B. I.; Collins, L. A. Phys. Rev. A. 1983, 28, 166.

[25] See Chapter 1 for references to much of this work.

[26] Nesbet, R. K. Phys. Rev. A 1979, 20, 58.

[27] Engelhardt, A. G.; Phelps, A. V. Phys. Rev. 1964, 133, A375.

[28] Haddad, G. N.; Crompton, R. W. Aust. J. Phys. 1980, 33, 975.

ELECTRON COLLISION CROSS-SECTIONS FOR MOLECULES DETERMINED FROM BEAM AND SWARM DATA

Makoto Hayashi

Nagoya Institute of Technology
Gokiso-cho, Showa-ku, Nagoya, 466, Japan

Elastic and inelastic electron collision cross-sections for molecules of interest in plasma processes, gas lasers, gaseous dielectrics, diffuse discharge opening switches, space science and radiation research have been determined from available electron beam and electron swarm data utilizing the Boltzmann equation and a Monte Carlo simulation method. The method is the same as the one used by Phelps, Crompton, Thomas, Tagashira, and their co-workers. The discussion will cover a number of molecular species, e.g., F_2, HCl, SO_2, CH_4, SiH_4, CF_4, CCl_2F_2, CCl_4, SF_6, C_2H_6, and Si_2H_6, in the 0 to 1000 eV energy range. In spite of several studies, inelastic collision cross-sections for N_2 are still not well-known. This matter will be briefly addressed. The discussions will then focus mainly on the cross-sections for CF_4. Some calculated results of electron swarm data for CF_4 and CF_4-inert gases mixtures will be presented.

The goal of electron collision studies is to provide absolute cross-sections for all processes involved: elastic scattering, excitation, dissociation, attachment and ionization, as a function of incident electron energy and scattering angle.

In this general invited lecture at the Dusseldorf ICPIG, entitled "Developments in the use of swarm experiments for investigating electron collision processes", R. W. Crompton discussed how swarm experiments can be used to derive cross-section data for low-energy electrons scattered by, or attached to atoms and molecules. With selected examples, Crampton illustrated how such techniques can be used, often in combination with the results of beam experiments and theory to obtain cross-section data. The molecules discussed were H_2, N_2, O_2, CO, CO_2 and CH_4.

During the past ten years, there has been a growing

Table 1. Areas of applications for various atoms and molecules. (** indicates "very important")

	Plasma Process	Gaseous Dielectrics	Discharge Switch	Gas Laser	Space Science	Radiation Research
He				**	*	*
Ne				**		*
Ar	*		*	**		*
Kr				**		*
Xe				**		*
H					**	
O	*				**	
F	**	*		*		
Cd				*		
H_2	*				**	*
N_2	*	*		**	**	*
O_2	*	*			**	*
F_2	**			**		*
Cl_2	**			*		*
CO				*	*	*
HCl				**		*
H_2O					*	**
CO_2				**	*	*
SO_2		*			*	
N_2O		*				*
COS		*		*		
NH_3				*	*	*
NF_3	*			*		
PH_3	*					
CH_4	**		*		*	*
SiH_4	**					
CF_4	**	*	*	*		*
CCl_2F_2	*	*			*	
CCl_4	*	*		*		
GeH_4	**					
SF_6	**	**		*		*
C_2H_6	*		*			*
Si_2H_6	**					
C_2F_6	*	*	*			*
C_3F_6		*				
C_3F_8		*	*			
C_4F_8		*				

interest in such systems as CH_4, SiH_4, CF_4, SF_6 and others as materials for the production of very large scale integrated circuits. Scientists require the values of cross-sections for these plasma-processing gases, for the analysis of glow discharges in plasma devices and in calculations of electron swarm data for these gases and their mixtures with inert gases, hydrogen, oxygen and others. The molecules F_2, HC_1, CO_2, OCS and others are important for various gas lasers. To analyze the mechanism of gas lasers, the values of electron-collision cross-sections for these gases are essential. Recent reviews are given in Applied Atomic Collision Physics, Vol. 3, 1982 [1]. For the gaseous insulators and diffuse discharge opening switches, the molecules SF_6, CCl_2F_2, SO_2, C_3F_6, C_2F_6 and their mixtures are of interest. L. G. Christophorou and his coworkers have published extensive experimental data in this area [2]. Cross-sections are also useful in the field of space science and radiation physics and chemistry. There are good reviews of this field [3][4].

The areas of applications of various gases are shown in Table 1. There are many other molecules for which electron collision cross-sections are yet to be determined.

We have determined the cross-sections for 24 atoms and molecules over the last ten years. These atoms and molecules are listed in Table 2.

Table 2. The gases for which the author has determined sets of cross-sections. The molecules in parentheses will be studied in the near future.

He	Ne	Ar	Kr	Xe			
H	O	F	Cd				
H_2	N_2	O_2	F_2		(Cl_2)	(Br_2)	
HCl					(CO)	(NO)	
H_2O	SO_2				(CO_2)	(N_2O)	(COS)
					(NH_3)	(NF_3)	
CH_4	SiH_4	CF_4	CCl_2F_2	CCl_4	(GeH_4)		
SF_6							
C_2H_6	Si_2H_6				(C_2F_6)	(C_3F_6)	(C_4F_8)

Table 3. Symbols of electron collision cross-sections and electron swarm parameters along with their typical errors.

Collision Cross-Sections (as a function of electron energy)

elastic (total)	q_t	10 %
momentum transfer	q_m	15
rotational excitation	q_r	30
vibrational excitation	q_v	30
electronic excitation	q_e	30
dissociation	q_d	30
ionization	q_i	10
attachment	q_a	20
grand total (sum of q, except q_m)	Q_T	5

Electron Swarm Parameters (as a function of E/N, E: electric field, N: gas number density)

drift velocity	W	2 %
diffusion coefficient/mobility	D_T/μ	5
(transverse and longitudinal)	D_L/μ	10
diffusion coefficient	D_TN	5
(transverse and longitudinal)	D_LN	10
mean energy	$<\varepsilon>$	10
ionization coefficient	α/N	2
excitation coefficient	α_e/N	20
attachment coefficient	η/N	5

These electron collision cross-sections for molecules have been determined from available data from electron beam and the electron swarm experiments via the Boltzmann equation and the Monte Carlo simulation method. The beam data was given highest priority. Theoretical values of cross-sections were sometimes used. The method is almost the same one that was used by Phelps and Crompton, and their co-workers. The conventional, isotropic two-term approximation to the Boltzmann equation was used. As shown in Table 3, the errors in the cross-sections q are from 10 to 30% or more. It has often been discussed that the two-term approximation is not adequate for the gases with Ramsauer minimum in q_m and large values of q_v, such as CH_4 and CF_4. Actually, calculated values of the electron energy distribution function $f^{(1)}$ are larger than the values of $F^{(0)}$ at medium E/N values. Moreover, their shapes are sometimes unusual. The errors of the calculated swarm parameters using a two-term approximation are not as large as expected. Errors of 10 to 20%

for electron swarm parameters are certainly not unusual. We plan to present very accurate values of electron swarm parameters for relevant gases as soon as possible. The present cross-section values are not final. When exact cross section values or very good, swarm data are available, we will refine these cross-section values.

I would like to emphasize that the values of electron drift velocity W and ionization coefficient α/N can be measured with 2% error, as shown in Table 3. Now, the errors of q from the beam experiments are about 10% or larger. The errors in Q_T are about 5%. The values of α/N at low E/N are very sensitive to the values of q_e and q_d. When the exact values of q_i are given, we can determine the absolute values of total electronic excitation cross-sections q_e. Exact values of α/N are important for the determination of absolute values of q_e.

Cross-sections were determined for F_2, HCl, SO_2, CH_4, SiH_4, CF_4, CCl_2F_2, CCl_2, CCl_4, SF_6, C_2H_6 and Si_2H_6. For F_2, preliminary results were discussed elsewhere [5][6]. For SF_6, we pointed out that the measured Q_T values were inconsistent with the sum of elastic and inelastic cross-sections used by many workers [7]. Our cross-section also have this same inconsistency for SF_6. For HCl, Penetrante and Bardsley [8] determined a set of cross-sections by the Monte Carlo simulation method. Independently, we also obtained the same cross-sections. The first determination of a set of cross-sections for CH_4 and SiH_4 was carried out by Kline [9] and Garscadden, et al. [10]. Recently, Tagashira and his co-workers also carried out similar studies for CH_4 and SiH_4. Very recently, Novak, et al. [11] have published cross-sections for CCl_2F_2. The other gases, SO_2, CF_4, CCl_4, C_2H_6 and Si_2H_6, have no cross-section data until recently. We have reported a set of cross-sections for these molecules and also their calculated electron swarm parameters at a recent meeting in Japan.

In this report, we begin discussing the cross-sections for N_2. Data on the cross-sections for collisions of electrons and photons with N_2 was conpiled last year by Itikawa et al. [12]. These results are summarized in Fig. 1. Cross-sections for N_2 were also determined by Phelps [13]. Cross-sections for CF_4 will be discussed in detail, and cross-sections for the other gases are shown in the appendix.

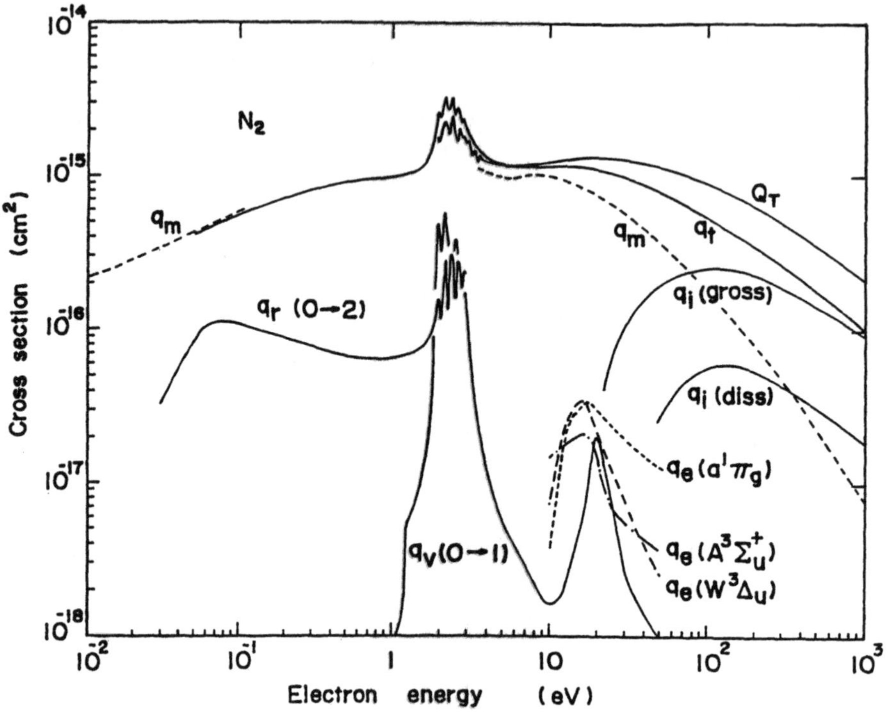

Fig. 1. Cross-sections of electrons in N_2 (1984).

There are no absolute elastic cross-sections (q_t and q_m) for CF_4. But, Hill and Woodcook [14] measured the relative DCS for electron energies from 4.1 to 39.7 eV. Recently, Tossell and Davenport [15] calculated the values of q_t for electron energies from 0.2 to 7.2 eV. Riley, et al's [16] theoretical data for q_t and q_m were used for electron energies up to 1000 eV. Field, et al. [17] reported experimental data, but we could not use their data. Important data are the values of W measured by Christophorou, et al. [18] and Naidu and Prasad [19], and values of D_T/μ measured by Naidu and Prasad [19] and Lakshminarashimha, et al. [20]. Values of q_m and q_v have been chosen so as to agree with the calculated and measured values of W and D_T/μ. Schatz and Hornig [21] measured the intensities of the infra-red absorption bands and Shimanouch [22] have summarized the molecular vibrational frequencies for CF_4. After many trials, we chose tentative values of q_m, q_{v24} and q_{v13}, on the basis of my experience. As shown in Fig. 2, the values of q_{v13} are very large. We hope to measure these q_{v13} and q_{v24} values by beam experiments.

Leiter, et al.[23] measured the values of q_i from

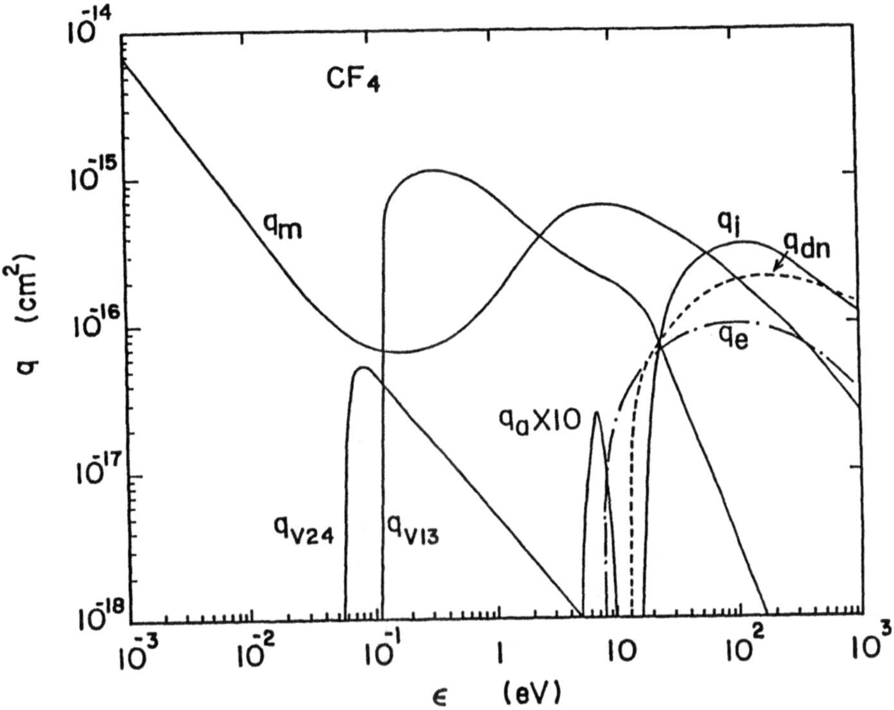

Fig. 2. Cross-sections of electrons in CF_4 (1984).

threshold to 180 eV. We have used these values, multiplied by 1.1 so as to fit the experimental values of α/N. The ionization energy is 16.25 eV which is the largest for molecules. Dissociation cross-sections q_d have been measured by Winters, et al. [24] [25]. Their q_d values probably include the contribution of dissociative ionization. We hence used the neutral q_d values $q_{dn} = q_d - q_i$. The dissociation energy is 12.5 eV. There are no electronic excitation cross-section q_e for CF_4. The values of q_e were hence determined by the same method given by Hayashi [26] from experimental values of α/N.

Values of q_a have been measured by Lifshitz and Grajower [27], Harland and Franklin [28], Spyrou, et al. [29] and Hunter and Christophorou [30]. The shape of the total q_a (sum of F^- and CF_3^-) are almost the same. The values of η/N were calculated and then compared with the experimental values of η/N, Finally, the values of q_a were increased with the maximum value for q_a of 2.52 x 10^{-16} cm^2 at 7.0 eV. Hunter's maximum value is 1.56 , 10^{-16} cm^2 at 7.5 eV.

The values of α/N and η/N were measured by Bozin and Goodyear [31], Bornik and Panov [32], Naidu and Prasad [33],

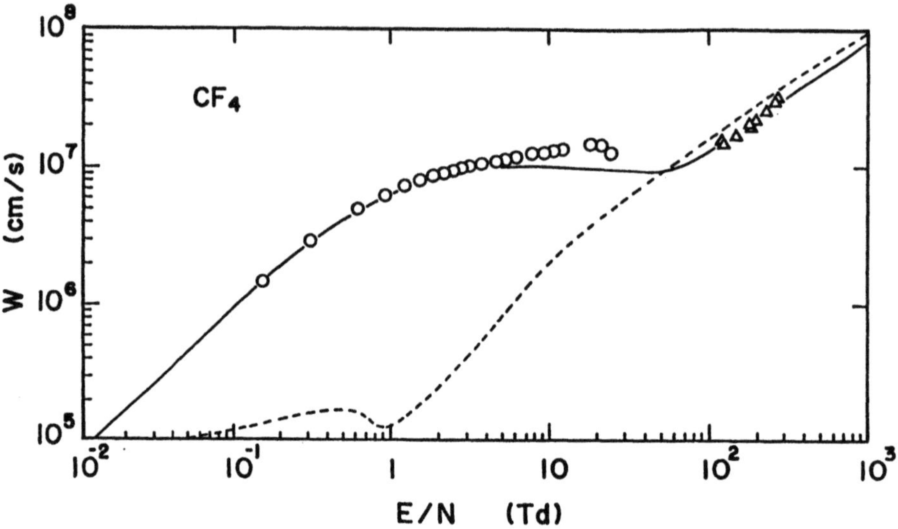

Fig. 3. Drift velocity of electrons W in CF_4. (---): calculated values of W for $q_v = 0$.

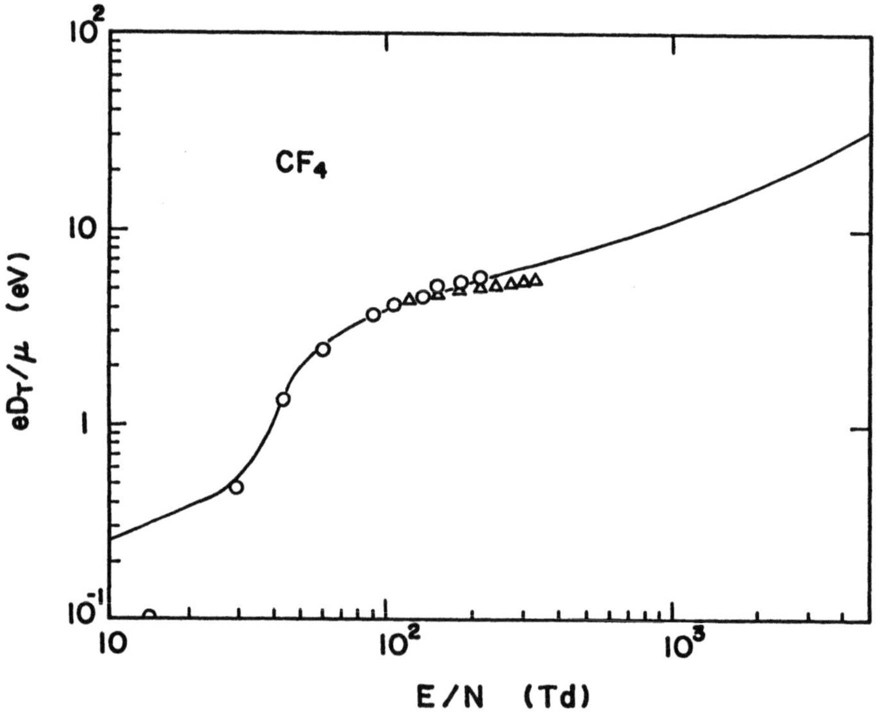

Fig. 4. Ratio of transverse diffusion coefficient to mobility D_T/μ in CF_4.

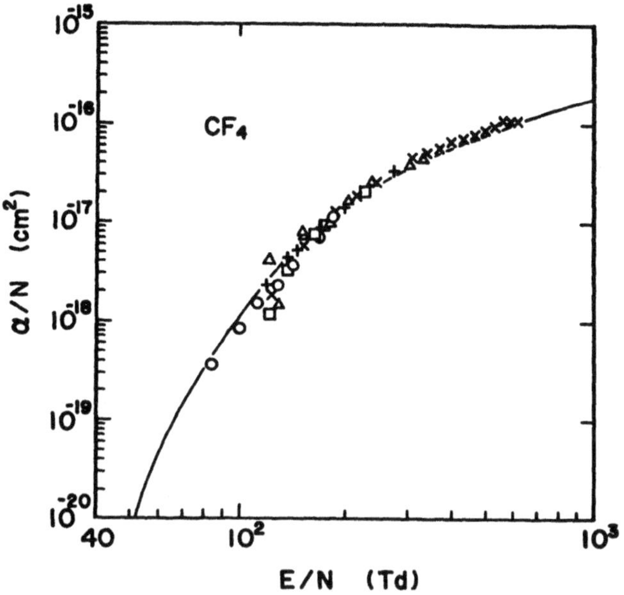

Fig. 5. Ionization coefficient α in CF_4.

Lakshiminarasimha, et al. [34] [35] and Shimozuma, et al. [36] for E/N from 120 to 624 Td.

Our calculated swarm parameters are shown in Figures 3, 4, 5 and 6 along with experimental values. In these figures, the solid lines are our calculated values,and points such as circle and cross are the experimental data given in the literatures.

Next, we calculated the values of W for the CF_4 - inert gas (all, but mainly in Ar and Xe) mixtures. Christophorou et al. [37][38] measured the values of W for these mixtures. An example of these calculated values is shown in Fig. 7 for the CF_4 - Xe mixtures. The calculated values of W are almost identical with the experimental values, except for Ar - 1%CF_4 and Xe - 1%CF_4 mixtures. I believe that the disagreements are due to errors of the assumed value of 1% presence of CF_4. The values of a set of cross-sections for the other nine gases are shown in Fig. 8 to 17. Work on other gases is in progress. The details will be published in the near future.

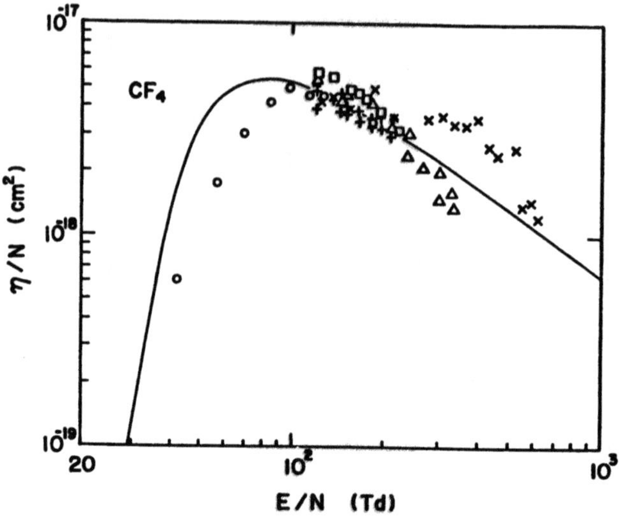

Fig. 6. Attachment coefficient η in CF_4.

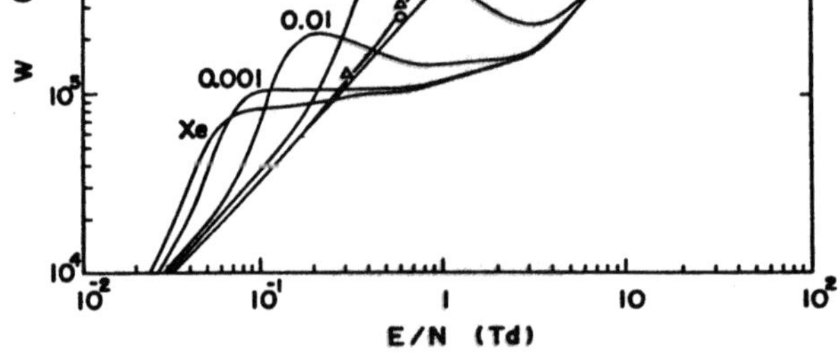

Fig. 7. Drift velocity of electrons W in Xe - CF_4.

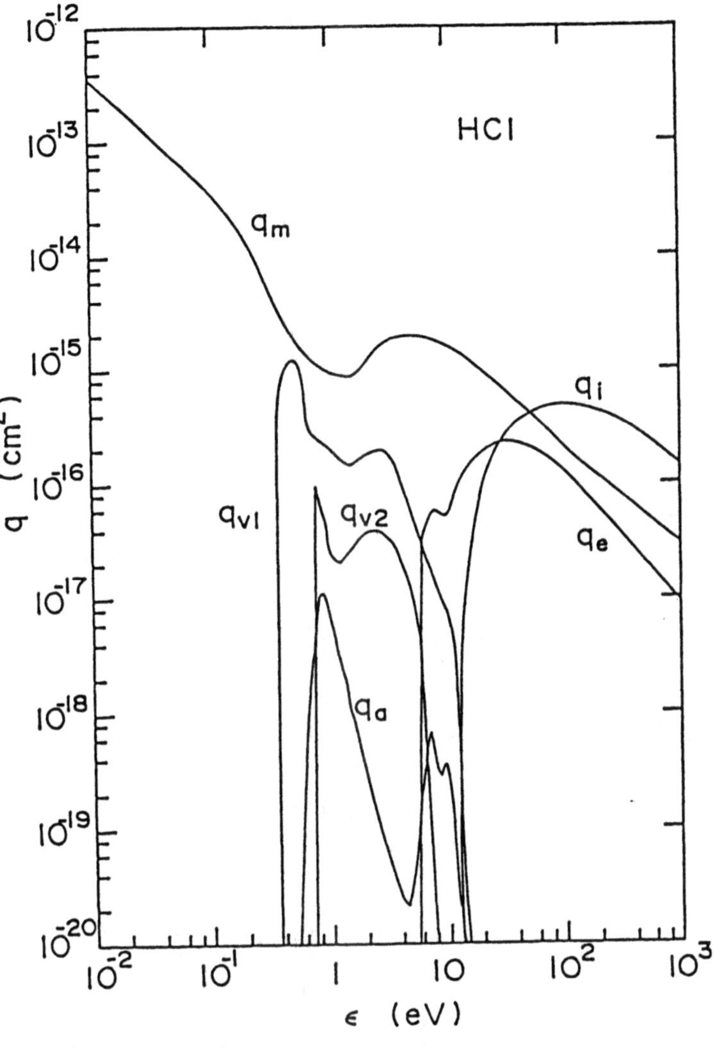

Fig. 8. Cross-sections of electrons in HCl(1983).

Important references for HCl:
N. T. Padial and D. W. Norcross, Phys. Rev. A27, 141 (1983)
B. M. Penetrante, et al., J. Appl. Phys. 54, 6150 (1983)
K. Rohr and F. Linder, J. Phys. B9, 2521 (1976)
F. Linder, Private Communication (1983) DCS of elast. scatt.
P. J. Chantry, Applied Atomic Collision Physics III, 35 (1982)
D. K. Davies, 33rd GEC, 84 (1980)
D. K. Davies, et al., Proc Phys. Soc. 80, 898 (1962)
K. T. Compton, et al., Phys. Rev. 26, 436 (1925)
V. A. Bailey and W. E. Duncanson, Phil. Mag. 10, 145 (1930)

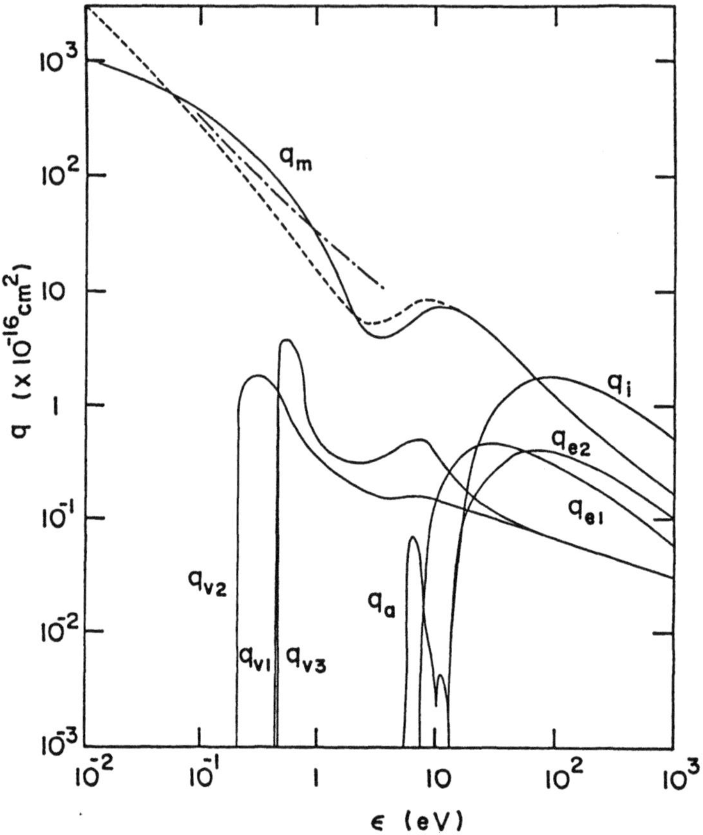

Fig. 9. Cross-sections of electrons in H_2O(1983).

Important references for H_2O:
M. Inokuti, ed., Proc. of the Workshop, ANL-84-28 (1984)
M. Zaider, et al., Radiation Res. 95, 231 (1983)
J. L. Pack and A. V. Phelps, Phys. Rev. 127, 2084 (1962)
G. Seng and F. Linder, J. Phys. B3, 1252 ('70), B9, 2539 ('76)
A. Danjo, et al., J. Phys. Soc. Japan, 54, 1224 (1985)
S. Trajmar, et al., J. Chem. Phys. 58, 2521 (1973)
A. E. S. Green, et al., J. Geophys. Res. 82, 5104 (1977), Proc. Biophys. Aspects Radiat. Qual. Symp. 79 (1971)
J. J. Olivero, et al., J. Geophys. Res. 77, 4797 (1972)
C. E. Melton, J. Chem. Phys. 57, 4218 (1972)
A. V. Risbud, et al., J. de Phys. 40, C7-77 (1979)
J. J. Lowke and J. A. Rees, Aust. J. Phys. 16, 447 (1963)
J. F. Wilson, et al., J. Chem. Phys. 62, 4204 (1975)
R. W. Crompton, et al., Aust. J. Phys. 18, 541 (1965)
F. E. Spencer,Jr., et al., Proc. 15th Sympo. Engineering Aspects of Magnetohydrodynamics, 103 (1976)
F. A. Gianturco and D. G. Thompson, J. Phys. B13, 613 (1980)

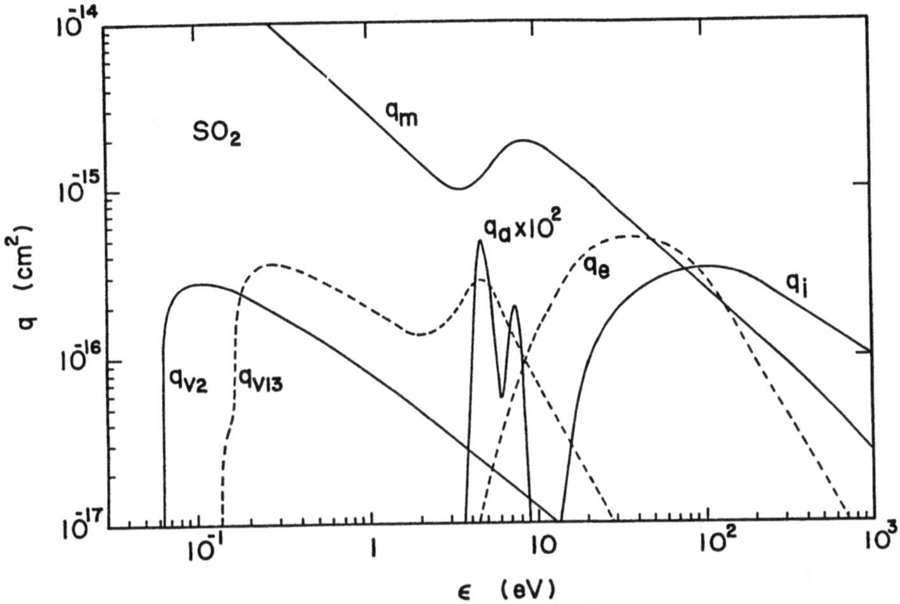

Fig. 10. Cross-sections of electrons in SO_2(1985)

Important references for SO_2:

O. J. Orient, et al., J. Chem. Phys. 77, 3523 (1982)
O. J. Orient, et al., J. Chem. Phys. 78, 2949 (1983)
M. Zubek, et al., Europ. Conf. Atomic Phys. Abst. 763 (1981)
F. E. Spencer, Jr. and A. V. Phelps, Proc. 15th Sympo. on Eng. Aspects of Magnetohydro. 103 (1976)
L. Vuskovic and S. Trajmar, J. Chem. Phys. 77, 5436 (1982)
O. I. Smith and J.S. Stevenson, J. Chem.Phys. 74, 6777 (1981)
I. M. Cadez, et al., J. Phys. D16, 305 (1983)
J. Rademacher, L. G. Christophorou and R. P. Blaunstein, J. Chem. Soc. Faraday Trans., II 71, 1212 (1975)
F. W. Wheatley, Phil. Mag. 26, 1034 (1913)
H. Schlumbohm, Z. f. Phys. 166, 192 (1962)
J. L. Moruzzi, et al., J. de Phys. C7, 11 (1979)
R. O. Jones, J. Chem. Phys. 82, 325 (1985) (1973)
C. Lifshitz, et al., Int. J. Mass Spect. Ion Phys. 11, 243
P. W. Harland, et al., J. Chem. Phys. 58, 1430 (1973)
V. Y. Foo and J. B. Hasted, Proc. Roy. Soc. A322, 535 (1971)
J. S. Bulger, et al., Can. J. Phys. 49, 1437 (1971)
J. Erickson, et al., J. Chem. Phys. 75, 1650 (1981)
T. Shimanouchi, J. Phys. Chem. Ref. Data 6, 993 (1977)
D. Simon, et al., J. Phys. B11, L561 (1978)
L. Andric, et al., J. Phys. B16, 1837 (1983)
K. Kraus, Z. Naturforsch. 16a, 1378 (1961)

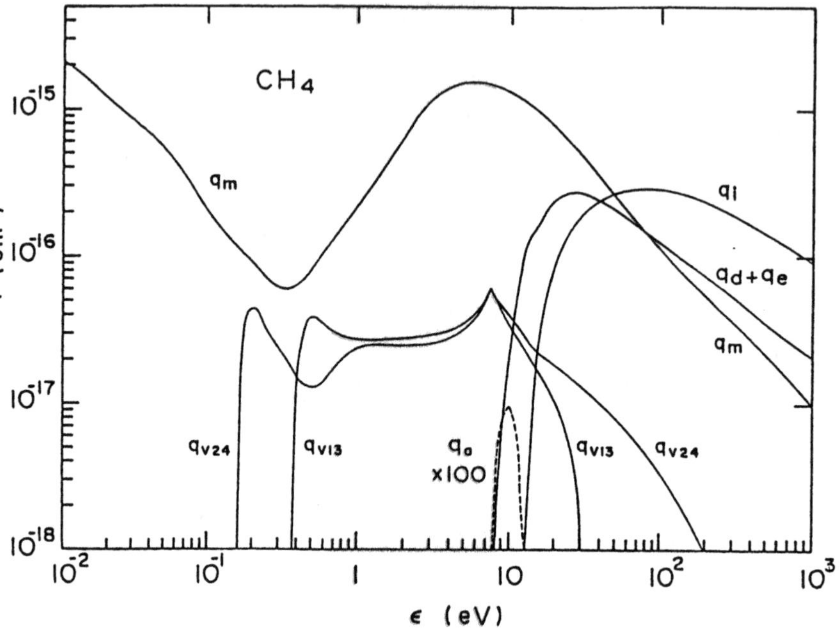

Fig. 11. Cross-sections of electrons in CH_4(1983).

Important references for CH_4:
L. E. Kline, IEEE Trans. PS-10, 224 (1982)
H. Tanaka, et al., J. Phys. B15, 3305 (1982), B16, 2861 (1983)
G. N. Haddad, from R.W. Crompton, XVI ICPIG, Dusseldorf (1983)
L. Vuskovic and S. Trajmar, J. Chem. Phys. 78, 4947 (1983)
W. Sohn, et al., J. Phys. B16, 891 (1983)
H. F. Winters, J. Chem. Phys. 63, 3462 (1975)
D. Rapp, et al., J. Chem. Phys. 43, 1464 (1965)
B. Adamczyk, et al., J. Chem.Phys. 44, 4640 (1966)
T. E. Sharp and J. T. Dowell, J. Chem. Phys. 46,1530 (1967)
X. Fink, et al., Helv. Phys. Acta, 38, 717 (1965)
Y. Nakamura, Private communication (1983) Values of W
T. E. Bortner, et al., Rev. Sci. Inst. 28, 103 (1957)
L. Frommhold, Z. f. Phys. 150, 172 (1958), 156, 144 (1959)
W. Franke, Z. f. Phys. 158, 96 (1960)
W. J. Pollock, Trans. Faraday Soc. 64, 2919 (1968)
N. Gee and G. R. Freeman, Phys. Rev. A20, 1152 (1979)
L. W. Cockran and D. W. Forester, Phys. Rev. 126, 1785 (1962)
C. W. Duncan and I. C. Walker, J. Chem. Soc. Faraday Trans.
II 68, 1514 (1972) Electronics 39, 653 (1975)
A. E. D. Heylen, J. Chem.Phys. 38, 765 (1963), Int. J.
C. S. Lakshminarasimha, et al., J. Phys. D10, 313 (1977)
R. K. Jones, J. Chem.Phys. 82, 5424 (1985)

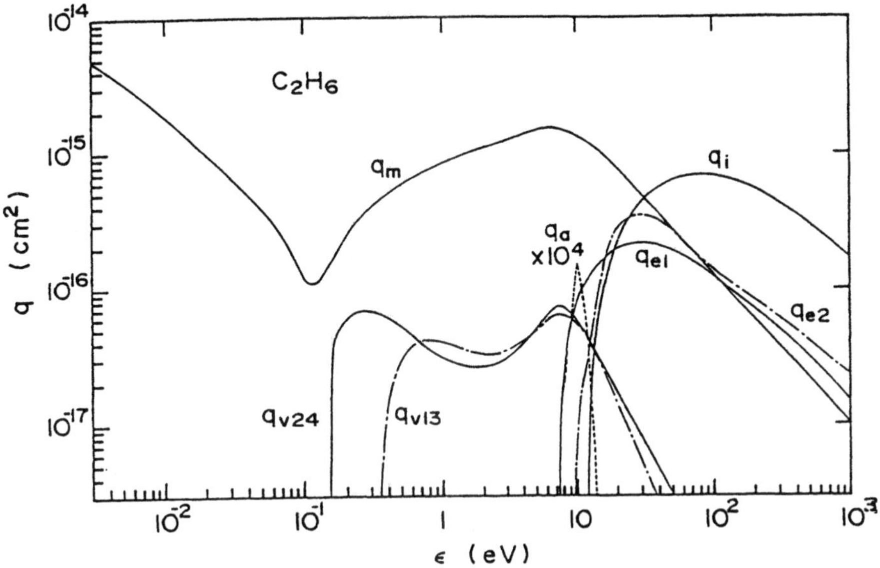

Fig. 12. Cross-sections of electrons in C_2H_6(1983).

Important references for C_2H_6:

C. R. Bowman and D.E. Gordon, J. Chem. Phys. 46, 1878 (1967)
J. L. Cottrell, et al., Trans. Faraday Soc. 61, 1585 (1965), 64, 2260 (1968)
C. W. Duncan and I. C. Walker, J. Chem. Soc. Faraday Trans. II 70, 577 (1974)
M. Fink, et al., J. Chem. Phys. 63, 1985 (1975)
D. L. McCorkle, et al., J. Phys. B11, 3067 (1978)
N. Gee and G. R. Freeman, Phys. Rev. A22, 301 (1980)
D. Matsunaga, et al., 12th ICPEAC, 358 (1981)
H. Tanaka, from S. Trajmar, et al., Phys. Rep. 97, 219 (1983)
C. R. Bowman and W. D. Miller, J. Chem. Phys. 42, 681 (1965)
H. H. Brongersma, et al., Chem. Phys. Lett. 3, 437 (1969)
L. V. Trepka and H. Neuert, Z. Naturforsch. 18a, 1295 (1963)
B. L. Schram, et al., J. Chem. Phys. 44, 49 (1966)
I. Suzuki, et al., Int. J. Mass Spect. Ion Phys. 24, 147 ('77)
H. Chatham, et al., J. Chem.Phys. 81, 1770 (1984)
J. Perrin, et al., Chem. Phys. 73, 383 (1982)
A. E. D. Heylen, J. Chem. Phys. 38, 765 (1963), Int. J. Electronics 39, 653 (1975)
O. H. LeBlanc, Jr. and J. C. Devins, Nature, 188, 219 (1960)
Y. Chatelus, et al., Nucl. Inst. Meth. 171, 127 (1980)
J. Ruthowsky, et al., Ann. d. Phys. 37, 259 (1980)
H. F. Winters, Chem.Phys. 36, 353 (1979)
J. H. Xu, et al., J. Phys. B16, 3863 (1983)
I. M. Nyquist, et al., J. Chem. Phys. 26, 552 (1957)
P. J. Curry, et al., J. Phys. B18, 2303 (1985)

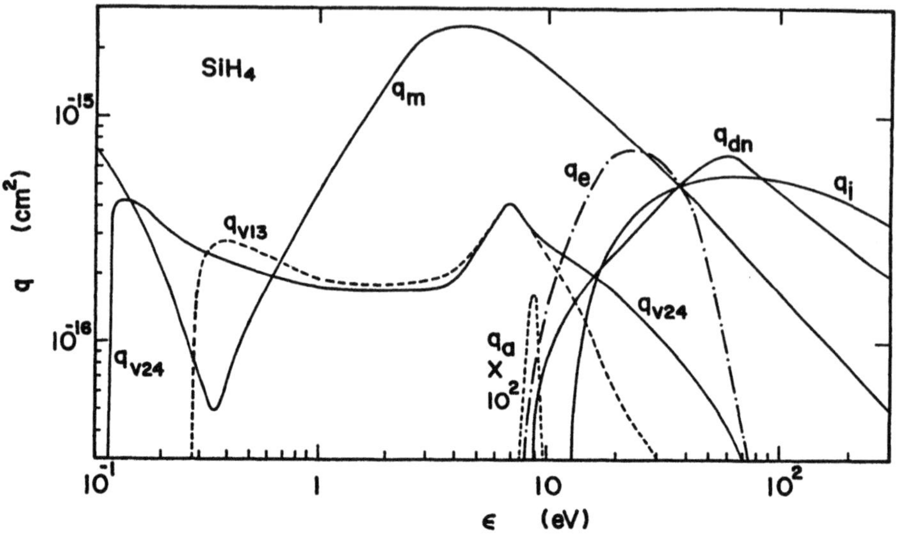

Fig. 13. Cross-sections of electrons in SiH_4(1984).

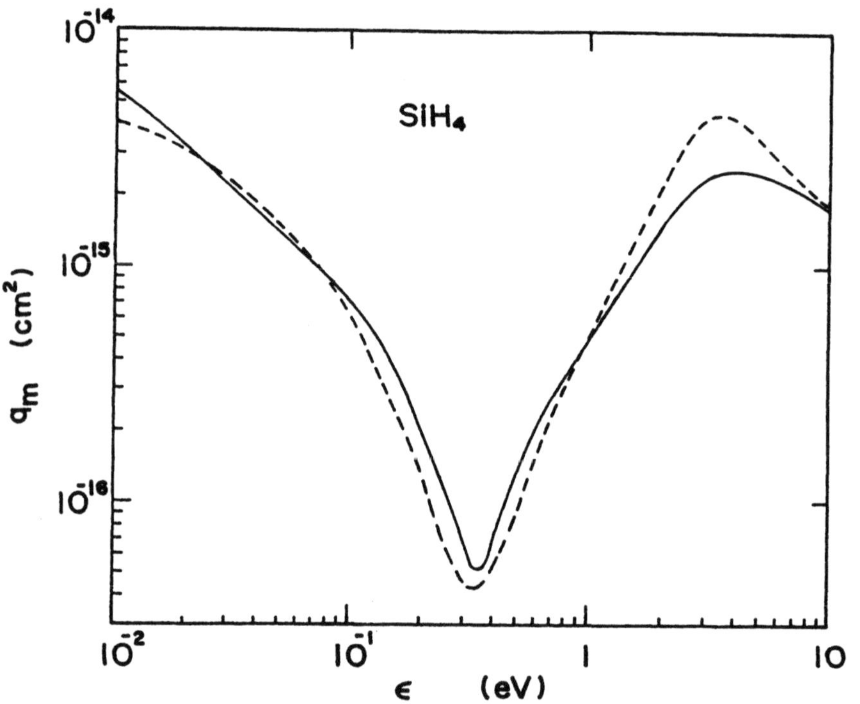

Fig. 14. Values of q_m for SiH_4 given by Jain. (---) in this figure and (——) in Fig. 13.

Important references for SiH_4:
C. A. DeJoseph, et al., Proc. Int. Conf. on Lasers, 738 ('82)
A. Garscadden, et al., Appl. Phys. Lett. 43, 1012 (1983)
M. Hayashi, Dry Process Symposium, Tokyo, V-7, 127 (1984)
J. Perrin, et al., Chem. Phys. 73, 383 (1982)
H. Chatham, et al., J. Chem. Phys. 81, 1770 (1984)
T. L. Cottrell, et al., Trans. Faraday Soc. 61, 1585 (1965)
W. L. Pollock, Trans. Faraday Soc. 64, 2919 (1968)
M. Shimozuma, et al., Tech. Papers of Electrical Discharge, IEE Japan, ED-83-86 (1983) (in Japanese) α and η
O. Sueoka, Private communication (1984) Exp. values of Q_T
I. C. Walker, Private comm. (1984) Exp. values of W and D_T/μ
A. Jain, Private communication (1985) Fig. 14 and others
J. Perin and J. F. M. Aarts, Chem. Phys. 80, 351 (1983)
W. B. Steward, et al., Phys. Rev. 47, 828 (1935)
C. H. Tindal, et al., Phys. Rev. 62, 151 (1942)
T. F. Deutsch, J. Chem. Phys. 70, 1187 (1979)

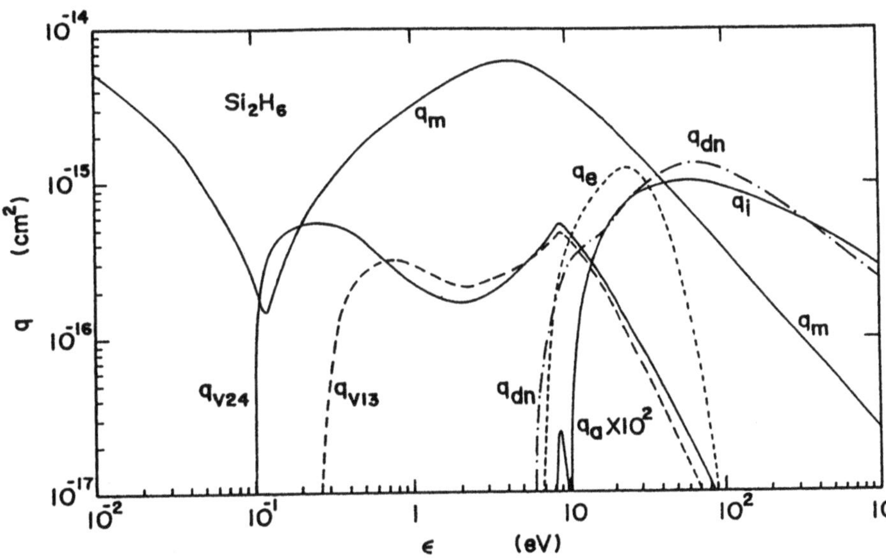

Fig. 15. Cross-sections of electrons in Si_2H_6(1985)

Important references for Si_2H_6:
P. Potzinger and F. W. Lampe, J. Phys. Chem. 73, 3912 (1969)
J. Perrin, et al., Chem. Phys. 73, 383 (1982)
M. Hayashi, Dry Process Symposium, Tokyo V-7, 127 (1984)
H. Chatham, et al., J. Chem. Phys. 81, 1770 (1984)
M. Shimozuma, et al., Meeting of IEE, Japan, 168 (1985) (in Japanese) Values of α and η

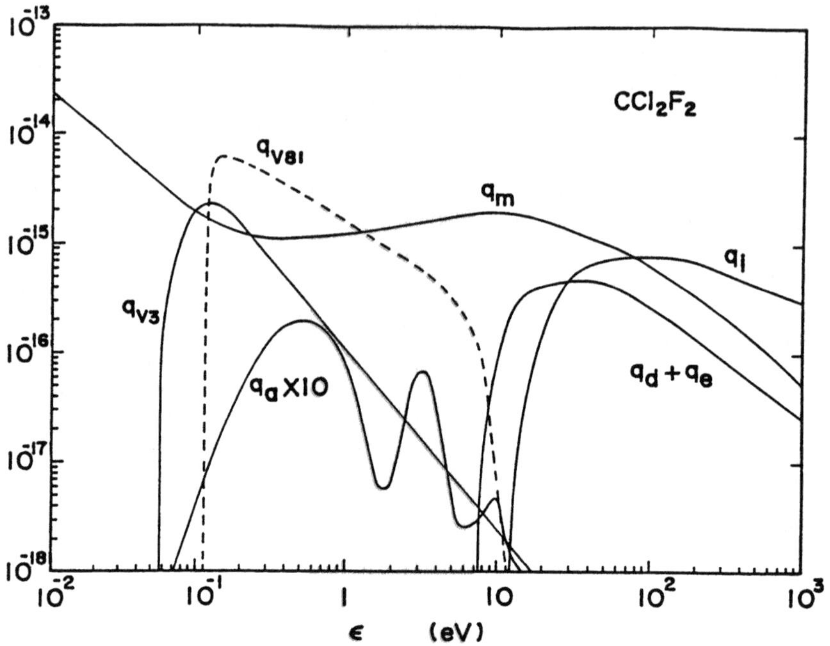

Fig. 16. Cross-sections of electrons in CCl_2F_2(1984).

Important references for CCl_2F_2:

T. Shimanouchi, J. Phys. Chem. Ref. Data 3, 269 (1974)
K. Rohr, XIth ICPEAC, Kyoto 322 (1979)
H. W. Jochims, et al., Ber. Bunsenges. Phys. Chem. 80, 130 (1976)
G. J. Verhaart, et al., Chem. Phys. 34, 161 (1978)
J. Doucet, et al., J. Chem. Phys. 58, 3708 (1973)
G. C. King and J. W. McConkey, J. Phys. B11, 1861 (1978)
V. M. Pejcev, et al., Chem. Phys. Lett. 63, 301 (1979)
D. L. McCorkle, et al., 2nd Int. Swarm Semi. ORNL I-3 (1981)
E. Illenberger, Ber. Bunsenges. Phys. Chem. 86, 252 (1982)
K. Stephan, Private communication (1984) Exp. values of q_i
H. Deutsch, et al., Beit. Plasmaphys. 24, 475 (1984)
M. S. Naidu and A. N. Prasad, J. Phys. D2, 1431 (1969)
V. N. Maller, IEEE Trans. IA-16, 724 (1980)
M. S. Siddagangappa, et al., Gaseous Dielectrics III, 40 ('82)
M. S. Siddagangappa, et al., J. Phys. D16, 1595 (1983)
M. A. Harrison and R. Geballe, Phys. Rev. 91, 1 (1953)
C. Raja Rao, et al., Int. J. Electronics 35, 49 (1973)
H. A. Boyd, et al., Proc. 1st Int.Conf. Gas Dis. 426 (1970)
V. N. Maller, et al., Proc. 3rd Int.Conf. Gas Dis. 409 (1974)
J. L. Moruzzi, Brit. J. Appl. Phys. 14, 938 (1963)
H. Schlumbohm, Z. f. Phys. 166, 192 (1962)
J. P. Novak, et al., J. Appl. Phys. 57, 4368 (1985)

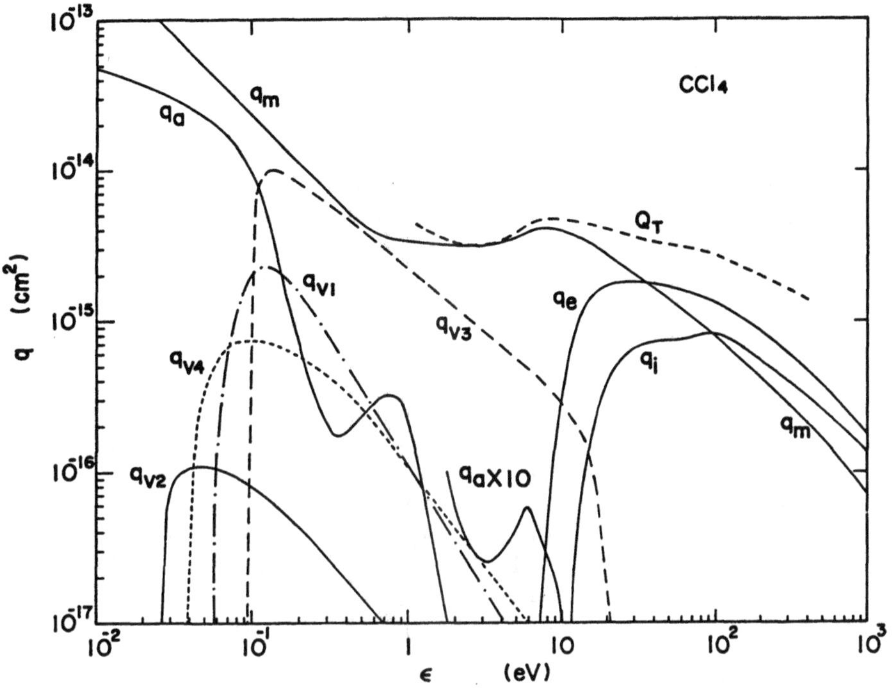

Fig. 17. Cross-sections of electrons in CCl_4(1984).

Important references for CCl_4:

H. Daimon, et al., J. Phys. Soc. Japan, 52, 84 (1983)
S. Hill, et al., Proc. Roy Soc. A155, 331 (1936)
J. A. Tossell, et al., J. Chem. Phys. 80, 816 (1984)
T. Shimanouchi, J. Phys. Chem. Ref. Data 3, 269 (1974)
C. R. Zobel, et al., J. Am. Chem. Soc. 77, 2611 (1955)
H. U. Scheunemann, E. Illenberger and H. Baumgartel, Ber. Bunsenges. Phys. Chem. 84, 580 (1980)
L. G. Christophorou, et al., J. Phys. B4, 1163 (1971)
A. A. Christodoulides, et al., J. Chem. Phys. 54, 4691 (1971)
K. Leiter, et al., Symp. Atomic and Surface Phys. (1984)
W. Holst, et al., K. Nor. Vidensk. Selsk, 4, 89 (1931)
O. Sueoka, Private communication (1984) Exp. values of Q_T
M. E. Riley, et al., Atomic Data Nucl.Data Table 5, 443 ('75)
R. Geballe and M. A. Harrison, from L. B. Loeb, Basic Processes of Gaseous Electronics, Univ. Calif. P., 414 (1955)
F. J. Davis, et al., J. Chem. Phys. 59, 2324 (1973)
J. Marriott, et al., Proc. Phys. Soc. B67, 437 (1954)
D. Spence and G. Schulz, J. Chem. Phys. 58, 1800 (1973)
T. G. Lee, J. Phys. Chem. 67, 360 (1963)
K. Watanabe, J. Chem. Phys. 26, 542 (1957)
A. Chutjian and S. H. Alajajian, Phys. Rev. A31, 2885 (1985)

The authors wish to thank A.V. Phelps, R.W. Crompton, K. Takayangi; I. Ogawa, S. Trajmar and V. McKoy for useful discussions and encouragement.

References

1 E. W. McDaniel and W. L. Nighan, ed., Applied Atomic Collision Physics, Vol.3, Gas Lasers, Academic Press, 1982
2 H. S. W. Massey, E. W. McDaniel and B. Bederson, ed., Applied Atomic Collision Physics, Vol. 5, Special Topics, Academic Press, 1982
3 H. S. W. Massey and D. R. Bates, ed., Applied Atomic Collision Physics, Vol. 1, Atmospheric Physics and Chemistry, Academic Press, 1982
4 M. Inokuti, ed., Proceedings of the Workshop on Electronic and Ionic Collision Cross-sections needed in the Modeling of Radiation Interactions with Matter, ANL-84-28, Argonne National Laboratory, 1983
5 M. Hayashi and T. Nimura, J. Appl. Phys. 54, 4879 (1983)
6 M. Hayashi, Jap. J. Appl. Phys. Lett. 22, 565 (1983)
7 M. Hayashi and T. Nimura, J. Phys. D17, 2215 (1984)
8 B. M. Penetrante and J. N. Bardsley, J. Appl. Phys. 54, 6150 (1983)
9 L. E. Kline, IEEE Trans., PS-10, 224 (1982)
10 A. Garscadden, G. L. Duke and W. F. Bailey, Appl. Phys. Lett. 43, 1012 (1983)
11 J. P. Novak and M. F. Frechette, J. Appl. Phys. 57, 4368 (1985)
12 Y. Itikawa, M. Hayashi, A. Ichimura, K. Onda, K. Sakimoto, K. Takayanagi, M. Nakamura, H. Nishimura and T. Takayanagi, ISAS Research Note, 291 (1985)
13 A. V. Phelps and L. C. Pitchford, Phys. Rev. A31, 2932 (1985); JILA Report No. 26 (1985)
14 S. Hill and A. H. Woodcock, Proc. Roy. Soc. A155, 331 ('36)
15 J. A. Tossell and J. W. Davenport, J. Chem. Phys. 80, 813 (1984)
16 M. E. Riley, C. J. MacCallum and F. Biggs, Atomic Data & Nucl. Data Tables 5, 443 (1975)
17 D. Field, J. P. Ziesel, P. M. Guyon and T. R. Govers, J. Phys. B17, 4565 (1984)
18 L. G. Christophorou, D. L. McCorkle, D. V. Maxey and J. G. Carter, Nucl. Instr. Meth. 163, 141 (1979)
19 M. S. Naidu and A. N. Prasad, J. Phys. D5, 983 (1972)
20 C. S. Lakshminarasimha, J. Lucas and D. A. Price, Proc. IEE, 120, 1044 (1973)
21 P. N. Schatz and D. F. Hornig, J. Chem. Phys. 21, 1516('53)
22 T. Shimanouch, NSRDS-NBS 39, 48 (1972)
23 K. Leiter, K. Stephan, H. Deutsch and T. D. Mark, Symp. Atomic and Surface Phys. (1984)

24 H. F. Winters, J. Appl. Phys. 48, 4973 (1977)
25 H. F. Winters and M. Inokuti, Phys. Rev. A25, 1420 (1982)
26 M. Hayashi, J. Phys. D16, 581 (1983)
27 C. Lifshitz and R. Grajower, Int. J. Mass Spect. Ion Phys. 10, 25 (1972)
28 P. W. Harland and J. L. Franklin, J. Chem.Phys. 61, 1621 (1974)
29 S. M. Spyrou, I. Sauers and L. G. Christophorou, J. Chem. Phys. 78, 7200 (1983)
30 S. R. Hunter and L. G. Christophorou, J. Chem. Phys. 80, 6150 (1984)
31 S. E. Bozin and C. C. Goodyear, J. Phys. D1, 327 (1968)
32 I. M. Bortnik and A. A. Panov, Sov. Phys. Tech. Phys. 16, 571 (1971)
33 same as 19
34 same as 20
35 C. S. Lakshminarasimha, J. Lucas and R. A. Snelson, Proc. IEE, 122, 1162 (1975)
36 M. Shimozuma, H. Tagashira and H. Hasegawa, J. Phys. D16, 971 (1983)
37 same as 18
38 L. G. Christophorou, D. V. Maxey, D. L. McCorkle and J. G. Carter, Nucl. Instr. Meth., 171, 491 (1980)

PART V

STUDIES OF SINGLE-COLLISION PHENOMENA

LOW-ENERGY, HIGH-RESOLUTION ELECTRON-MOLECULE COLLISION STUDIES

H. Ehrhardt

Fachbereich Physik der Universität, D-6750 Kaiserslautern
Federal Republic of Germany

Abstract

This paper deals with recent developments of

i) the production and detection of low-energy (i.e. 0.1 eV to ca. 10 eV) and high-resolution (ca. 10 meV to 30 meV) electrons in crossed beam experiments,
ii) rotational excitations of molecules by electrons in the ground electronic states of molecules, and
iii) vibrational excitations of molecules close to their thresholds. Non-resonant and resonant interaction mechanisms are discussed.

Developments of experimental techniques in the last few years have made it feasible to perform reliable crossed beam electron-molecule collision studies with high energy and angular resolution ($\Delta E, \Delta\vartheta$) at very low impact energies (E_o). The optimum numbers which have been reached are of the order of

$\Delta E \sim 10$ meV for the overall energy uncertainty due to the energy spread in the electron gun and the electron detectors,

$\Delta\vartheta \sim \pm 2°$ for the uncertainty of the scattering angle due to the angular spreads in the primary electron beam, the beam intersection, and of the electron optics of the detectors,

$E_o \sim 50$-100 meV minimum energy of the electrons in the electron beam with reasonable beam currents of ca. 10^{-10} A and $E \sim 10$ meV for the minimum energy of the scattered electrons.

These achievements extend the well-developed electron impact spectroscopy [1] with crossed beams into the regime of swarm techniques with their high degree of accuracy of the determination of total cross-sections, but with weaknesses with respect to quantitative statements on selective cross-sections, if several reaction channels are open.

In this decade of impact energies from 0.08 eV to about 2 eV, which is now accessible to reliable beam experiments, rotational and vibrational excitations close to their thresholds can be studied. Here, direct, i.e. non-resonant and resonant interaction mechanisms can play quite different roles for different molecules and/or reaction channels. Since such interactions have different energy and angular momentum behaviour, the experiment can make statements on their contributions through energy and angular dependence measurements.

Recent Experimental Developments

Possibly the most important progress in high resolution electron impact spectrometry with crossed beams has been achieved in the last few years by the use of position-sensitive detectors. J. Comer et al. [2] have done most of the developing work. They have reached a lowest overall energy spread of 10 to 25 meV in the range of impact energies from about 2 eV to some 100 eV. In comparison to conventional spectrometers the time for measurements is reduced by a factor of 1000, which means a much higher reproducibility and therefore reliability of the data and an enormous increase of sensitivity. Lately, they have made measurements for N_2, CO, CO_2, HCl and other molecules.

In the last 10 years more and more experimental groups [1] use double monochromators and analysers, often called tandem energy selectors. As in optical spectrometers, they do not increase the resolution, but they reduce the background due to stray electrons, i.e., tails on the high and low-energy side of peaks are reduced considerably, so that small signals close to large peaks can be measured. Figure 1 shows schematically a somewhat unusual system [3] of this type, since it uses 127° selectors, which are electrically separated (in the tandem configuration) by a small lens system. This lens system can be used to shift the two transmission curves of the two selectors slightly with respect to each other, so that the overall resolution is electrically tunable. The exit lens system of the gun and the entrance lens system of the detector are quite open in

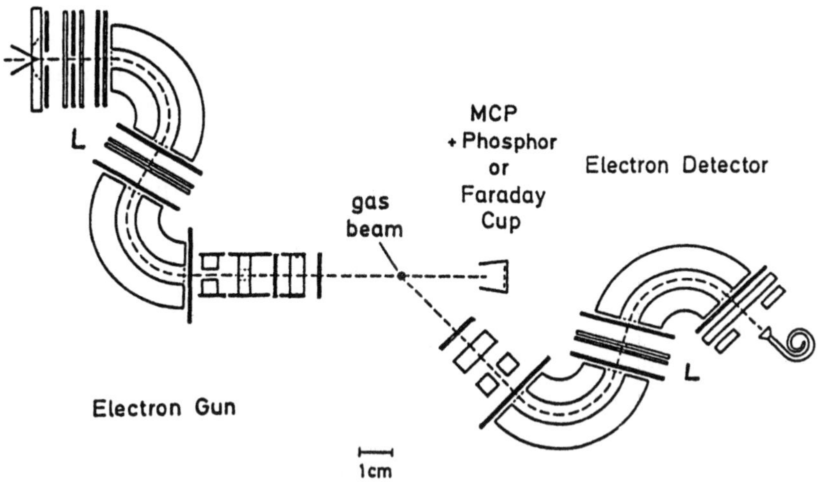

Figure 1. A double tandem electron impact spectrometer with electrically separated energy selectors. The lense systems L allow to tune the resolution from outside, shifting the transmission curve of each selector. The impact energy can be chosen as low as 50 meV, $\Delta E \geq 12$ meV FWHM, beam current $I_o \geq 1 \times 10^{-10}$A, scattering angle 0°...150°.

order to reduce electron impact contaminations on metal surfaces. This is important for low energies of the electrons before and after the collision. The shape of the electron beam is visualised by multichannel plates followed by a phosphor.

Besides these improvements of more conventional systems, new techniques are coming up and important progress can be expected in electron-molecule collision physics. One innovation concerns the preparation of molecules in a special state either by laser pumping or by selecting a special rotational state in a quadrupole field. After this special state has been bombarded with electrons or heavy particles, the new population of states is probed by laser fluorescence. In this way state-to-state inelastic cross-sections have already been measured for NO, Na_2, CaCl, S_2, NH_2, and other molecules after collisions with heavy particles, but the same methods could be used also for electron collisions.

Another rapidly developing technique concerns the use of polarized electrons and target beams [4] and collision studies of molecules bombarded with low energy and quite monochromatic positron beams [5]. By both techniques quite specific data are obtained which are mostly of great value for the study of exchange effects, short-range polarization interactions, shape resonances, virtual states etc. These data will be of importance for the further improvement of theoretical approximations in electron-molecule collisions.

Rotational Excitations of Molecules by Low-Energy Electrons in Beam Experiments [6-21]

It is known by now that pure rotational excitation of molecules by low-energy electrons can have cross-sections which differ by approximately five orders of magnitude depending on the interaction mechanism. For a molecule with a weak permanent quadrupole and weak polarizability, e.g. N_2 [14], rotational excitation cross-sections are ca. $10^{-18}cm^2$, whereas for the same molecule rotational excitation via a shape resonance [6,7,13,14] easily has cross-sections of ca. $10^{-16}cm^2$, and for molecules with a strong permanent dipole moment cross-sections of ca. 10^{-13} cm^2 can be obtained [8-12,15]. It is therefore interesting to make a systematic study on electron impact rotational excitations with specific considerations of the interaction mechanisms. This can only be done in a beam experiment with angular dependence measurements. On the other hand, the energy resolution even of the best electron spectrometers is still too low to separate state-to-state transitions, except for H_2, HD, and D_2 [6,7]. Therefore, the only measurable quantity which contains information on the cross-sections for rotational transitions is the line width and line shape of the "elastic" scattering peak, if it is measured in a spectrometer with low enough energy spread (ca. 10 to 20 meV) and stable conditions [13,14]. Figure 2 shows a few examples of line-shape measurements and results of rotational analysis. The rotationally broadened lines are found to have half-width between the width of the apparatus profile (true elastic scattering) and 2 to 4 times this value due to excessive rotational transitions. Important are the asymmetries in the line wings, which contain information on the energy transfer processes from the free electron to the molecule (energy loss) and transfer of rotational energy from the molecule to the free electron (energy gain).

For the evaluation of line shapes in terms of rotational cross-sections two methods have been worked out, i) the branch construction method, which depends on the high-J-approximation [17] and ii) a cross-section formula for transitions of a spherical symmetric rotator by Shimamura [18].

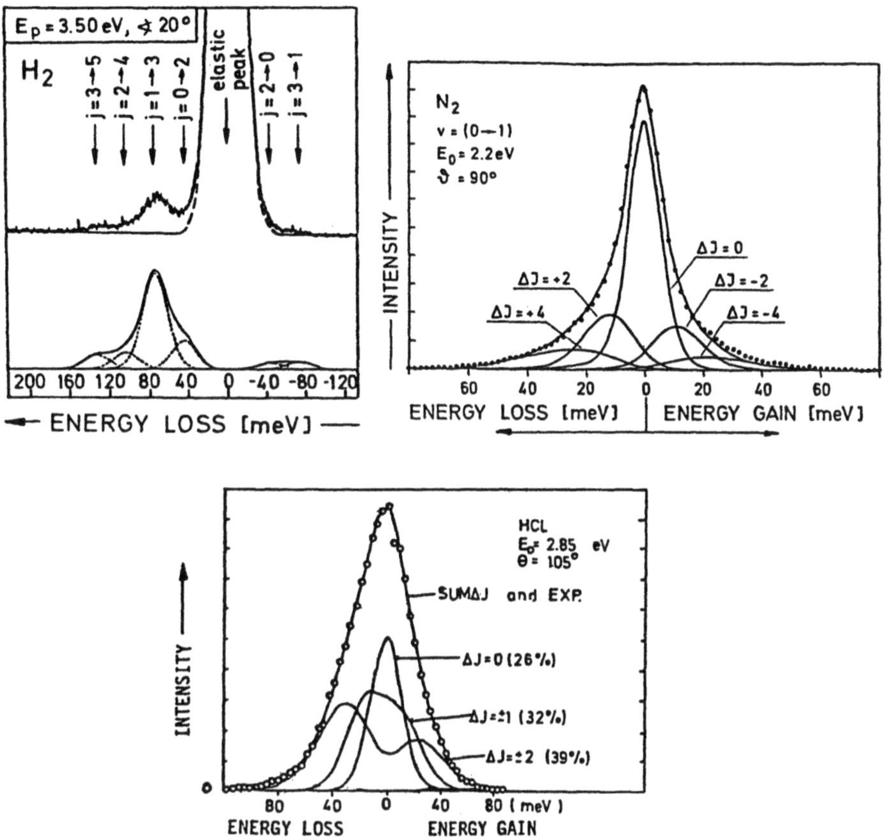

Figure 2. Rotational line broadening by electron impact on H_2, N_2 and HCl [6,14,21].

i) Branch Construction

This method starts from the assumption that all cross sections $\sigma(J \rightarrow J + \Delta J)$ have the same value independent of the initial J, i.e. a branch for each ΔJ exists and within one ΔJ-branch all transition probabilities are equal. For high J quantum numbers ($J \gtrsim 4$) this assumption is very good.as can be seen [17] from the Clebsch-Gordon coefficients. The evaluation of a line shape proceeds via the following steps:

1. Calculation of the initial rotational distribution in the molecular beam due to the measured beam temperature, which is typically 500 K in our experiments, in order to ensure high initial J.

2. For each ΔJ a branch shape is calculated using the initial rotational distribution, the new energy position of a line and the apparatus profile measured by using helium, neon or argon.
3. The individual lines of the energy gain and energy loss branches are multiplied by the detailed balance factors

$$\frac{2(J+\Delta J)+1}{2J+1} \qquad\qquad \frac{2(J-\Delta J)+1}{2J+1}$$

   Selection rules may limit the number of possible branches.
4. A least square fit of the branch intensities with $\Delta J = 0, \pm 1, \pm 2, \ldots$ is made to the measured line shape. The number of free parameters is equal to half the number of branches +1, since the truly elastic peak height and only the loss or the gain branches have to be considered (because of the detailed balance factors).

ii) The Shimamura formula [18]

$$\sigma(J' \leftarrow J) = \sum_{J_t=|J-J'|}^{J+J'} \frac{2J'+1}{(2J+1)(2J_t+1)} \cdot \sigma(J_t \leftarrow 0)$$

is a high-energy approximation, which essentially contains the assumption that the rotational transition is adiabatic, which is true practically for all energies used in the experiments. The formula has also only a few free parameters ($\sigma(J' \leftarrow 0)$) depending on the selection rule, and it can provide state-to-state transition cross-sections ($\sigma(J' \leftarrow J)$).

Table 1 lists the molecules for which electron impact rotational excitation cross-sections have been measured in low-energy beam experiments. The cross-sections in this table are only order-of-magnitude numbers.

Figure 3 shows integrated cross-sections from recent measurements of the energy and angular dependence of the rotational transition $J = 1 \rightarrow 3$ for H_2 in the range of impact energies 0.2 eV to 0.6 eV. The high impact energy is ca. 12 times the threshold energy of the process, and the angular dependence is found to be constant from the lowest angle (12°) to 135° as expected if the quadrupole is the most important interaction. The low impact energy is only 4 times the threshold energy, and the cross-section decreases by a factor of 6 in the angular range from 40° to 12°, since the process is no longer adiabatic, i.e., a fixed nuclei approximation cannot be used [19]. The integrated cross-sections are in excellent agreement with the swarm results of Crompton et al. [20].

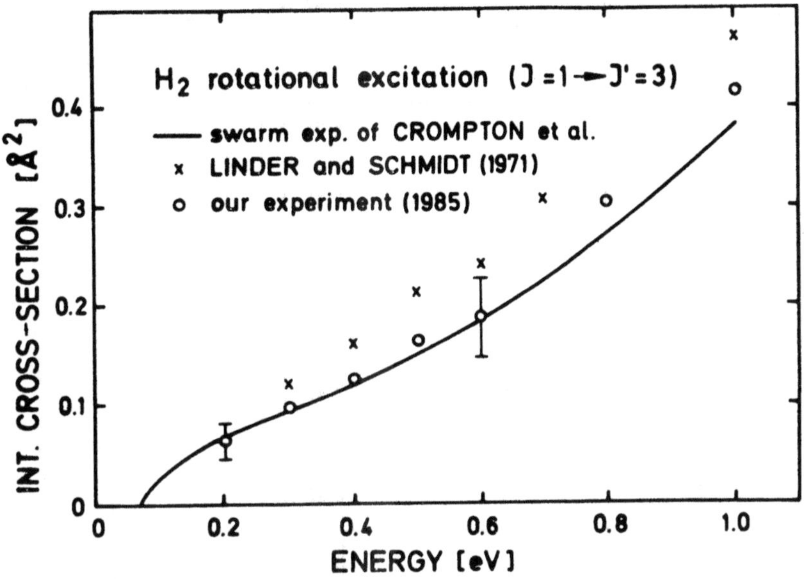

Figure 3. Integrated cross-sections for the rotational excitation J = 1 → 3 in H_2.

For the sake of simplicity in this paper, only pure rotational transitions are discussed and not simultaneous rotational-vibrational transitions. Data for such processes are given, for example, in the papers [6,7,14,21].

Table 1. Integrated Rotational Cross-Sections

Molecule	non-resonant interaction: main interaction	ΔJ	$\sigma(cm^2)$	transition via shape-resonance: ΔJ	$\sigma(cm^2)$	publication
H_2	weak quadrupole polarizability	0,±2	$\sim 10^{-17}$	0,±2	0.8×10^{-16}	[6,7,21]
N_2	weak quadrupole polarizability	0,±2	1.5×10^{-18}	0,±2,±4	2.8×10^{-16}	[13,14]
CO	weak dipole	0,±1	1.7×10^{-17}	0,±1,...,±4	1×10^{-16}	[14,16]
H_2O	strong dipole	0,±1	2.5×10^{-16}	0,±1	$> 0.3 \times 10^{-16}$	[14]
HCl	strong dipole	0,±1	$\sim 10^{-16}$	0,±1,±2	$\sim 10^{-16}$	[21]
CH_4	octopole polarizability	0 3 4	6×10^{-18}	0,±1,...,±4	$\sim 10^{-16}$	[21]
CO_2	strong quadrupole polarizability	0 2 4	$\sim 10^{-17}$	0,±2,...	6×10^{-17}	[21]

The table contains only order-of-magnitude numbers and the interactions and ΔJ-values, which are believed to be most important. For quantitative details see original publications. The impact energies are in the range from 0.1 eV to about 10 eV. Molecules with very large dipole moments such as CsF, ScCl, KI, LiF, CsBr (see ref. 8,9,10,11,12,15) may have cross-sections up to 10^{-13} cm^2.

Vibrational Excitations by Electron Impact in the Ground Electronic States of Molecules

For the vibrational excitation of a molecule by colliding electrons, it is necessary that the electron cloud of the molecule changes, and consequently the motions of the nuclei are altered. Therefore, vibrational transitions occur simultaneously with electronic transitions. It is also well known that strong vibrational excitation may occur via short-lived negative ion states such as shape resonances. In this case the electronic cloud of the molecule is changed for a short time (10^{-12}-10^{-15} sec), but after autoionization of the resonance the molecule is back in its electronic groundstate, and the only effect after the collision is the vibrational excitation. This mechanism is well understood now. All molecules seem to have such shape resonances in energy positions between their electronic groundstate and their first electronic excited state.

Another mechanism could be a virtual state, a "shape" resonance at zero energy and in the s-wave channel. The colliding electron with nearly zero energy is not really bound, but has a large amplitude close to the molecule. It is known [22] that especially polar molecules have such virtual states, but also others as SF_6 [23]. They produce a large cross-section close to the threshold. The question is, of course, what the effect is of a virtual state on the excitation of vibrational modes.

Little is known experimentally about other mechanisms which are also capable of exciting vibrational transitions in the electronic groundstate of the molecule. In principle, all direct interactions between the colliding electron and the molecule could induce such transitions, leading to more or less large cross-sections. Permanent dipoles, induced dipoles, higher multipoles, short and long-range polarizabilities have to be considered.

Although vibrational excitation via direct interactions was discussed already as early as 1935 by H.S.W. Massey [24] and in evaluations of swarm experiments, practically all beam experiments have dealt with vibrational excitation via a shape resonance or a virtual state. Only few exceptions are known, for example the strong non-resonant excitation of the bending and asymmetric stretch mode in CO_2 in the energy region below the shape resonance [25].

In the last few years in the Physics Department in Kaiserslautern we have made a series of measurements on (mostly) non-resonant vibrational excitations in the energy region from threshold to a few eV and in the angular range from close to zero to about 135° scattering angle. The list of molecules investigated is given in Table 2, together with their properties, which we think might be most important for vibrational excitations. We are aware that this is of course only an order-of-magnitude-approach for the discussion.

Table 2. Vibrational excitation via direct interactions

A. Molecules without a virtual state

H_2	weak quadrupole	[26]
N_2	weak quadrupole	[26]
CO	weak dipole	[27]
C_2H_2	weak polarizability	[28]

B. Molecules with a virtual state

CO_2	strong quadrupole	[29]
COS	intermediate strong dipole	[30]
CS_2	large polarizability	[30]
CH_4	octopole, polarizability	[31]

All molecules have shape resonances between ca. 1 - 2 eV and 6 eV collision energy.

This semi-quantitative line of argumentation gives the following picture for low-energy vibrational excitations.

1. Vibrational excitations induced by a shape resonance have for most molecules integrated cross-sections in the order of 10^{-16}cm². Transitions with high Δv up to 10 have been observed, but with lower cross-sections (10^{-17}-10^{-19}cm²). In polyatomic molecules, the shape resonances act often only on selected vibrational modes, namely on those with a symmetry in accordance with the symmetry of the resonance.

2. Vibrational excitations induced by a virtual state have large cross-sections only very close to the thresholds of the excitation functions. Here the cross-sections may easily be much larger than those from a shape resonance. The half-width of such a threshold peak is of the order of 0.3 eV. Mostly only $\Delta v = 1$ or 2 is induced. The angular dependence is constant, since for a virtual state only the s-wave shows an amplitude enhancement.

3. Vibrational excitation induced by dipole interaction has cross-sections, which depend on the dipole moment. For a molecule with a weak dipole moment (e.g. CO) the integrated cross-section is of the order of $10^{-17} cm^2$, while for molecules with large dipole moments the cross-sections are the order of $10^{-16} cm^2$ to $10^{-15} cm^2$, with a maximum at an impact energy around 3 to 5 times the threshold energy. The angular dependence is peaked in the forward direction. The Born Dipole Approximation describes the cross-section quite well at all impact energies and scattering angles from 0° to ca. 30°, but at higher angles the BDA gives low values. Only $\Delta v = 1$ has been observed in accordance with the well-known selection rule.

   In polyatomic molecules with and without a permanent dipole moment, vibrational transitions of infrared active modes are also excited by low-energy electron impact, following the same rules, i.e. $\Delta v = 1$, with strong forward scattering in agreement with the Born Dipole Approximation, and cross-sections of the order of $10^{-17} cm^2$ to $10^{-15} cm^2$. These facts have been seen for all polyatomic molecules of table 2.

4. Vibrational excitation, induced by quadrupole or octopole interaction, seems to be very small, e.g. in the order of $10^{-18} cm^2$. Test molecules have been H_2, N_2, and possibly the (1,0,0) mode in triatomic molecules outside the shape resonance region. These findings, however, may be obscured by the effects of the polarizability.

5. The contributions to the vibrational excitation through the polarization can be quite large; in a molecule like CS_2 the integrated cross-section may reach the order of $10^{-16} cm^2$, if the long-range polarizability is considered [30]. In general, the cross-sections for the excitation for one or two vibrational quanta are in the order of 10^{-18} to $10^{-16} cm^2$ close to the thresholds. The effects of the short-range polarizability are well visible in the backward scattering, as has been shown in the case of CO ($v = 0 \rightarrow 1$). The effect of the polarizability is best seen in the excitation of Raman active vibrational modes.

The above-mentioned experimental findings have been extrapolated from areas in the impact energy and the scattering angles, for which only one mechanism is predominant over others. Such situations can easily be found for shape resonances, virtual states and for direct dipole excitation, but not so for the effects of the polarizability and higher multipoles. In order to clear up their role, more theoretical work is needed. Coherent scattering due to different simultaneous interactions has also been found.

Acknowledgement

This work was supported by the Deutsche Forschungsgemeinschaft under Sonderforschungsbereich 91.

References

[1] Trajmar, S. and Register, D.F. In "Electron-Molecule Collisions"; Shimamura, I. and Takayanagi, K., Eds.; Plenum Press: New York 1984.

[2] Hick, P.J.; Daviel, S.; Wallbank, B.; and Comer, J. J. Phys. E 1980, 13, 713.
York, T.A. and Comer, J. J. Phys. B 1983, 16, 3627.
York, T.A. and Comer, J. J. Phys. B 1984, 17, 2563.

[3] Kochem, K.H.; Sohn, W.; Jung, K.; and Ehrhardt, H. to be published.

[4] see for example: XIV ICPEAC 1985, Book of invited papers, Lecture by U. Heinzmann, and Book of contributed papers, Abstracts by J. Kessler et al.

[5] 3. International Workshop on Positron(Electron)-Gas Scattering, Wayne State University, July 1985, Book of invited papers.

[6] Ehrhardt, H. and Linder, F. Phys. Rev. Lett. 1968, 21, 419.

[7] Linder, F. and Schmidt, H. Zeitschrift für Naturforschung 1971, 26a, 1603.

[8] Slater, R.C.; Fickes, M.G.; Becker, W.G.; and Stern, R.C. Journ. Chem. Phys. 1974, 60, 4697.

[9] Becker, W.G.; Fickes, M.G.; Slater, R.C.; and Stern, R.C. Journ. Chem. Phys. 1974, 61, 2283.

[10] Slater, R.C.; Fickes, M.G.; Becker, W.G.; and Stern, R.C. Journ. Chem. Phys. 1974, 61, 2290.

[11] Rudge, M.R.H.; Trajmar, S.; Williams, W. Phys. Rev. 1976, A13, 2074.

[12] Vušković, L.; Srivastava, S.K.; and Trajmar, S. J. Phys. B 1978, B11, 1643.

[13] Wong, S.F. and Dubé, L. Phys. Rev. A 1978,17, 570.

[14] Jung, K.; Antoni, Th.; Müller, R.; Kochem, K.-H.; and Ehrhardt, H. J. Phys. B 1982,15, 3535.

[15] Anduszliwer, B.; Tino, A.; Weiss, P.; and Bederson, B. Phys. Rev. Lett. 1983, 18, 1644.

[16] Chang, E.S.; Antoni, Th.; Jung, K.; and Ehrhardt, H. Phys. Rev. A 1984, 30, 2086.

[17] Read, F.H. J. Phys. B 1972, 5, 255.

[18] Shimamura, J. Symposium on Electron-Molecule Collisions, Tokyo, Invited papers 1979.

[19] Feldt, A.N. and Morrison, M.A. J. Phys. B 1982, 15, 301.
Morrison, M.A.; Feldt, A.N.; and Austin, D. Phys. Rev. A 1984, 29, 2518.
Varracchio, E.F. and Lamanna, U.T. Phys. Rev. B 1984, 17, 4395.

[20] Crompton, R.W.; Gison, D.K., and McIntosh, A.I. Aust. J. Phys. 1969, 22, 715.

[21] Ehrhardt, H. et al., to be published
[22] Rohr, K. and Linder, F. J. Phys. B 1976, 9, 2521.
[23] Rohr, K. J. Phys. B 1977, 10, 1175.
[24] Massey, H.S.W. Trans. Far. Soc. 1935, 31, 556.
[25] Andrick, A.; Danner, D.; and Ehrhardt, H. Phys. Lett. 1969, 29A, 346.
[26] Ehrhardt, H. et al., to be published.
[27] Sohn, W.; Kochem, K.-H.; Jung, K.; Ehrhardt, H.; and Chang, E.S. J. Phys. B 1985, 18, 2049.
[28] Kochem, K.-H.; Sohn, W.; Jung, K.; Ehrhardt, H.; and Chang, E.S. J. Phys. B 1985, 18, 1253.
[29] Antoni, Th.; Jung, K.; Ehrhardt, H.; and Chang, E.S. to be published.
[30] Kochem, K.-H. et al. , to be published.
[31] Sohn, W.; Jung, K.; and Ehrhardt, H. J. Phys. B 1983, 16, 891.

THRESHOLD PHENOMENA IN ELECTRON-MOLECULE COLLISIONS

W. Domcke

Theoretische Chemie, Physikalisch-Chemisches Institut,
Universität Heidelberg,
D-6900 Heidelberg, Federal Republic of Germany

INTRODUCTION

Threshold phenomena are ubiquitous in electron-molecule scattering as there is a multitude of excitation and reaction thresholds associated with processes such as rotational excitation, vibrational excitation, dissociative attachment, ionization and dissociation. In this brief account we shall confine ourselves to a discussion of vibrational excitation thresholds in electronically elastic electron-molecule scattering.

Cross-sections for vibrational excitation by low-energy electrons have been measured for many molecules using swarm or beam techniques [1]. For e + H_2, which is the simplest system from a theoretical point of view, extensive ab initio rotational-vibrational close-coupling calculations have been performed [2] and good agreement with experiment has been achieved, in particular for rotational excitation. For many-electron targets, in particular polyatomics, the ab initio close-coupling treatment of vibrational excitation is not feasible at present, and simple approximate theories and phenomenological models are required. A starting point for a simple description is the fact that the long-range part of the electron-molecule interaction potential should dominate the threshold behavior of the cross-sections [3]. The excitation of infrared-active vibrational modes by low-energy electron impact, for example, is often quantitatively described by the simple Born-dipole approximation [4].

High-resolution electron-molecule scattering experiments have revealed, over the past decade, distinct threshold structures in vibrational excitation functions, of a number of molecules which have attracted considerable theoretical interest. The most pronounced features are the threshold peaks observed in the hydrogen halides, in particular HCl [5]. In these cases the vibrational excitation cross-sections exceed the estimates based on the Born approximation by orders of magnitude. To explain this enhancement, the participation of either a resonance or a virtual state has been proposed [6,7].

The basic problem is that the standard approximations and models such as the adiabatic-nuclei-approximation [8] (for nonresonant scattering) and the local-complex-potential model [9] (for resonant scattering) are not applicable near threshold. A number of theoretical formulations have been put forward recently to fill this obvious gap, among them the Feshbach projection-operator method [10], the zero-range-potential model [11,12], the R-matrix method [13] and the vibrational frame-transformation theory [14]. We outline here the approach based on the projection-operator formalism, and illustrate the ideas by application to the virtual-state effect in electron-CO_2 scattering.

DESCRIPTION OF THE FORMALISM

Assume that the electron-molecule collision system under consideration possesses, in the fixed-nuclei approximation, a bound state for a certain range of internuclear distances. If a weakly-bound state exists at the equilibrium geometry R_o of the target molecule, or if a bound state is introduced by a small variation of the internuclear distance R around R_o, this may result in a strong enhancement of the scattering cross-section at low energies. From general scattering theory [15] we know, at least for short-range interaction potentials, that a bound state which disappears as a function of R becomes either a virtual state or a resonance. In the presence of longe-range interaction potentials the situation is more complex (see [16]). In any such case the adiabatic-nuclear-vibration approximation [8] is certainly inapplicable, since the fixed-nuclei T-matrix varies strongly with energy and internuclear distance. The basic idea of the projection-operator approach is to extract the rapid variations of the T-matrix caused by resonances, virtual states or bound states such that the remaining "background" T-matrix is weakly dependent on energy and internuclear distance, and can be treated in the adiabatic-nuclei approximation. The formalism yields a simple prescription to treat the "resonant" part of the T-matrix essentially exactly, thus taking account of non-Born-Oppenheimer and threshold effects.

Let $|\phi_d\rangle$ be a discrete (square-integrable) electronic state which approximates the bound state or resonance under consideration. We require that $|\phi_d\rangle$ is a quasi-diabatic state, i.e., that the electronic wavefunction changes slowly when the internuclear distance is varied. In particular, $|\phi_d\rangle$ should not exhibit abrupt changes when the bound state approaches or crosses the continuum threshold. We then define the electronic projectors

$$Q = |\phi_d\rangle\langle\phi_d| \quad , \quad P = \int k dk d\Omega_k |\phi_k\rangle\langle\phi_k| \tag{1}$$

where the $|\phi_k\rangle$ are continuum states orthogonal to $|\phi_d\rangle$ such that the orthogonality and completeness relations in electronic Hilbert space

$$PQ = 0 \quad , \quad P + Q = 1 \tag{2}$$

are fulfilled. Given these projectors, the fixed-nuclei electron-molecule scattering T-matrix decomposes into a resonant and a background term [17]

$$T(k',k) = T_{bg}(k',k) + T_{res}(k',k) \tag{3a}$$

where T_{bg} is the T-matrix associated with the scattering states $|\phi_k\rangle$ and T_{res} is given by the explicit expression [17]

$$T_{res}(k',k) = V_{k'}(R)[E-\varepsilon_d(R)-\Delta(R,E) + i\Gamma(R,E)/2]^{-1}V_k(R)^* \tag{3b}$$

with the definitions (H_{el} denotes the electronic Hamiltonian)

$$V_k(R) = \langle\phi_k|H_{el}|\phi_d\rangle \tag{4}$$

$$\varepsilon_d(R) = \langle\phi_d|H_{el}|\phi_d\rangle \tag{5}$$

$$\Gamma(R,E) = 2\pi\int d\Omega_k |V_k(R)|^2 \tag{6}$$

$$\Delta(R,E) = (2\pi)^{-1} P\int dE'\Gamma(R,E')/(E-E') \tag{7}$$

Methods for the direct ab initio calculation of the basic quantities ε_d, Γ and Δ have been developed by Hazi [18] and Berman and Domcke [19] and applied to shape and Feshbach resonances. Here, we wish to emphasize the usefulness of the projection-operator approach as a simple phenomenological tool to describe resonance and threshold effects. The phenomenological approach is based on the fact that Eq. (3a) is equivalent to a decomposition of the eigenphase sum $\delta = (2i)^{-1} \ln \det \underline{S}$, where $\underline{S}$ is the fixed-nuclei S-matrix, into a resonant and a background term [19]

$$\delta = \delta_{bg} + \delta_{res} \tag{8}$$

with

$$\delta_{res}(R,E) = -\tan^{-1} \{\frac{1}{2}\, \Gamma(R,E)/[E-\varepsilon_d(R)-\Delta(R,E)]\} \tag{9}$$

It is thus possible, in principle, to extract information on the functions $\varepsilon_d(R)$, $\Gamma(R,E)$, $\Delta(R,E)$ from calculated fixed-nuclei eigenphase sums.

When the nuclear degrees of freedom are included (we omit the rotational motion for simplicity) the electron-molecule scattering problem becomes a multi-channel problem, the channels being the vibrational states of the target molecule. The clue of the projection-operator formalism is that the multi-channel resonant T-matrix is (within the assumption of diabatic $|\phi_d\rangle$ and $|\phi_k\rangle$) <u>exactly</u> given by the formal operator expression [20]

$$T_{res}(k_f,v';k_i,v) = \langle v'|V_{k_f}(R)(E-\mathcal{H})^{-1}\, V^*_{k_i}(R)|v\rangle \tag{10}$$

$$\mathcal{H} = H_o + \varepsilon_d(R) + \Delta(R,E-H_o) - i\Gamma(R,E-H_o)/2 \tag{11}$$

Here H_o is the vibrational Hamiltonian of the target molecule with eigenstates $|v\rangle$ and energies ε_v. k_i and k_f are the initial and final momenta of the scattered electron, and $E = k_i{}^2/2 + \varepsilon_v$ is the total energy which is conserved in the scattering process. Since the expression (10) is exact, it includes all threshold effects, provided the energy-dependence of V_k, Γ, Δ is fully taken into account. Eqs. (10,11) are equivalent to the nonlocal-complex-potential theory as formulated by O'Malley [21], Bardsley [22] and others. Although formally a resonance theory, the approach is not confined to the description of resonances. In fact, Eqs. (10,11) provide a general prescription to deduce multi-channel electron-molecule scattering cross-sections from fixed-nuclei scattering data. Threshold effects caused by bound states or virtual states, for example, are naturally included [16], as will be illustrated below.

The explicit evaluation of the operator expression (10) is only briefly sketched here. More detailed accounts can be found in the literature [23-25]. The simplest approach is to introduce the representation of $\mathcal{H}$ in the basis of target vibrational states

$$\mathcal{H}_{nm} = \langle n|\mathcal{H}|m\rangle \tag{12}$$

and to evaluate the matrix elements of the resolvent operator

$$G = (E - \mathcal{H})^{-1} \tag{13}$$

in Eq. (10) by inversion of the complex and energy-dependent matrix $(E\delta_{nm} - \mathcal{H}_{nm})$ [23,24]. This procedure is not appropriate when the dissociative attachment channel is open. In this case we may start from the operator identity

$$G = G_d + G_d(\Delta - i\Gamma/2)G \tag{14}$$

where

$$G_d = (E - H_o - \varepsilon_d)^{-1} \tag{15}$$

is the propagator for vibrational motion in the discrete state, which can be determined by standard numerical methods. Eq. (14) can be solved with the use of separable expansions of the nonlocal level-shift operator $\Delta - i\Gamma/2$ [25].

As mentioned in the Introduction, a variety of computational methods and phenomenological models have been developed to describe threshold phenomena in vibrationally inelastic electron-molecule scattering. The most obvious approach is the close-coupling expansion, which represents, however, a considerable computational challenge in cases of strong vibrational excitation. The advantage of the projection-operator approach in comparison with the close-coupling expansion consists in the reduction of the nuclear dynamical problem to a one-dimensional subspace of the electronic Hilbert space; the nuclear dynamics can thus be treated more efficiently than by solving a large number of coupled equations. In the R-matrix theory [13] electronic R-matrix states are introduced instead of the projected electronic states $|\phi_d\rangle$ and $|\phi_k\rangle$. The nuclear dynamics is treated explicitly in a small subset of R-matrix states, taking their non-Born-Oppenheimer coupling into account if necessary, while the contribution of the remaining states is neglected or treated in the adiabatic-nuclei approximation. The general strategy of the R-matrix method is thus similar to the projection-operator approach. Few applications of this method have been reported so far [26]. A related formalism, which connects fixed-nuclei scattering phaseshifts to the multi-channel S-matrix, is the vibrational frame-transformation theory [14]. The zero-range-potential or effective-range models [6,11,12,27] avoid the introduction of projected diabatic states; these models are parametrized in terms of a short-range boundary condition which is related to the adiabatic bound or virtual states of the collision system. Finally, we mention exactly solvable models discussed by Kazansky [28], where the coupling between electrons and nuclei is described by a rank-one separable interaction.

EXAMPLE: VIRTUAL-STATE EFFECT IN ELECTRON-CO_2 SCATTERING

The formalism outlined in the preceding section has been applied to describe a variety of threshold phenomena, among them shape resonances near threshold [23], threshold peaks in polar molecules [7,16,29] and threshold effects in electron-ion scattering [16]. Here we discuss briefly as a simple example the threshold peak [30] in the excitation of the symmetric stretching mode of CO_2 (more details can be found in [31]).

The possible existence of a virtual state in fixed-nuclei electron-CO_2 scattering has been pointed out by Morrison and Lane [32,33]. The dominant eigenphase shift at low energy is of ${}^2\Sigma_g^+$ symmetry and the scattering length is large and negative [33]. To apply the projection-operator formalism to this problem in a simple phenomenological manner, we parametrize the width function as

$$\Gamma(E) = A\, E^{1/2}\, e^{-BE} \tag{16}$$

thus taking account of the threshold law for s-wave scattering [3]. The parameters A and B and the discrete-state energy ε_d can then be determined by fitting the fixed-nuclei eigenphase sum given by Eqs. (8,9) to the ab initio electron-CO_2 scattering data at the equilibrium geometry [33]. The unknown background phaseshift in Eq. (8) is parametrized by its threshold expansion

$$\delta_{bg}(E) = C\, E^{1/2}. \tag{17}$$

The fit of the ab initio data and the decomposition of δ into a background and a resonant term is shown in Figure 1. Figure 2 shows the resulting width and level-shift functions $\Gamma(E)$, $\Delta(E)$. The level-shift $\Delta(E)$ exhibits

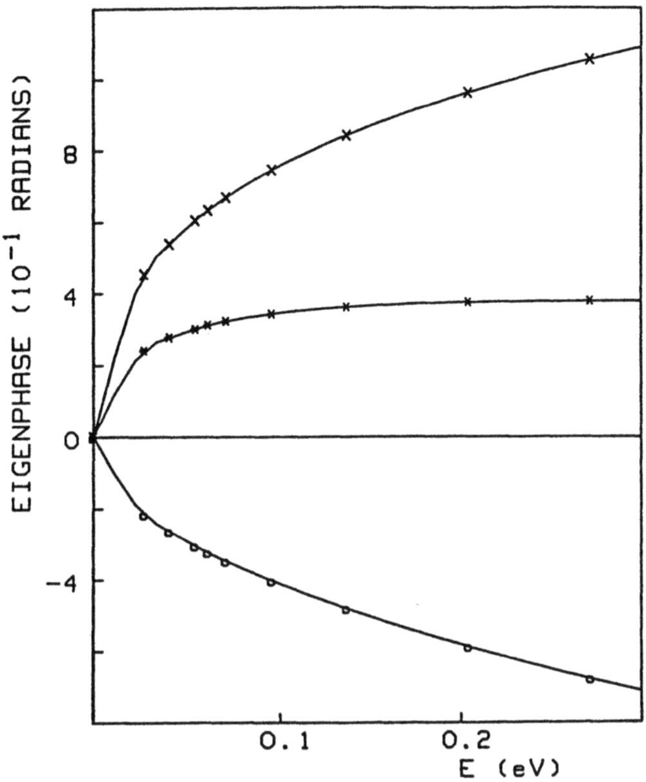

Figure 1. Fixed-nuclei s-wave eigenphase δ for e-CO_2 scattering [33] (stars) together with the fit obtained with the present model (full curve through the stars). The decomposition of δ into a resonant term (upper curve, with crosses) and a background term (lower curve, with circles) is shown.

a discontinuous derivative at E = 0 as a consequence of the square-root onset of $\Gamma(E)$ [16]. Given $\Gamma(E)$, $\Delta(E)$ and ε_d, the singularities of the fixed-nuclei scattering matrix in the complex energy or momentum plane can be obtained by direct analytic continuation [16]. It turns out that in the present case the S-matrix possesses a pole on the negative imaginary axis of the momentum plane at k_p = -0.13i [31]. Such a pole is called a virtual state [15] and causes an enhancement of the fixed-nuclei scattering cross-section near threshold.

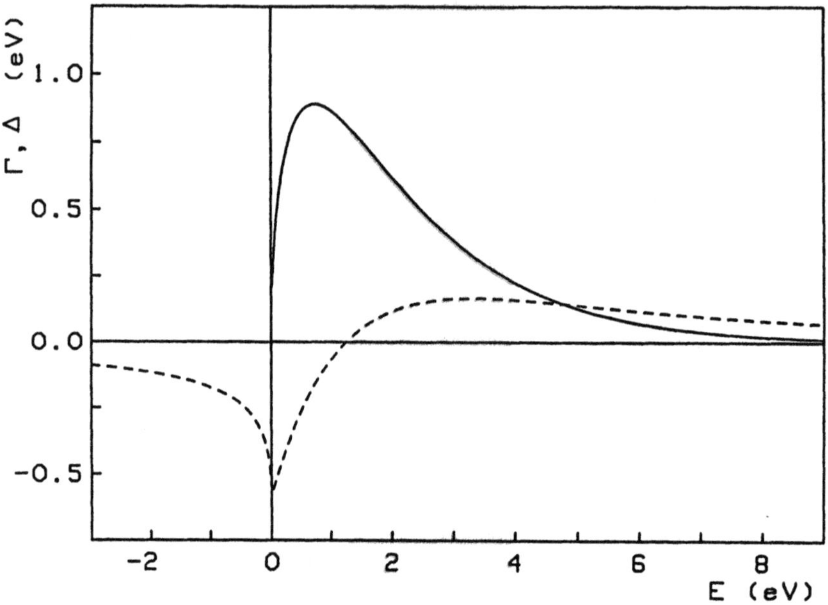

Figure 2. Width function Γ(E) and level-shift function Δ(E) (broken curve) of the present virtual-state model.

To complete the model we have to define the target potential-energy curve $V_o(R)$ and the discrete-state energy $\varepsilon_d(R)$, where R denotes the symmetric-stretching coordinate of CO_2. Since only weak vibrational excitation is involved [30], we may adopt the harmonic approximation for the former and may expand $\varepsilon_d(R)$ up to first order about the target equilibrium geometry R_o

$$\varepsilon_d(R) = \varepsilon_d(R_o) + \kappa(R - R_o) \tag{18}$$

If we assume, for simplicity, that A, B, C in Eqs. (16,17) are independent of R, all relevant parameters of the model except κ in Eq. (18) are determined by the fitting of the ab initio scattering data at $R = R_0$. The multi-channel T-matrix for $0 \rightarrow v$ vibrational excitation is now determined from Eq. (10) (note that T_{bg} vanishes for inelastic scattering as a consequence of our assumption that δ_{bg} is independent of R) and the single undetermined parameter κ is fixed by adjusting the absolute magnitude of the $0 \rightarrow 1$ vibrational excitation cross-section to experiment [30]. As shown in Figure 3, the energy dependence of the $0 \rightarrow 1$ excitation cross-section is very well reproduced by the calculation. Also shown in Figure 3 are the results of a two-state close-coupling model calculation of Whitten and Lane [34]. It is seen that the two methodically different theoretical approaches yield results in good agreement with each other and with the experimental data.

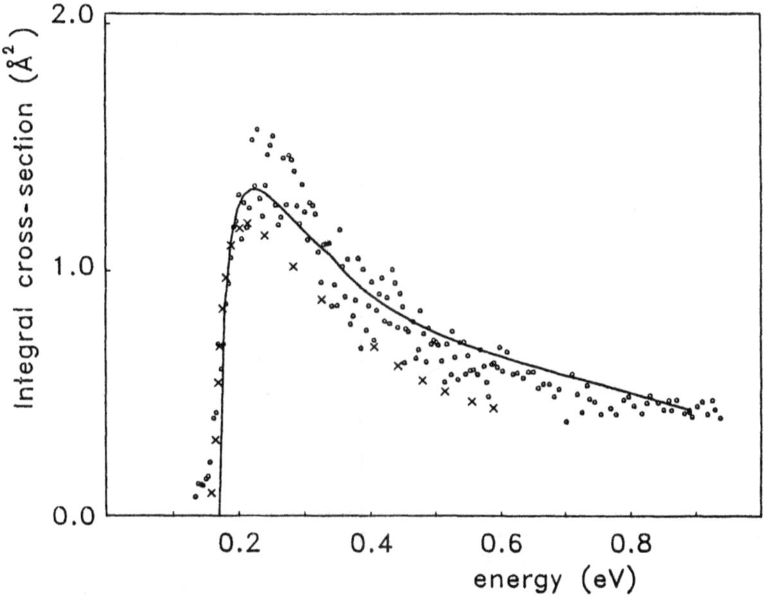

Figure 3. Integral cross-section for symmetric-stretch vibrational excitation of CO_2. Circles: experimental data [30]; crosses: close-coupling model calculation of Whitten and Lane [34]; full curve: present model.

The model calculation based on the projection-operator formalism exhibits more directly than the close-coupling calculation [34] the connection between the threshold peak in vibrational excitation and the virtual-state pole of the fixed-nuclei S-matrix. The present analysis results, for example, in a potential-energy curve for the virtual state which is shown in Figure 4. The full curve in Figure 4 represents the harmonic symmetric-stretch potential-energy curve $V_o(R)$ of the $X^1\Sigma_g^+$ electronic ground state of CO_2.

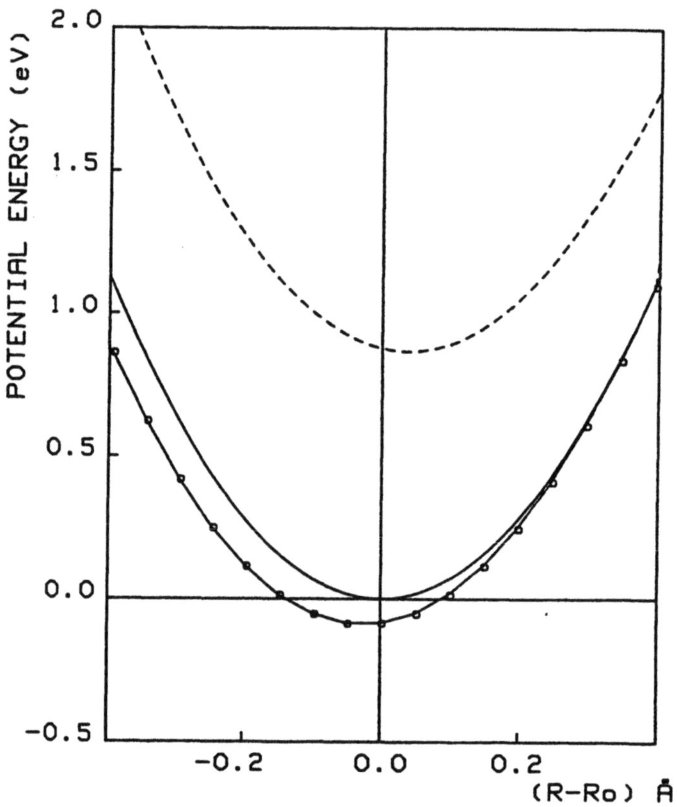

Figure 4. Symmetric-stretch potential-energy curves of the present model. Full curve: electronic ground state of CO_2; broken curve: discrete state of CO_2^-; full curve with circles: virtual state on the unphysical sheet of the complex energy surface.

The broken curve is the potential-energy curve of the discrete state, defined by $V_d(R) = V_o(R) + \varepsilon_d(R)$. The full curve with circles represents the potential-energy curve of the virtual state in CO_2, obtained by analytic continuation. It should be emphasized that the virtual-state pole is situated on the unphysical sheet of the complex energy surface and is thus not directly observable.

ACKNOWLEDGEMENTS

The work surveyed here has been supported for many years by the Deutsche Forschungsgemeinschaft through SFB 91. The results presented in Figures 1-4 have been obtained in collaboration with H. Estrada at Universidad Nacional, Bogotá, Colombia.

REFERENCES

[1] S. Trajmar, D.F. Register and A. Chutjian, Phys. Rep. 97, 219 (1983).
[2] M.A. Morrison, A.N. Feldt and D. Austin, Phys. Rev.A 29, 2518 (1984); M.A. Morrison, A.N. Feldt and B.C. Saha, Phys. Rev. A 30, 2811 (1984).
[3] R.G. Newton, "Scattering Theory of Waves and Particles", McGraw-Hill, New York, 1966, § 17.2.
[4] See, for example, W. Sohn, K.-H. Kochem, K. Jung, H. Ehrhardt and E.S. Chang, J. Phys. B 18, 2049 (1985).
[5] K. Rohr and F. Linder, J. Phys. B 9, 2521 (1976); K. Rohr, J. Phys. B 11, 1849 (1978).
[6] L. Dubé and A. Herzenberg, Phys. Rev. Lett. 38, 820 (1977).
[7] W. Domcke and L.S. Cederbaum, J. Phys. B 14, 149 (1981).
[8] F.H.M. Faisal and A. Temkin, Phys. Rev. Lett. 28, 203 (1972); M. Shugard and A.U. Hazi, Phys. Rev. A 12, 1895 (1975).
[9] A. Herzenberg, J. Phys. B 1, 548 (1968).
[10] W. Domcke and L.S. Cederbaum, in "Physics of Electronic and Atomic Collisions", S. Datz, Ed., North-Holland, Amsterdam, 1982, p. 569; in "Electron-Atom and Electron-Molecule Collisions", J. Hinze, Ed., Plenum Press, New York, 1983, p. 255.
[11] A.K. Kazansky, Teor. Mat. Fyz. 36, 414 (1978); 37, 84 (1978).
[12] J.P. Gauyacq, J. Phys. B 16, 4049 (1983); D. Teillet-Billy and J.P. Gauyacq, J. Phys. B 17, 4041 (1984).

[13] B.I. Schneider, M. Le Dourneuf and P.G. Burke, J. Phys. B 12, L365 (1979).
[14] C.H. Greene and C. Jungen, Phys. Rev. Lett. 55, 1066 (1985).
[15] see ref. [3], § 12.
[16] W. Domcke, J. Phys. B 14, 4889 (1981); 16, 359 (1983).
[17] H. Feshbach, Ann. Phys. (NY) 19, 287 (1962).
[18] A.U. Hazi, in "Electron-Atom and Electron-Molecule Collisions", J. Hinze, Ed., Plenum Press, New York, 1983, p. 103.
[19] M. Berman and W. Domcke, Phys. Rev. A 29, 2485 (1984).
[20] W. Domcke and L.S. Cederbaum, Phys. Rev. A 16, 1465 (1977).
[21] T.F. O'Malley, Phys. Rev. 150, 14 (1966).
[22] J.N. Bardsley, J. Phys. B 1, 349 (1968).
[23] W. Domcke and L.S. Cederbaum, J. Phys. B 13, 2829 (1980).
[24] M. Berman, H. Estrada, L.S. Cederbaum and W. Domcke, Phys. Rev. A 28, 1363 (1983).
[25] C. Mündel and W. Domcke, J. Phys. B 17, 3593 (1984).
[26] B.I. Schneider, M. Le Dourneuf and Vo Ky Lan, Phys. Rev. Lett. 43, 1927 (1979); M. Le Dourneuf and J.M. Launay, in "Wavefunctions and Mechanisms from Electron-Scattering Processes", F.A. Gianturco and G. Stefani, Eds., Springer, Berlin, 1984, p. 69.
[27] I.I. Fabrikant, Zh. Eksp. Teor. Fiz. 73, 1317 (1977); J. Phys. B 11, 3621 (1978).
[28] A.K. Kazansky, Zh. Eksp. Teor. Fiz. 82, 1422 (1982); J. Phys. B 16, 2427 (1983).
[29] W. Domcke and C. Mündel, J. Phys. B., in press.
[30] K.-H. Kochem, W. Sohn, N. Hebel, K. Jung and H. Ehrhardt, J. Phys. B, in press.
[31] H. Estrada and W. Domcke, J. Phys. B, in press.
[32] M.A. Morrison and N.F. Lane, Chem. Phys. Lett. 66, 527 (1979).
[33] M.A. Morrison, Phys. Rev. A 25, 1445 (1982).
[34] B.L. Whitten and N.F. Lane, Phys. Rev. A 26, 3170 (1982).

ROTATIONAL AND VIBRATIONAL EXCITATION OF MOLECULES BY LOW-ENERGY ELECTRONS

David W. Norcross*

Joint Institute for Laboratory Astrophysics,
National Bureau of Standards and University of Colorado,
Boulder, Colorado 80309-0440

My goal in this paper is to review and summarize recent advances (since about 1982) in theoretical and computational techniques, and to illustrate the application of some of these developments in practical calculations. Important features of recent work are increased interest in polar molecules and complex polyatomics, the development of more realistic and complete representations of the interaction at short as well as long range, and important progress in theory and calculations for the treatment of nuclear dynamics in general, and vibrational excitation in particular.

Not all of the papers to be cited include specific results for rotational and/or vibrational excitation, but rather present developments that promise or could lead to detailed calculations. References to experimental work must, unfortunately, be limited to those papers directly pertinent to the discussion of theoretical developments. Points of contact with swarm measurements will be emphasized.

This paper is intended to supplement, rather than duplicate, the comprehensive outlines of the fundamental theory and current state-of-the-art in recent reviews: of rotational excitation by Shimamura [1]; of vibrational excitation by Chang, by Herzenberg, and by Thompson [2]; of general electron-molecule collision theory by Morrison, and by Kazansky and Fabrikant [3]; and of scattering by polyatomics by Thompson and Gianturco, and by Gianturco and Jain [4]; all of which the serious reader is strongly encouraged to consult.

Targets of Opportunity

Four molecules -- H_2, N_2, CO, and CO_2 -- have probably received more attention from both theorists and experimentalists over the years than all other molecules combined. Significant advances in calculations of rotational and vibrational excitation of H_2 are discussed in detail by Crompton and Morrison elsewhere in this volume. Owing to the implications of much of the recent work on H_2 [5-20] for other molecules, some of it will also be touched upon here. Given all the work on H_2, it should be no surprise that a bit of new information on HD [17] and D_2 [17,20] has also recently appeared.

N_2 continues to receive attention [8,9,16,17,21-34] in roughly equal measure, primarily because the infamous 2.3 eV resonance provides an ideal vehicle for testing new dynamical theories and approaches to treatments of correlation and charge polarization. Recent work on CO by our group and others [17,21,35-39] continues long-standing interest in this molecule, both theoretical and experimental. Recent work on CO_2 has also been quite substantial [16,40-44].

Theoretical work on nonpolar diatomics other than H_2 and N_2 has been very limited, only F_2 [45] and Li_2 [46] having been recently studied. Polar diatomics, on the other hand, have received enormous attention. Most effort has been devoted to HF [47-51] and $HC\ell$ [22,52-59]; but LiH [36], LiF [48,49], and BeO [48] have also been studied.

The only linear triatomic other than CO_2 recently treated is HCN [48,60,61]. There has also been some work on other triatomics -- the ever-popular H_2O [48,62-64], and H_2S [65]. More complicated polyatomics that have come under study recently include: NH_3 [66]; CH_4 [62,64,67,68]; the complexes CX_4 and SiX_4 (X=H,F,Cl) [69]; C_2H_2 [38,43,70]; and even C_2H_4, $C_2H_3C\ell$, and $C_6H_3C\ell_3$ [71].

The Interaction Potential

It is helpful, if not completely defensible, to consider the interaction potential as composed of four parts: direct (a.k.a. static), exchange, (long-range) polarization, and (short-range) correlation. Progress is measured, of course, by growth in our ability to include some or all of these effects completely and rigorously. Just as important, perhaps, is growth in our ability to judge when it is necessary to do so. When the goal is physics other than quantitative accuracy in calculated cross-sections, a local model potential may be perfectly acceptable.

The simplest representations of even the static interaction potential are adequate for some of the most fundamental physics. For example, noting the essential separability of the long- and short-range contributions to the eigenphase sum for scattering by polar molecules, Clark [72] derived closed-form expressions for its threshold behavior for diatomics and simple polyatomics. In the same spirit, Herrick and Engelking derived the cross-section threshold law for rotational excitation of a polar symmetric top [73], and studied the effect of rotational doubling on photodetachment threshold behavior for polar molecules [74].

Very simple models of the interaction of an electron with a polar molecule have been used [75,76] in detailed investigations of the analytic properties of the scattering matrix and its singularities. Knowledge of the short-range interaction was unnecessary, as the energy dependence of threshold behavior depends only on the long-range interaction. These studies showed that while the concept of virtual states is strictly invalid for polar molecules, it can still be of practical value for weakly polar systems. Still left open is the question of whether virtual states are the correct interpretation of the provocative threshold resonances observed in vibrational excitation of many systems [2].

In effective range and quantum defect theories the potential is partially and very simply represented, the rest deduced semiempirically. Expressions have been developed [16, 38,48] for cross-sections that are valid for low energies and involve, beyond multipole moments and polarizabilities, only a very few unknown parameters that can be deduced using scattering or electron affinity data. This permits, for example: prediction of scattering lengths [16], analysis of differential cross-sections for dominant interactions [38], and extrapolations into unknown energy regions [48].

Examples of similarly productive use of local model exchange potentials are: a recent study of potential-adapted scattering wave functions for $HC\ell$ and N_2 [22], development of a polarization potential for N_2 [25], and an attempt to test the effect of correlation in the target wave function on scattering [29].

When collision energy and/or the angular momentum barrier are such that exchange is a weak effect, use of local models of exchange to supplement results obtained using an exact treatment can yield much greater computational efficiency at no sacrifice in accuracy [39,51,59,60]. Similarly, the use of the distorted-wave approximation with a very simple asymptotic representation of the static potential [15] can provide a useful and accurate bridge between sophisticated variational calculations for low partial waves and the Born approximation for high partial waves.

Exchange. The treatment of the exchange interaction has advanced considerably since simple prescriptions based on orthogonalization of bound and continuum orbitals, and/or on free-electron-gas (FEG) theory, e.g., the Slater, Hara, and semiclassical [see 67] potentials, were the best we could do. While for relatively simple molecules it is now possible to do much better, these approximations will continue to be quite useful in several situations.

For example, the FEG model has often been found, sometimes by tuning, to agree well with an exact treatment at the static-exchange level in a few test cases, providing a justification for its use in production calculations [5,7,11-13, 41,54]. The same kinds of argument have been made to justify advances into a new regime, for which an exact treatment of exchange is not universally possible, on the basis of results obtained with a model exchange potential, e.g., from vibrationally elastic scattering to vibrational excitation [14,56].

Exact treatments of exchange are now the rule in several computational programs based on the R-matrix method [8,28,36, 45], the linear-algebraic method [6,9,31,41], the Schwinger-variational method [18-20,24,30,32,68,77], and the integral-equations approach [12,39,46,47,51,59-61]. This year has seen the first application of exact exchange in a calculation of scattering by a linear, polar molecule [60] and by a nonlinear system [68]. New methods continue to be developed and applied: another version of the linear-algebraic approach [10]; a noniterative partial differential equations technique [33]; and a generalization [59] of the separable approach [78] used in earlier fixed-nuclei calculations to an integral-equations treatment of full vibrational close-coupling in the molecular body frame.

Much, however, remains to be done. There have been very few [20,47,51,58,59] recent calculations of vibrational excitation in which exchange was treated exactly, and only two [20,58] of these were not based on an approximate (adiabatic) treatment of the nuclear dynamics.

As exact treatments of exchange have become more common and widely used, there has been less detailed study of the importance of an exact treatment. Indeed, it is difficult, if not impossible, within the contexts of the R-matrix, linear algebraic, and Schwinger methods to isolate or approximate exchange as a separate effect. Comparisons have been made between the results of these and different, independent programs, e.g., model exchange in the integral-equations method and exact exchange in the linear-algebraic method for H_2 [12] and CO_2 [41]. These supported earlier conclusions that the orthogonalized Hara FEG model is usually quite good for total and vibrationally elastic cross-sections for heavier systems, auguring well for its use in more complex systems.

Caution is still advised -- exchange can be much more important for some processes, e.g., vibrational and some rotational transitions, and for some energy regions, e.g., near resonances, than for others. In one recent comparison [59] it was shown that the consequence of exact vs. model exchange is quite large in vibrational excitation of HCℓ -- as much as a factor of two in the cross-section for v-v'=0-1 and five in that for v-v'=0-2 (Fig. 1).

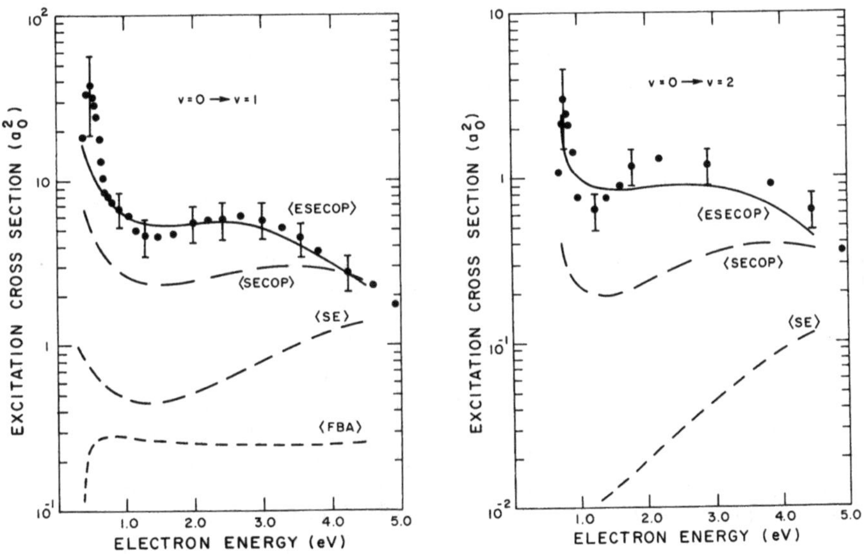

Figure 1. Cross-sections for vibrational excitation of HCℓ from calculations [59] and measurements [79] (renormalized by a factor of 0.6). Notations are: FBA, first Born approximation; SE, static-model exchange including orthogonalization; SECOP, SE plus correlation and polarization effects (COP); ESECOP, static-exact exchange plus COP.

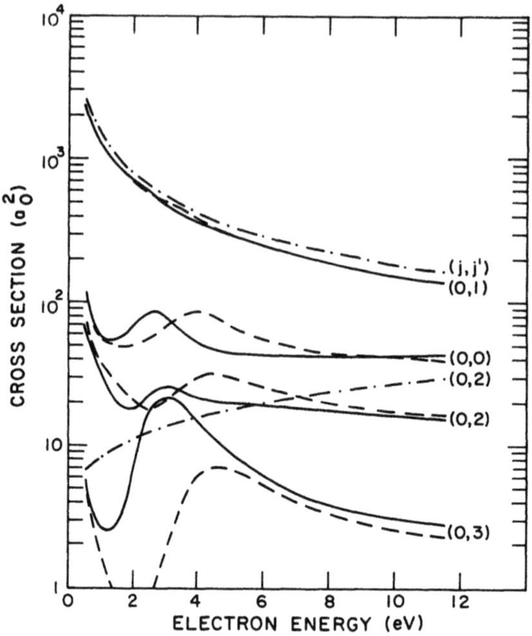

Figure 2. Calculated [60] cross-sections for rotational transitions in HCN: —•—, FBA; - - -, SECOP; ———, ESECOP.

Another illustration is provided in Fig. 2. Here, only rotational transitions with $\Delta j \neq 1$ are significantly affected, and then only in the energy range 1-5 eV where a prominent resonance exists. In both of these examples the Hara FEG model of the exchange potential was employed.

Fortunately, there will probably always be pioneers [e.g., 62-67,69-71] willing to attack ever more complex polyatomic molecules for which information is desperately needed, but for which high rigor is not yet possible or unacceptably expensive, using this kind of local model exchange potential. For the most complex polyatomics, single-center approaches and exact treatments of exchange will probably always be prohibitive, and the more qualitative techniques such as the continuum-multiple-scattering method [e.g., 27,69-71] will remain essential tools.

Polarization and correlation. One could not wish a more striking demonstration of the importance of the correlation-polarization effect than the comparison of the SE and SECOP curves in Fig. 1. A similar, though somewhat smaller, effect was found in vibrational excitation of HF [51]. Not too long ago [e.g., 7,54] our only recourse was to a semiempirical approach, whereby a cutoff function with one or more adjustable parameters served double duty - to soften the r^{-4} singularity of the long-range polarization potential and to incorporate missing physics, e.g., correlation, by tuning the potential to match measurement or some more rigorous calculation.

Consider, for example, the results shown in Fig. 3. HCℓ has a resonance at about 2.5 eV, which is visible in Fig. 1 and which a calculation [54] with an appropriately chosen cutoff polarization potential and FEG model exchange potential did reproduce. In the later ESECOP calculation [59] both of these approximations were avoided, but the results for the momentum transfer cross-section were little changed (the bumps in the solid curve in Fig. 3 are an estimated contribution for vibrational excitation, not included in the dashed curve). Thus while much better than a simple parameterized polarization potential, now often possible thanks to recent advances as will be discussed below, less elegant techniques will no doubt continue to be of considerable value.

Several ab initio polarization potentials have been developed for special molecular cases: for H_2 [5,12], N_2 [25], and several polyatomics [62,65,66]. These are all based in one way or another on the seminal polarized orbital theory of Temkin, and rely on a somewhat ad hoc truncation of

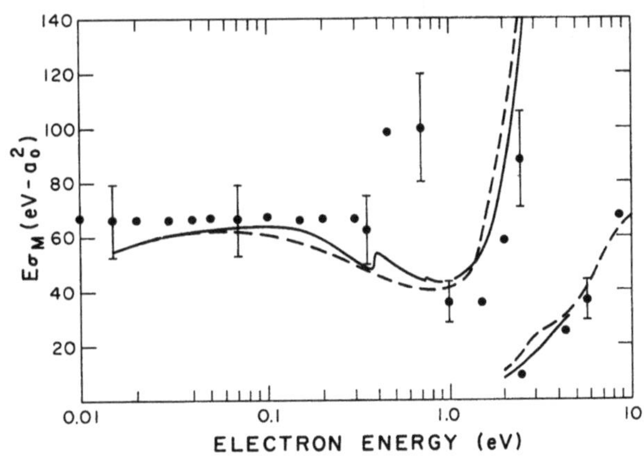

Figure 3. Momentum transfer cross-section for HCℓ: - - -, model potential calculations [54]; ——, ESECOP calculations [59]; •, swarm measurements [80]. The data are divided by 10 above 2 eV.

certain terms and integrals at small distances, i.e., the nonpenetrating approximation) to incorporate nonadiabatic effects. In extensive applications to H_2 [10-14] this model seems to work quite well. The local kinetic energy semiclassical model [81] used by Abusalbi et al. in their CH_4 calculations [67] is similar in spirit.

There has been much recent activity devoted to the development [6,9,18,24,28,30] and application [19,20,31,32, 36,45] of optical potential theory. These methods offer the possibility of including both short- and long-range effects, i.e., correlation as well as polarization, rigorously and without semiempiricism.

The effect of correlation in the target itself, or more precisely ensuring balance between the recorrelation in the N-particle target and that in the N+1 particle collision system, has been of some concern in the CI approaches to generating the optical potential [28,31,36]. One of the alleged advantages of the two-particle-hole Tamm-Dancoff approximation (TDA) used by Berman et al. [24,30] is that balanced correlation is automatically achieved. The effect of target correlation has been shown to be substantial, for example shifting the position of the Π_g resonance in N_2 by about 0.5 eV [24,29], in addition to the shift of about 1.0 eV obtained from the second-order optical potential.

It is also difficult to simultaneously include both long-range polarization and short-range correlation effects in the CI approaches [31,36], and perhaps also in the TDA. In N_2, for example, long-range effects dominate scattering in Σ symmetry, while short-range effects dominate Π_g (resonant) scattering [31]. A technique in which both are easily included is essential if cross-sections over a wide range of energy are to be routinely obtainable.

An alternative technique, at an intermediate level of sophistication and difficulty, has recently been suggested [82]. It is based on a hybrid combination of a short-range correlation potential derived from free-electron-gas theory and the (long-range) asymptotic form of the polarization potential, the two being joined by a physically motivated but ad hoc procedure. The correlation potential is determined, eschewing semiempiricism, solely from the theoretical molecular charge density; the latter requires knowledge of experimentally or theoretically determined values for the molecular polarizabilities.

This method has been applied in both test and production calculations for a large variety of molecules: H_2, N_2, and CO_2 [82]; CO [39]; Li_2 [46]; HF [51]; HCℓ [56,59]; and HCN [60,61]. This is the model designated "COP" in the captions to Figs. 1 and 2. So far it has been remarkably, indeed surprisingly, successful. For example, the position of the Π_g resonance in N_2 was calculated to be 2.17 eV, in excellent agreement with the results [24,30-32] of much more rigorous calculations.

Results [39] for the analogous resonance in CO, illustrated in Fig. 4, are also excellent -- the calculated resonance position (width) is 1.85 (0.95) eV, compared with 1.72 (0.75) eV from recent calculations [36] using an ab initio optical potential, and 1.9 eV from recent crossed-beam measurements [83,84]. Just as for N_2, the resonance position is primarily determined by short-range correlation, and is shifted down by about 1.5 eV from its position obtained in the static-exchange approximation. It seems that this method does, somehow, achieve a balanced treatment of correlation effects, as well as the inclusion of polarization.

We may conclude from the last two sections that there has been very substantial recent progress in production-level calculations in which the exchange, correlation, and polarization interactions are all treated accurately and ab

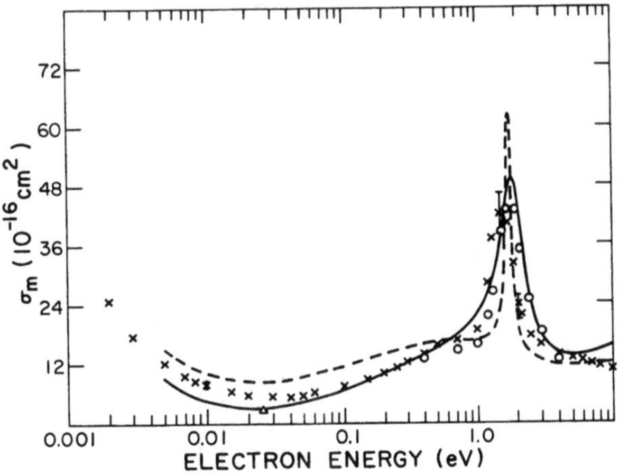

Figure 4. Momentum transfer cross-section for CO: - - -, model potential calculations [85]; ———, ESECOP calculations [39]; × and o, swarm measurements [86,87]; Δ, cyclotron resonance measurements [88].

initio; indeed the past two years have seen the first such calculations for any molecules except H_2 and N_2.

Treatment of Nuclear Dynamics

The most straightforward solution to the problem of treating nuclear motion is known as full laboratory-frame close coupling. While occasionally used [11,13,67], it can be prohibitively expensive, if not totally impractical, for almost any molecule except H_2. There has yet to appear any calculation in which both rotational and vibrational motion were included ab initio and simultaneously with a sophisticated treatment of short-range interactions.

The major difficulty arises from the very complicated couplings that follow from the multicenter nature of the interaction at short distances. Ways of dealing with this problem can be roughly grouped into three (overlapping) classes: i) simplify and/or semiempiricise the short-range interaction, ii) use adiabatic approximations in the treatment of the long-range interactions, iii) retain rigor throughout, and employ frame transformation theories.

Semiempirical and parameterized theories. A great deal of physics can be learned by purely formal studies using simple parameterized potentials. For example, a simple model potential with a few adjustable parameters has been used in vibrational close-coupling calculations to provide fascinating insight into the problem of threshold vibrational excitation in CO_2 [42,44]. Inclusion of coupling to the bending mode in the more recent work [44] appears to rob a significant amount of flux from excitation of the first symmetric stretch, but the total cross-section is relatively little changed.

Fabrikant [49] generalized the effective range theory to treat energies of order of rotational spacings and near threshold for vibrational excitation. Another work in the same spirit is the development [89] of an exact solution to the problem of electron scattering by a rigid rotor, in which the interaction is described by two zero-range potentials whose strengths can be adjusted to mimic certain physically interesting situations.

The next level of complexity involves parameterizing a boundary condition at short range, e.g., the log derivative of the scattering wave function, and then integrating the scattering equations outward as usual with the simplest possible forms of the asymptotic interaction potential. The functional form of the boundary condition may be energy dependent, but depends on the internuclear coordinate; its value may be determined by reference to other calculations or by fitting the results of the calculations to experiment. This approach was used in calculations of vibrational excitation of HF [50] and HCℓ [57], based on calculated potential energy curves for HF^- and $HC\ell^-$, and equilibrium fixed-nuclei scattering calculations for HCℓ.

A related approach is the use of parameterized potentials, a recent example of which is that proposed by Pozdneev [27] for calculating cross-sections for a variety of processes in electron collisions with N_2, including elastic scattering, vibrational excitation, and dissociative attachment. He employs a multiple-scattering approach with the short-range interactions treated as local, separable, and determinable semiempirically. The scattering equations are then solved using the Faddeev method.

Also in the category of exactly solvable models with parameterized potentials is that developed by Kazansky [52, 55] for vibrational excitation of diatomic molecules, based on a parameterized antibonding intermediate state. Applied to vibrational excitation and dissociative attachment in electron collisions with HCℓ, the qualitatively correct results supported the physical hypothesis of the virtual state. When the nuclear vibrations are taken into account, these can be interpreted as nuclear-excited Feshbach resonances [53].

Formalisms based on the concept of a resonance have obvious appeal for processes such as vibrational excitation that are often dominated by resonances. Berman et al. [90] studied parameterized local complex potentials often used to describe nuclear motion in resonance states, casting "serious doubt on the uniqueness and reliability of empirical local complex potentials obtained by fitting experimental scattering cross-sections" in some cases. Nevertheless, this approach has a long and honorable history, and a guaranteed future. A new wrinkle has been recently added [18] -- a semiclassical treatment that yields a simplifying factorization of the cross-sections.

Berman et al. [26] went beyond the local approximation in treating nuclear dynamics in resonant electron-molecule scattering, employing basis-set methods to treat a nonlocal, complex, and energy-dependent potential. Remarkably good results for the vibrational structure in the 2.3 eV resonance in N_2 were obtained by using a combination of earlier ab initio calculations and empirical fitting; the results were also used [23] to unfold the population of the vibrational levels from experimental electron transmission spectra.

In another recent application [58] of the nonlocal complex potential formalism -- to $HC\ell$ and $DC\ell$ -- calculated eigenphases for electron-$HC\ell$ scattering as a function of internuclear distance [59] were used to determine the parameters in the model. It is interesting to note that good results have now been obtained for the same processes using both a nonresonant (virtual state) [55] and resonant [58] formalism, suggesting that the physical concepts behind the models should perhaps not be taken too seriously, i.e., lack uniqueness, and that the distinction between resonant and nonresonant methods is beginning to blur as they become more flexible and powerful.

Adiabatics -- rotation. For most electron energies, and for most molecules of interest, the collision energy is a great deal larger than the rotational spacing. When this is true, the dynamics and kinematics of the collision process can be usefully considered separately. The collision may be considered classically sudden, i.e., the electron has come, transferred a certain amount of angular momentum to the molecule, and gone before the molecule has rotated significantly. This leads to a variety of sum rules and formulas valid in the limit of high initial rotational quantum number that are independent of the details of the target molecule, interaction potential, or transition mechanism. For example, Chang [21] has deduced a simple limiting expression for the relative magnitude of various rotational branches in the limit of high initial rotational quantum number.

Many of these sum rules and limiting forms, which relate to energy loss as well as conventional rotational cross-sections, can be carried over in a rigorous quantum mechanical approach to rotational or ro-vibrational excitation. Much of the algebra relevant to rotational excitation of molecules more complex than linear (spherical, symmetric and asymmetric tops) has only recently been completely worked out [1,4,64, 67,91]. The review by Shimamura [1] provides a particularly clear and comprehensive discussion of this whole topic, and a summary of relevant literature through 1983. His own original contributions, cited therein, have been particularly notable.

As an example of the usefulness of such concepts, consider the results for rotationally state-selected transitions in CO shown in Fig. 5. The measured results [92] were obtained by unfolding rotationally broadened energy-loss spectra from

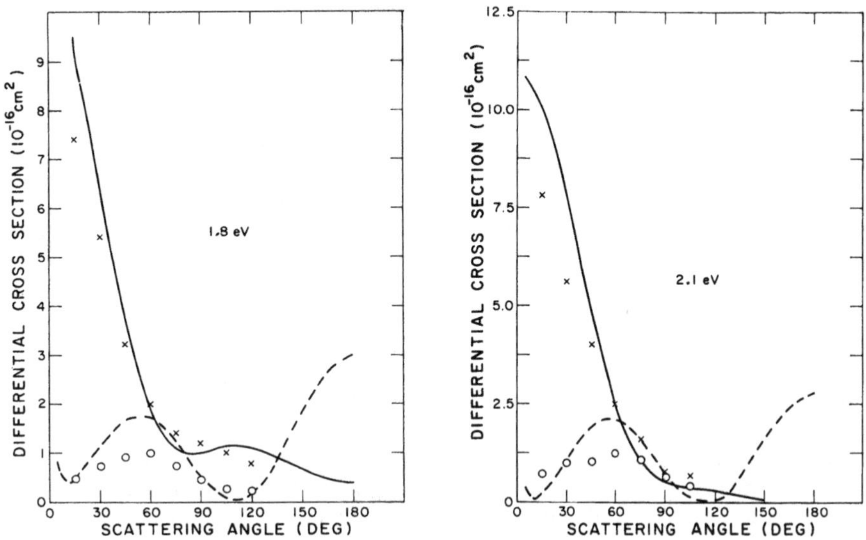

Figure 5. Cross-sections for rotational transitions in CO. The measured [92] and calculated [39] results are, respectively, × and ——— for $\Delta j=0$, and o and - - -, for $\Delta j=\pm 1$.

crossed-beam measurements using a high-j approximation. This provides an extraordinarily valuable test of calculations, obtainable in no other way. In this case, a factor-of-two discrepancy between the calculated and measured results was discovered, and has yet to be explained.

The assumption that the collision can be considered rotationally sudden has been central to the vast majority of calculations of electron-molecule scattering to date. When cross-sections for individual rotational transitions are desired, they are usually obtained using the so-called adiabatic-nuclear-rotation (ANR) approximation [see 1,3]. At the relatively low energies of interest in swarm measurements, collision energies approach rotational spacings and this approximation must obviously break down. The ANR approximation breaks down at all energies for forward scattering and total cross-sections for polar molecules, a pathological consequence of the very long-range nature of the interaction for scattering from a fixed dipole. This approximation will, however, die hard, because it still offers the only practical way to obtain results for scattering of very low energy electrons by the highly rotationally excited moleclules typical of a thermal gas.

Recent years have seen many different attempts both to more precisely clarify the regime and degree of breakdown of the ANR approximation, and to develop techniques both practical and accurate for obtaining results at low collision energies that retain at least some of its simplicity. Morrison and co-workers [7,13] have carried out extremely careful, detailed, and critical examinations of the ANR approximation for H_2, concluding that "the breakdown of the AN theory for rotation is found to be more severe at energies a few times threshold than was anticipated by rough qualitative arguments made heretofore."

They also developed a new scaling procedure [11], derived from the fact that most of the important lab-frame and ANR scattering matrix elements reduce to their Born approximation values at threshold, and introduced simple kinematic corrections accordingly. This is similar in spirit to the so-called MEAN approximation developed [93] specifically for polar molecules. In the latter is exploited the fact that forward scattering and total cross-sections for polar molecules are dominated by contributions for which the Born approximation is valid in the laboratory frame.

Both techniques extend the range of validity of the ANR approximation for individual rotational excitation cross-sections closer to threshold. The MEAN approximation also precludes divergence in forward scattering and integrated cross-sections for transitions with $\Delta j=1$ (usually dominant) in polar molecules. The integrated cross-sections in Fig. 2 are an example of its use, in this case to provide a relatively small correction to the lab-frame FBA value of the cross-section for $j-j'=0-1$. Another example is the differential cross-sections shown in Fig. 6, which also illustrates the fact that transitions with $\Delta j\neq 1$ can be important for scattering out of the forward direction and for vibrational excitation even for highly polar molecules such as HF.

There has also been a significant advance beyond these simple procedures, involving fewer approximations but still without resort to a complete lab-frame representation. A new "off-shell" formalism [10] involves T-matrix elements all defined in the molecular body frame, and therefore parametrically dependent on the internuclear distance, but still explicitly involves the distinct momenta of the incoming and outgoing scattering channels. The results of tests for both total and differential cross-sections for H_2 are very encouraging. Frame transformation approaches, to be discussed below, are a further improvement.

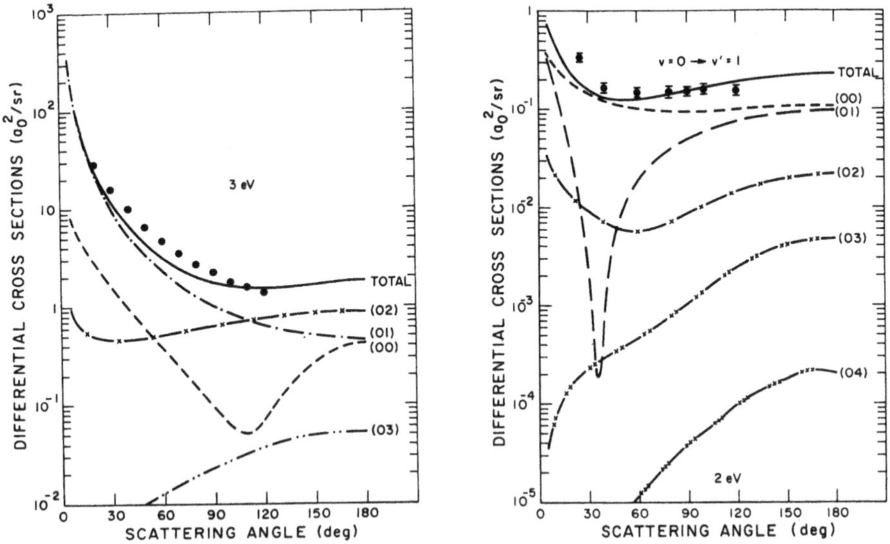

Figure 6. Differential cross-sections for rotational transitions in vibrationally elastic and inelastic scattering in HF. The calculated results [51] are given as curves and the points (●) are from measurements [79].

Adiabatics - vibration. Another attack in the spirit of adiabatic theory is the recently suggested algebraic-eikonal approach [94]. It combines an algebraic description of molecular motion (i.e., the vibron model) with the Glauber (or eikonal) and adiabatic approximations. Explored to date for only dipolar diatomics, it should easily generalize to more complex molecules and interactions. Its major attraction is the more-or-less equal footing on which it places rotational and vibrational excitation. It remains, however, strictly valid only well away from thresholds.

Just as for rotation, there is an adiabatic approximation for vibration, the simplest form of which amounts to taking the integral between initial and final vibrational states of transition matrices calculated as a function of a series of fixed internuclear distances. Likewise, its range of validity has only recently been seriously scrutinized. Morrison et al. [14] showed that it leads to serious errors up to several times threshold for excitation of H_2, and that carrying out vibrational close-coupling calculations in the molecular body frame, in which rotation is still neglected, yields significant improvement.

Gauyacq [50] carried out a model study for excitation of HF with and without this approximation, finding much less serious, but still significant, errors near threshold. Our results for excitation of HF [51] and HCℓ [59], when compared with experiment as in Fig. 1, confirm this conclusion. Domcke and Mündle [58] carried out an analysis for HCℓ analogous to that of Gauyacq for HF, but with a much more accurate basis. The results were nevertheless similar -- error of a factor of two or more very near threshold, minimal error at twice threshold.

Frame transformation theories. These are designed to exploit the fact that, in certain regimes, the neglect of otherwise important parts of the interaction introduces no (or negligible) error. For example, molecular rotation and perhaps also vibration can be neglected when the electron is very close to the nuclei; and exchange and the difficulties associated with a multicenter interaction vanish at large distances. This accounts for the appeal of computational schemes designed to work in the molecular body frame with fixed nuclei at short range, coupled at some intermediate point to a scheme that includes the rotational and/or vibrational Hamiltonians at large range.

As usually conceived, this approach is based on a transformation of the scattering wave function itself and numerical integration in the outer region, e.g., the application of effective range theory to polar molecules [49]. The function taken on the inner boundary is, in fact, more often the fixed-nuclei R matrix, i.e., essentially the log derivative of the wave function, parameterized by the internuclear distance [50,57]. With appropriate vibrational integration over this quantity, the integrations can then be continued in the outer region. When the outer region is dominated by a single interaction that is, or can be assumed, central, the outer region can often be treated analytically. Such has been done for the dipole interaction in the contexts of quite elaborate close-coupling calculations for CO [35].

Chang and coworkers have extended frame transformation theory for vibrational excitation to include dipole coupling between vibronic levels [37], and to treat more complicated linear polyatomics [43]. The former recognizes the competing importance of resonance (short-range) and the long-range (dipole) effects, yielding a relatively simple parameterization of the ro-vibrational excitation cross-sections. The same concern to incorporate short- and long-range effects in a coherent theory of vibrational excitation stimulated Greene and Jungen [34] to develop a frame transformation scheme that unifies the usual resonant and nonresonant points of view.

While not usually thought of in this category, models of resonant vibrational excitation based on the definition of a complex potential do qualify. The rigid-rotor, fixed-nuclei approximation is implicit in internuclear-distance dependence of the essential parameters, but the adiabatic approximation need not be made in the course of extracting vibrational excitation cross-sections. Such calculations are usually critically dependent on fitting to experimental data and/or parameters provided by other accurate ab initio calculations, as discussed above.

More recently, it has become more common to compute directly and ab initio the resonance energy, width, and shift functions [19] that are the necessary input to the nonlocal complex potential models. Mündel and Domcke [95] recently extended the nonlocal complex potential formalism to include dissociative attachment channels, and applied the technique in very impressive calculations of cross-sections for vibrational excitation and dissociative attachment in H_2 and D_2 [20]. Thus so-called parameterized methods are breaking out of the box of semiempiricism.

Concluding Remarks

Progress in the theory and practice of rotational and vibrational excitation of molecules by low energy electrons has clearly been substantial in recent years, but is still perhaps more virtual than real. We have summarized numerous theoretical and computational advances that will ultimately lead to the production of a large amount of high-quality data, but the number of such calculations for molecules other than H_2 and N_2 is still quite small. Several theoretical challenges nevertheless remain, in particular the development of a more general capability for accurately treating excitation near thresholds and in polyatomic and nonlinear systems.

Perhaps most important in the context of this volume is improving the productivity of the connection between detailed calculations and swarm measurements. It is only a slight exaggeration to say that currently, with regard to rotational excitation, results near threshold are either easily generated but of poor reliability, or of high quality but rare and not very useful. The lack of reliability may derive from their parentage, often some variant of the adiabatic approximation. This approximation facilitates, however, the generation of data for practically any desired rotational states. High-quality lab-frame close-coupling calculations are, on the other hand, not easily made compatible with the generation of data for the highly excited rotational states typical of the room temperature gas, e.g., most-probable j values of 3 and 7 for the HCℓ and CO data shown in Figs. 3 and 4, much higher for heavier systems.

We must therefore remain aware of the risk that as our ability to carry out more accurate calculations grows, our usefulness to the swarm community may be reduced, at least for the short term. Further work is also required if we are to extend the applicability of the results of high-quality calculations and measurements. To do this we must develop more powerful techniques for separating the essential dynamical quantities characterizing the interaction from those characterizing the kinematics. The former might be sets of coefficients for moments of differential cross-sections expressed in terms of angular momentum transferred during the collision, from which cross-sections for arbitrary rotational transitions might be deduced. Such an approach has been suggested [93], but has yet to be seriously exploited and certainly might be improved upon.

Acknowledgments

I would like to acknowledge numerous helpful discussions with A. V. Phelps and L. C. Pitchford on the subject of swarms, and the contributions of N. T. Padial and A. Jain to the code development and calculations leading to the results for HF, HCℓ, CO and HCN used to illustrate this paper. This work was supported by the U. S. Department of Energy (Division of Chemical Sciences).

References

*Staff Member, Quantum Physics Division, National Bureau of Standards.

[1] Shimamura, I. In "Electron-Molecule Collisions"; Shimamura, I.; Takayanagi, K., Eds.; Plenum Press: New York and London, 1984; Chapter 2.
[2] Herzenberg, A. ibid.; Chapter 3; Thompson, D.G. Adv. At. Mol. Phys. 1983, 19, 309; Chang, E.S. Comments At. Molec. Phys. 1985, 16, 293.
[3] Morrison, M.A. Aust. J. Phys. 1983, 36, 239; Kazansky, A.K.; Fabrikant, I.I. Sov. Phys. Usp. 1984, 27, 607.

[4] Thompson, D.G.; Gianturco, F.A. Comments At. Mol. Phys. 1985, 16, 307; Gianturco, F.A.; Jain, A. Phys. Repts., in press.
[5] Gibson, T.L.; Morrison, M.A. J. Phys. B 1982, 15, L221.
[6] Schneider, B.I.; Collins, L.A. ibid., p. L335.
[7] Feldt, A.N.; Morrison, M.A. ibid., p. 301.
[8] Noble, C.J.; Burke, P.G.; Salvini, S. ibid., p. 3779.
[9] Schneider, B.I.; Collins, L.A. Phys. Rev. 1983, A27, 2847.
[10] Ficocelli Varracchio, E.; Lamanna, U.T. Chem. Phys. Lett. 1983, 101, 38; J. Phys. B 1984, 17, 4395.
[11] Feldt, A.N.; Morrison, M.A. Phys. Rev. 1984, A29, 401.
[12] Gibson, T.L.; Morrison, M.A.; ibid., p. 2497.
[13] Morrison, M.A.; Feldt, A.N.; Austin, D. ibid., p. 2518.
[14] Morrison, M.A.; Feldt, A.N.; Saha, B.C. ibid., A30, 2811.
[15] Nesbet, R.K. J. Phys. B 1984, 17, L897.
[16] Fabrikant, I.I. ibid., p. 4223.
[17] Kazansky, A.K.; Yelets, I.S. ibid., p. 4767.
[18] Gibson, T.A.; Lima, M.A.P.; Takatsuka, K.; McKoy, V. Phys. Rev. 1984, A30, 3005.
[19] Berman, M.; Mündel, C.; Domcke, W. ibid., 1985, A31, 641.
[20] Mündel, C.; Berman, M.; Domcke, W. ibid., A32, 181.
[21] Chang, E.S. J. Phys. B 1982, 15, L873.
[22] Le Dourneuf, M.; Lan, Vo Ky; Launay, J.M. ibid., 1982, 15, L685.
[23] Estrada, H.; Berman, M.; Cederbaum, L.S.; Domcke, W. Chem. Phys. Lett. 1983, 97, 352.
[24] Berman, M.; Walter, O.; Cederbaum, L.S. Phys. Rev. Lett. 1983, 50, 1979.
[25] Onda, K.; Temkin, A. Phys. Rev. 1983, A28, 621.
[26] Berman, M.; Estrada, H.; Cederbaum, L.S.; Domcke, W. ibid., p. 1363.
[27] Pozdneev, S. J. Phys. B 1983, 16, 867.
[28] Burke, P.G.; Noble, C.J.; Salvini, S. ibid., p. L113.
[29] Rumble, J.R.Jr.; Stevens, W.J.; Truhlar, D.G. ibid., 1984, 17, 3151.
[30] Berman, W.; Domcke, W. Phys. Rev. 1984, A29, 2485.
[31] Schneider, B.I.; Collins, L.A. ibid., A30, 95.
[32] Berman, M.; Domcke, W. J. Phys. B 1984, 17, L453.
[33] Weatherford, C.A.; Onda, K.; Temkin, A. Phys. Rev. 1985, A31, 3620.
[34] Greene, C.H.; Jungen, Ch. Phys. Rev. Lett. 1985, 55, 1066.
[35] Chandra, N. J. Phys. B 1982, 15, 4465.

[36] Salvini, S.; Burke, P.G.; Noble, C.J. ibid., 1984, 17, 2549.
[37] Chang, E.S.; Antoni, Th.; Jung, K.; Erhardt, H. Phys. Rev. 1984, A30, 2086.
[38] Sohn, W.; Kochem, K.-H.; Jung, K.; Erhardt, H.; Chang, E.S. J. Phys. B 1985, 18, 2049.
[39] Jain, A.; Norcross, D.W. Phys. Rev. 1986, in press.
[40] Morrison, M.A. ibid., 1982, A25, 1445.
[41] Collins, L.A.; Morrison, M.A. ibid., p. 1764.
[42] Whitten, B.L.; Lane, N.F. ibid., p. 3170.
[43] Chang, E.S. J. Phys. B 1984, 17, 3341.
[44] Kimura, M.; Lane, N.F. Phys. Rev. 1986, in press.
[45] Morgan, L.A.; Noble, C.J. J. Phys. B 1984, 17, L369.
[46] Padial, N.T. Phys. Rev. 1985, A32, 1379.
[47] Resigno, T.N.; Orel, A.E.; Hazi, A.U. ibid., 1982, A26, 690.
[48] Fabrikant, I.I. J. Phys. B 1983, 16, 1253.
[49] Ibid., p. 1269.
[50] Gauyacq, J.P. ibid., p. 4049.
[51] Jain, A.; Norcross, D.W. Phys. Rev. 1986, in press.
[52] Kazansky, A.K. Sov. Phys. JETP 1982, 55, 824.
[53] Gauyacq, J.P.; Herzenberg, A. Phys. Rev. 1982, A25, 2959.
[54] Padial, N.T.; Norcross, D.W.; Collins, L.A. ibid., 1983, A27, 141.
[55] Kazansky, A.K. J. Phys. B 1983, 16, 2427.
[56] Padial, N.T.; Norcross, D.W. Phys. Rev. 1984, A29, 1590.
[57] Fabrikant, I.I. J. Phys. B 1985, 18, 1873.
[58] Domcke, W.; Mündel, C. ibid., in press.
[59] Padial, N.T.; Norcross, D.W. Phys. Rev. 1986, in press.
[60] Jain, A.; Norcross, D.W. ibid., 1985, A32, 134.
[61] Jain, A.; Norcross, D.W. J. Chem. Phys. 1986, 84, in press.
[62] Jain, A.; Thompson, D.G. J. Phys. B 1982, 15, L631.
[63] Jain, A.; Thompson, D.G. ibid., 1983, 16, L347.
[64] Jain, A.; Thompson, D.G. ibid., p. 3077; ibid., 1984, 17, 1930.
[65] Jain, A.; Thompson, D.G. ibid., 1984, 17, 443.
[66] Jain, A.; Thompson, D.G. ibid., 1983, 16, 2593.
[67] Abusalbi, N.; Eades, R.A.; Nam, T.; Thirumalai, T.; Dixon, D.A.; Truhlar, D.G. J. Chem. Phys. 1983, 78, 1213.
[68] Lima, M.A.; Gibson, T.L.; Huo, W.M.; McKoy V. Phys. Rev. 1985, A32, 2696.
[69] Tossell, J.A.; Davenport, J.W. J. Chem. Phys. 1984, 80, 813.
[70] Tossell, J.A. J. Phys. B 1985, 18, 387.

[71] Bloor, J.E.; Sherrod, R.E. Chem. Phys. Lett. 1982, 88, 389; Bloor, J.E. Int. J. Quant. Chem. Symp. 1983, 17, 101.
[72] Clark, C.W. Phys. Rev. 1984, A30, 750.
[73] Herrick, D.R.; Engelking, P.C. ibid., A29, 2421.
[74] Herrick, D.R.; Engelking, P.C. ibid., p. 2425.
[75] Estrada, H.; Domcke, W. J. Phys. B 1984, 17, 279.
[76] Herzenberg, A. ibid., p. 4213.
[77] Takatsuka, K.; McKoy, V. Phys. Rev. 1984, A30, 1734.
[78] Schneider, B.I.; Collins, L.A. ibid., 1981, A24, 1264; Rescigno, T.N.; Orel, A.E. ibid., p. 1267.
[79] Rohr, K.; Linder, F. J. Phys. B 1975, 8, L200; 1976, 9, 2521.
[80] Davies, D.K. Bull. Am. Phys. Soc. 1981, 26, 726; 1982, 27, 110; AFWAL Report TR-82-2083.
[81] Valone, S.M.; Truhlar, D.G.; Thirumalai, D. Phys. Rev. 1982, A25, 3003.
[82] Padial, N.T.; Norcross, D.W. ibid., 1984, A29, 1742.
[83] Kwan, C.K.; Hsieh, Y.F.; Kaupilla, S.J.; Smith, S.J.; Stein, T.S.; Uddin, M.N.; Dababneh, M.S. Phys. Rev. 1983, A27, 1328.
[84] Buckman, S.J. private communication.
[85] Chandra, N. Phys. Rev. 1977, A16, 80.
[86] Haddad, G.N.; Milloy, M.B. Aust. J. Phys. 1983, 36, 473.
[87] Land, J.E. J. Appl. Phys. 1978, 49, 5716.
[88] Tice, R.; Kivelson, D. J. Chem. Phys. 1967, 46, 4748.
[89] Ostrovsky, V.N.; Ustimov, V.I. J. Phys. B 1983, 16, 991.
[90] Berman, M.; Cederbaum, L.S.; Domcke, W. ibid., p. 875.
[91] Natanson, G.A. J. Phys. B 1985, 18, 4481.
[92] Jung, K.; Antoni, Th.; Muller, R.; Kochem, K.-H.; Erhardt, H. J. Phys. B 1982, 15, 3535.
[93] Norcross, S.W.; Padial, N.T. Phys. Rev. 1982, 25, 226.
[94] Bijker, R.; Amado, R.D.; Sparrow, D.A. private communication.
[95] Mündel, C.; Domcke, W. ibid., 1984, 17, 3593.

STUDIES OF ELASTIC AND ELECTRONICALLY INELASTIC ELECTRON-MOLECULE COLLISIONS

Marco A. P. Lima,[a] Thomas L. Gibson,[b]
Luiz M. Brescansin,[c] and Vincent McKoy

California Institute of Technology, Pasadena, California 91125

and

Winifred M. Huo

University of Notre Dame, Notre Dame, Indiana 46556

Introduction

Cross-sections for the scattering of low-energy electrons by molecules play an important role in the modeling of swarm and plasma etching systems, gas lasers, and planetary atmospheres. In contrast to the related atomic problem, the progress to date in both theoretical and experimental studies of electron-molecule scattering cross-sections has been limited [1]. On the theoretical side, this situation is primarily due to the additional complexities arising from the nonspherical potential fields of molecular targets. Most studies of electronic excitation of molecules by low-energy electrons have hence been carried out using low-order theories. These include plane-wave theories such as the Born Ochkur-Rudge approximations [2,3], the impact-parameter method [4], and distorted-wave theories [5,6]. Several studies of elastic scattering by molecules have also used local approximations to the nonlocal exchange potentials [7]. Although such theories and approximations can be computationally easy to apply, they do not contain enough of the collision physics to yield consistently reliable differential and integral cross

[a]Permanent address: Instituto de Estudos Avançados, Centro Técnico Aeroespacial, 12200 São José dos Campos, São Paulo, Brazil

[b]Present address: Department of Physics, Texas Tech University, Lubbock, Texas 79409

[c]Present address: Instituto de Física "Gleb Wataghin" Universidade Estadual de Campinas, 13100 Campinas, São Paulo, Brazil

sections, particularly at low and intermediate energies [8]. What is clearly needed are theoretical methods which can provide quantitatively reliable cross-sections for the elastic and inelastic scattering of low-energy electrons by molecules.

In my talk I will present some results of our studies of elastic and electronically inelastic electron-molecule collisions which we have obtained using a multichannel extension of the Schwinger variational principle. We have formulated this multichannel extension of the Schwinger principle so that it is capable of dealing with important aspects of electron-molecule collisions such as polarization effects, which are included through closed channels, electronically inelastic scattering with several open and closed channels, and, very importantly, nonlinear targets. Some important features of the method are as follows. As in the original Schwinger principle [9], the trial scattering functions need not satisfy any specific boundary conditions, and hence can be expanded in an L^2 basis. The method avoids the explicit construction of the closed-channel Green's function. Furthermore, with the use of an insertion-like quadrature to evaluate the second-Born-like terms and with an expansion of the trial function in a Gaussian basis, all matrix elements in the variational expression can be evaluated analytically.

I will begin by outlining the development of our multichannel extension of the Schwinger principle and discussing essential aspects of its implementation. I will then present some results of recent applications of this formulation to low-energy electron-molecule collisions. These results will include cross-sections for excitation of the $X^1\Sigma_g^+ \rightarrow b^3\Sigma_u^+$, $a^3\Sigma_g^+$, $c^3\Pi_u$, and $B^1\Sigma_u^+$ transition in H_2 and for elastic scattering of electrons by H_2O and CH_4. The excitation cross-section for each transition is obtained with two open channels only, while the elastic cross-sections for H_2O and CH_4 have been studied at the static-exchange level. We conclude with a brief outline of studies of electron-molecule collisions which are under way or planned.

Theoretical Developments

The Hamiltonian for the collision system can be written as

$$H = (H_N + T_{N+1}) + V = H_0 + V \tag{1}$$

where H_N is the target Hamiltonian, T_{N+1} is the kinetic energy operator for the incident electron, and V is the interaction potential between the incident electron and the target, i.e.,

$$V = \sum_{i=1}^{N} \frac{1}{r_{i,N+1}} - \sum_{\alpha} \frac{Z_\alpha}{R_{\alpha,N+1}} \quad . \tag{2}$$

In Eq.(2) the first and second terms are the electron repulsion and electron-nuclei attraction, respectively. Our goal is to obtain a multichannel Schwinger variational principle for the scattering matrix associated with the Hamiltonian of Eq.(1). One could begin by writing the Lippmann-Schwinger equation for the Hamiltonian $H = H_0 + V$, i.e.,

$$\Psi_m^{(+)} = S_m + G_0^{(+)} V \Psi_m^{(+)} \tag{3}$$

where $G_0^{(+)}$ is the Green's function associated with $E-H_0$ and S_m is the regular solution of $E-H_0$. Based on this equation it is straightforward to construct the Schwinger variational functional for the scattering amplitude, i.e.,

$$f_{mn} = -\frac{1}{2\pi} \frac{< \Psi_m^{(-)} |V| S_n > < S_m |V| \Psi_n^{(+)} >}{< \Psi_m^{(-)} |V - V G_0^{(+)} V| \Psi_n^{(+)} >} \quad . \tag{4}$$

Formally this variational principle is applicable to the problems of interest and has no major drawback at least for the collisions of nonidentical particles. However, as pointed out by Geltman [10], the continuum states of the target molecule must be included in the total Green's function $G_0^{(+)}$ in order to make the wave function on the left-hand side of Eq.(3) antisymmetric. This certainly suggests that it would be difficult to construct and treat this Green's function exactly in the collision of identical particles. For this reason Eq.(3) is usually avoided, and instead the Schwinger principle is normally applied to the coupled equations which are more manageable [11].

To obtain a multichannel Schwinger variational principle based on the total (N+1)-particle wave function we proceed as follows. We begin by introducing a projection operator P which defines the open-channel space in terms of the eigenfunctions of H_N

$$P = \sum_{m=1}^{open} |\Phi_m(1,2,...,N) > < \Phi_m(1,2,...N)| \tag{5}$$

and

$$H_N\Phi_m = E_m\Phi_m \quad , \quad E-E_m > 0 \tag{6}$$

where E is the total energy of the (N+1) particle system. Note that the projector of Eq.(5) is different from the P operator of Feshbach formalism [12]. With this operator we obtain a projected Lippmann-Schwinger equation for the mth scattering state

$$P\Psi_m^{(+)} = S_m + G_P^{(+)}V\,\Psi_m^{(+)} \tag{7}$$

where $\Psi_m^{(+)}$ is the total scattering wave function with plane-wave plus outgoing-wave boundary conditions. S_m is hence the product of the target wave function Φ_m and an incident plane wave, i.e.,

$$S_m = \Phi_m(1,2,...,N)\; e^{i\underset{\sim}{k}_m\cdot\underset{\sim}{r}_{N+1}} \; . \tag{8}$$

The outgoing-wave Green's function $G_P^{(+)}$, which is defined only in the open-channel space, is given by

$$G_P^{(+)} = \sum_{m=1}^{open} |\Phi_m > g_m^{(+)}(\underset{\sim}{r}_{N+1}, \underset{\sim}{r}'_{N+1}) < \Phi_m| \tag{9}$$

with

$$g_m^{(+)}(r,r') = -\frac{1}{2\pi}\frac{e^{ik_m|\underset{\sim}{r}-\underset{\sim}{r}'|}}{|\underset{\sim}{r}-\underset{\sim}{r}'|} \; . \tag{10}$$

Multiplication of Eq.(7) from the left by V leads to

$$(VP - VG_P^{(+)}V)\Psi_m = VS_m \; . \tag{11}$$

Since the operator in parentheses in Eq.(11) is not Hermitian we cannot transform this equation directly into a variational principle.

To recover the unprojected component of the Green's function $G_0^{(+)}$ of Eq.(3) we consider the following projected Schrödinger equation

$$[\hat{H} - a(P\hat{H} + \hat{H}P)]\Psi_m^{(+)} = a(VP - PV)\Psi_m^{(+)} \tag{12}$$

where $\hat{H} = E - H_{N+1}$ and a is a parameter. This equation contains information about the closed channels without defining the closed-channel Green's function. A simple combination of Eqs.(11) and (12) together with a = (N+1)/2 defines a complete equation for $\Psi_m^{(+)}$, i.e.,

$$[\tfrac{1}{2}(PV + VP) - VG_P^{(+)}V + \frac{1}{(N+1)}(\hat{H} - \frac{(N+1)}{2}(P\hat{H} + \hat{H}P))]\Psi_m^{(+)} = VS_m . \tag{13}$$

Based on Eq.(13) we can write a multichannel variational functional for the scattering amplitude

$$[f_{mn}] = -\frac{1}{2\pi}\frac{< \Psi_m^{(-)}|V|S_n > < S_m|V|\Psi_n^{(+)} >}{< \Psi_m^{(-)}|A^{(+)}|\Psi_n^{(+)} >} \tag{14a}$$

where

$$A^{(+)} = \frac{1}{2}(PV + VP) - VG_P^{(+)}V + \frac{1}{(N+1)}[\hat{H} - \frac{(N+1)}{2}(P\hat{H} + \hat{H}P)]. \tag{14b}$$

This is an extension of the usual Schwinger principle. If we expand the scattering wave function in a basis of (N+1)-electron Slater determinants, this variational expression can be written as

$$\tilde{f}_{mn} = -\frac{1}{2\pi}\sum_{i,j} < S_m|V|\psi_i > \Delta_{ij}^{-1} < \psi_j|V|S_n > . \tag{15}$$

The Δ in this equation is the matrix representation of the operator of the denominator of Eq.(14) in this basis of

Slater determinants. Note that in this formulation all configurations representing both open and closed channels are treated equivalently. This variational formulation is obviously based on the variation of the total wave function rather than that of a specific scattering orbital.

Some important features of the variational principle in Eq.(15) are as follows. As in the original Schwinger principle, an L^2expansion of the trial wave function is possible. Furthermore, if the orbitals appearing in the (N+1)-electron Slater determinants in the expansion of the scattering wave function are chosen to be Cartesian Gaussian functions, all the matrix elements arising in Eq.(15), except those of $VG_P^{(+)}V$, can be evaluated analytically for a general molecular target. The matrix elements associated with the $VG_P^{(+)}V$ term can also be obtained analytically with an insertion-like quadrature around $G_P^{(+)}$ again using Cartesian Gaussian functions. Furthermore, the "residue" contribution to these matrix elements of $VG_P^{(+)}V$ can be obtained essentially exactly via insertion of a complete set of plane waves around $G_P^{(+)}$. This procedure results in an S-matrix that is very nearly unitary without resorting to large Cartesian Gaussian insertion basis sets. Details of this plane wave insertion technique will be presented elsewhere [13].

Finally, we note that Eq.(15) provides an analytical approximation to the *full scattering amplitude* in the body frame. To obtain the physical cross-section one needs the laboratory-frame scattering amplitude. The partial wave decompositions arising in this frame transformation are carried out via a Gauss-Legendre quadrature [14]. This simply requires that the body-frame amplitudes be determined at an appropriate set of angles.

Applications

We now summarize the results of several applications of this theory to elastic and electronically inelastic scattering of low-energy electrons by several molecules.

1. *Cross-sections for elastic e-H_2 collisions:* We begin by presenting some results for elastic e-H_2 scattering with and without polarization effects. This system provides a convenient test of how effectively our formulation can represent the effects of polarization since accurate experimental and calculated cross-sections are available for comparison. We stress that our discussion of these cross-sections and particularly their comparison with the results of other calculations

are only meant to be illustrative and do not review the results of many other studies of these cross-sections [15,16]. *Figure 1* shows our calculated differential cross-sections at 3 eV in the static-exchange and static-exchange-polarization approximation. These and other calculated results include

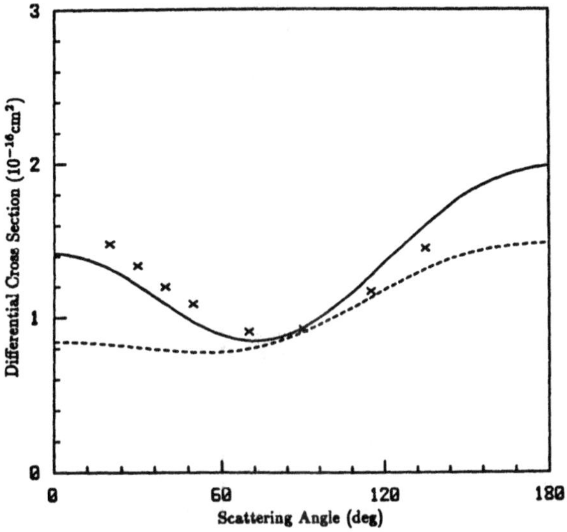

Figure 1. Elastic e-H_2 differential cross-sections at 3 eV: - - -, static-exchange: ——, with polarization; x, experimental data of Ref. 18.

contributions from $^2\Sigma_g$, $^2\Sigma_u$, $^2\Pi_u$, and $^2\Pi_g$ symmetries and a modest number of pseudostates in the closed channels, e.g., 10 pseudostates in Σ symmetry. Details of these calculations have been discussed elsewhere [17]. These differential cross-sections at 3 eV and those at 10 eV [17] show that the major features in these cross-sections due to polarization are reproduced by our method. *Figure 2* shows our calculated integral elastic cross-sections at the static-exchange level and including polarization along with the results of Gibson and Morrison [19], whose treatment of polarization was based on a polarized orbital approach. *Figure 2* also includes the measured values of Jones and Bonham [20]. A comparison of these results shows that most of the closed-channel effects are properly accounted for by our procedure. In our calculated cross-sections of *Fig. 2* we did not include pseudostates of π symmetry in the closed-channel expansion for the $^2\Sigma$ symmetries. Thus, configurations of the type $1\sigma n\pi m\pi$, which can be important at lower energies, were not included. To indicate the importance of such configurations *Fig. 2* also shows the elastic cross-section at 1 eV obtained including these $^2\Sigma(1\sigma n\pi m\pi)$ contributions.

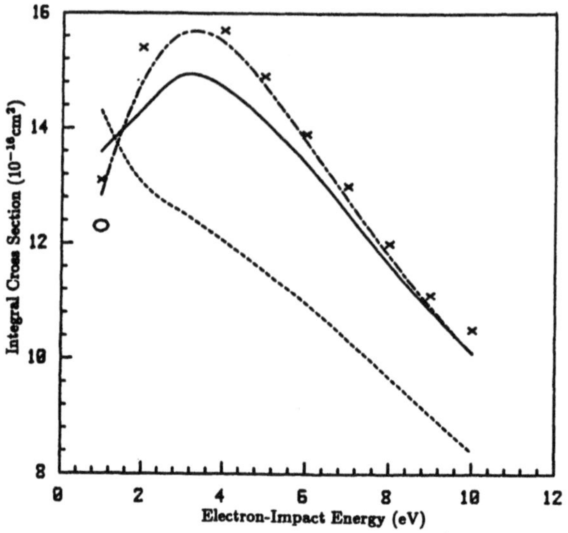

Figure 2. Elastic e-H_2 integral cross-sections: ---, present static-exchange results; ——, present results including polarization; — - —, results of Gibson and Morrison (Ref.19); x, experimental data of Jones and Bonham (Ref.20); ○ , present cross-section at 1 eV with polarization including $^2\Sigma(1\sigma n\pi m\pi)$ contributions.

2. Cross-sections for excitation of the $b^3\Sigma_u^+$, $a^3\Sigma_g^+$, $c^3\Pi_u$, and $B^1\Sigma_u^+$ states of H_2: We have used our multichannel extension of Schwinger's variational principle to study the cross-sections for excitation of the $b^3\Sigma_u^+$, $a^3\Sigma_g^+$, $c^3\Pi_u^+$, and $B^1\Sigma_u^+$ states of H_2 for collision energies from near threshold to 30 eV. These cross-sections are also being studied experimentally by Trajmar and his collaborators [21,22]. Our calculated $b^3\Sigma_u^+$ cross-sections, obtained at the two-state level, assume a special significance in view of the independent studies of these cross-sections by Schneider and Collins [23] and by Baluja et al [24] using the linear algebraic and R-matrix methods, respectively, and of the experimental studies by Hall and Andrić [25] and Nishimura [26].

Figure 3 shows our calculated differential excitation cross-section for the $X^1\Sigma_g^+ \rightarrow b^3\Sigma_u^+$ transition at 10.5 eV along with the experimental data of Hall and Andrić [25]. These cross-sections have been calculated with the two open channels only. Other details of these studies are given elsewhere [27]. *Figure 4* shows the cross-sections for this transition at 12 eV. At this energy we also compare our two-state results with the distorted-wave cross-sections of Fliflet and McKoy [6]. The agreement between these two-state cross-sections and the experimental results at impact energies 0.5 eV and 1.5 eV

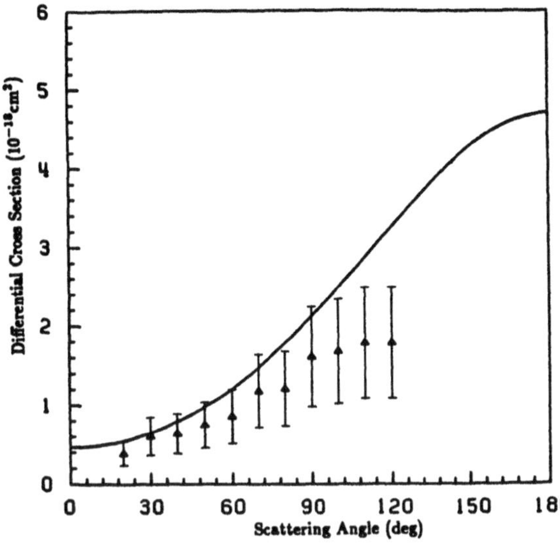

Figure 3. Differential inelastic cross-section for the $X^1\Sigma_g^+ \rightarrow b^3\Sigma_u^+$ transition in H_2 at 10.5 eV: ——, present two-state results; Δ, experimental data of Hall and Andrić (Ref. 25).

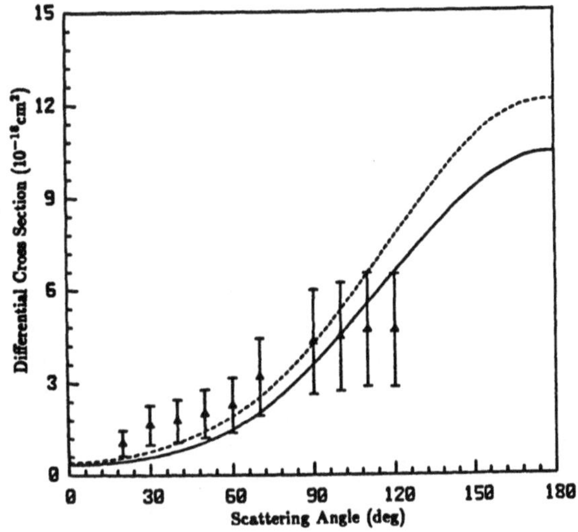

Figure 4. Differential inelastic cross-section for the $X^1\Sigma_g^+ \rightarrow b^3\Sigma_u^+$ transition in H_2 at 12 eV: ——, present two-state results; ---, distorted-wave cross-sections of Fliflet and McKoy (Ref.6); Δ, experimental data of Hall and Andrić (Ref.25).

above threshold is very encouraging. Our calculated inelastic differential cross-sections at 20 eV in *Fig. 5* also agree quite well with the experimental data of both Nishimura [26] and of Khakoo et al [21]. Why these two-state cross-sections agree as well as they do with the measured values at 20 eV, and also at 30 eV [27], is not obvious, and may be specific to the intravalence nature of this transition. Furthermore, the distorted-wave differential cross-sections are also quite simular to the measured cross-sections.

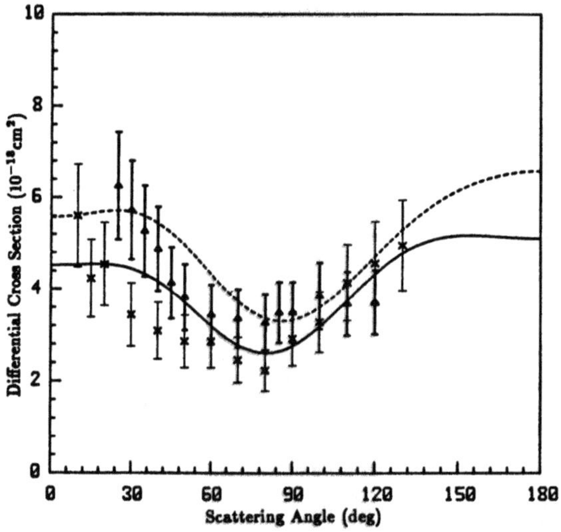

Figure 5. Differential cross-sections for the $X^1\Sigma_g^+ \rightarrow b^3\Sigma_u^+$ transition in H_2 at 20 eV: ——, present two-state results; ----, distorted-wave results of Ref.6; ▲, experimental data of Khakoo et al (Ref.21); ×, experimental data of Nishimura (Ref.26).

Figure 6 shows our calculated integral cross-sections for the $X^1\Sigma_g^+ \rightarrow b^3\Sigma_u^+$ excitation along with the experimental data of Hall and Andrić [25], Khakoo et al [21] and Nishimura [26]. In *Fig. 6* we also include the results of two-state calculations of Schneider and Collins [23], Baluja et al [24], and Chung and Lin [28]. The agreement between our calculated cross-sections, and also those of Schneider and Collins [23] and Baluja et al [24], and the experimental data of Hall and Andrić [25], of Khakoo et al [21], and Nishimura [26] is good at all energies. There are, however, significant differences between the two-state cross-sections of Chung and Lin [28] and the other two-state calculations shown in *Fig. 6*. These differences are due to the orthogonality imposed on the scattering function and the bound state orbital in the calculations of Chung and Lin [28]. In our studies this orthogonality

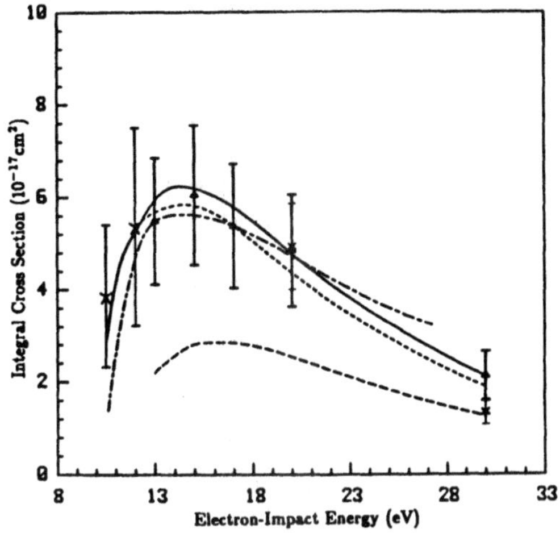

Figure 6: Integral cross-sections for excitation of the $b^3\Sigma_u^+$ state of H_2: ——, present results; — — — ; results of Schneider and Collins (Ref.23);— - —, results of Baluja et al (Ref.24); — — — , close-coupling results of Chung and Lin (Ref.28); ⚲, experimental data of Hall and Andrić Ref.25); Δ, experimental data of Nishimura (Ref.26); ×, measurements of Khakoo et al (Ref.21).

condition is relaxed by including the correlation terms in the expansion of the (N+1)-particle wave function [23,24].

We have also studied the cross-sections for excitation of the $a^3\Sigma_g^+$ state for collision energies from near threshold to 30 eV. These cross-sections were again obtained using only the $X^1\Sigma_g^+$ and $a^3\Sigma_g^+$ open channels. Our calculated differential inelastic cross-sections at 20 and 30 eV are compared with the experimental data of Trajmar et al [22] in *Fig. 7*. The agreement between these cross-sections at 20 eV is very good. Note that the measured cross-sections do not extend beyond 120° due to geometrical limitations in the apparatus. There are, however, significant differences between the calculated and measured differential inelastic cross-sections at 30 eV, particularly between 30° and 80°. Our calculated integral cross-sections for the $X^1\Sigma_g^+ \rightarrow a^3\Sigma_g^+$ transition are shown in *Fig. 8* along with the close-coupling results of Chung and Lin [28] and the experimental data of Khakoo and Trajmar [22]. The agreement between the calculated two-state integral cross-sections and the available experimental data [22] at 20 eV and 30 eV is reasonable. We note that these integral cross-sections are predicted to rise very sharply just above threshold, suggesting that the Franck-Condon approximation must be used carefully in the analysis of experimental data in this region.

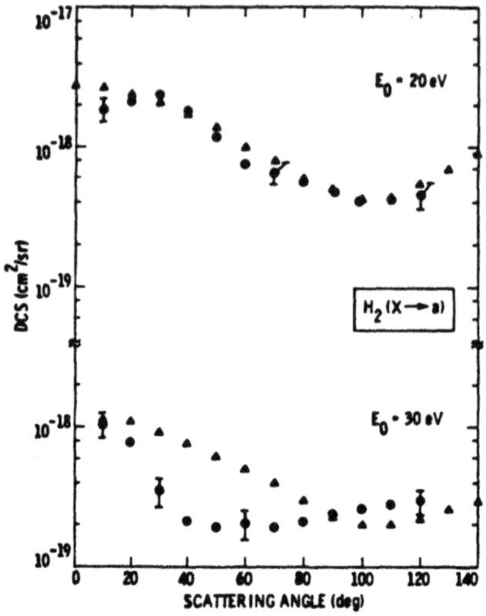

Figure 7. Differential cross-sections for excitation of the $a^3\Sigma_g^+$ state of H_2 of 20 eV and 30 eV: ●, experimental data of Ref.22; ▲, present two-state results

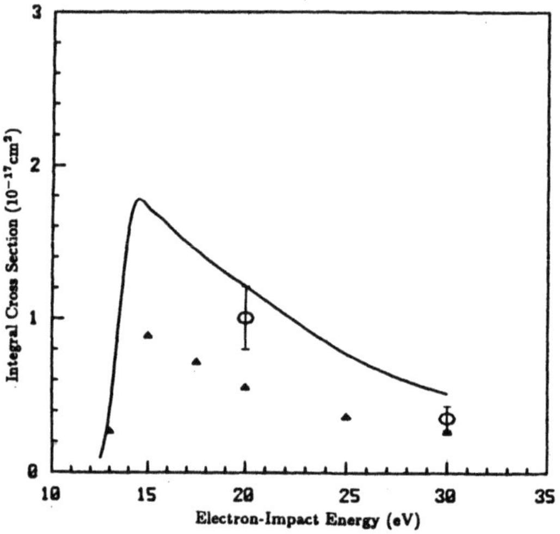

Figure 8. Integral cross-sections for the $X^1\Sigma_g^+ \rightarrow a^3\Sigma_g^+$ transition in H_2: ——, present two-state results; Δ, two-state close-coupling results of Chung and Lin (Ref.28); ○, experimental data of Ref.22.

We have also carried out calculations of the cross-sections for excitation of the $c^3\Pi_u$ state of H_2. *Figure 9* shows our differential inelastic cross-sections at 20 eV along with the

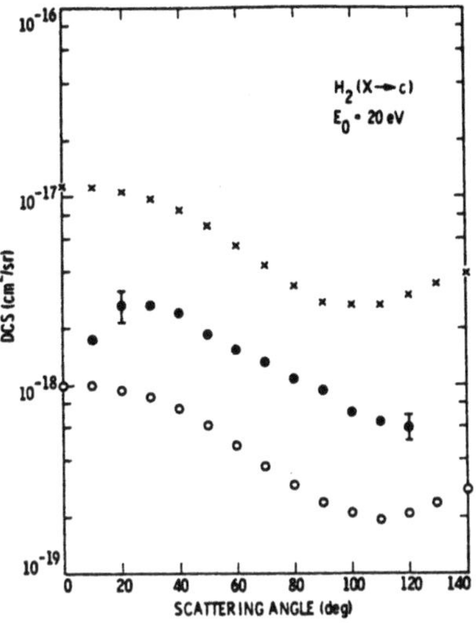

Figure 9. Differential cross-sections for excitation of the $c^3\Pi_u$ state of H_2 at 20 eV: ɸ, experimental data of Ref.22; ×, present results; ○, distorted-wave results of Ref.29.

experimental data of Trajmar et al [22] and the distorted-wave results obtained previously [29]. Although the shapes of our calculated and the experimental cross-sections are very similar, there are substantial differences in their magnitudes. The differences in the magnitudes of these two-state cross-sections and the distorted-wave results [29] are even larger. Further multichannel calculations are clearly required to determine the influence of additional open channels on these cross-sections. The decrease in the measured cross-sections going from 20° to 10° is almost certainly due to the larger experimental uncertainties closer to the forward direction.

In addition to these studies of cross-sections for the $X^1\Sigma_g^+ \rightarrow b^3\Sigma_u^+$, $a^3\Sigma_g^+$, and $c^3\Pi_u$ transitions with their short-range exchange interaction, we have also obtained cross-sections for the $X^1\Sigma_g^+ \rightarrow B^1\Sigma_u^+$ excitation. This is a dipole-allowed transition and the excitation mechanism has long-range dipolar character. The trial scattering wave function in such dipole-allowed transitions requires a more diffuse basis and higher angular momentum functions than for spin- or dipole-forbidden transitions. These higher ℓ-components in the wave function primarily influence the scattering cross-section at low angles and should be Born-like. Our strategy, therefore, is to use a fairly large basis to obtain the lower partial wave contributions to the inelastic differential cross-sections from the variational principle of Eq.(15) and to obtain the higher partial wave components, which contribute primarily in the forward direction, from the Born approximation. *Figure 10* shows the differential cross-section for the $X^1\Sigma_g^+ \rightarrow B^1\Sigma_u^+$ transition at 20 eV calculated with our multichannel variational principle using only a discrete basis along with the data of Trajmar et al [22] and of Srivastava and Jensen [30] and the distorted-wave results of Fliflet and McKoy [6]. The agreement between the calculated two-state cross-sections and the experimental data beyond 40° is, on average, quite good. This comparison beyond 105° is particularly interesting. As expected, the two-state cross-sections

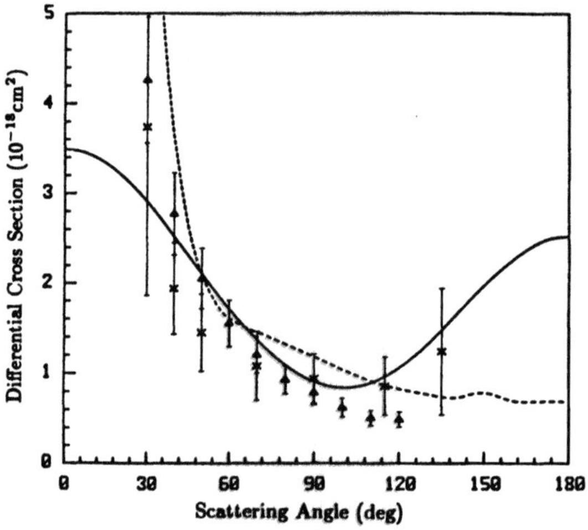

Figure 10. Differential cross-sections for excitation of the $B^1\Sigma_u^+$ state in H_2 at 20 eV: ▲, experimental data of Ref.22; ×, experimental data of Ref.30; ——, present results; ----, distorted-wave cross-sections (Ref.6).

obtained with a discrete basis only do not behave correctly in the forward direction. The contribution from the higher ℓ-waves is important in this region. Inclusion of these higher ℓ-waves in the Born approximation should lead to overall cross-sections in good agreement with the experimental data since, in the distorted-wave results in *Fig. 10*, the low-angle contributions were, in fact, obtained in the Born approximation [6]. On the other hand, the distorted-wave model does quite poorly at higher angles.

3. *Elastic scattering of electrons by* CH_4*:* We have also used our formulation to study the elastic cross-sections for e-CH_4 collisions. These cross-sections have been obtained at the static-exchange level for collision energies from 3 eV to 20 eV. However, we will also discuss some very preliminary results including polarization at 5 eV. To our knowledge, these are the first studies of electron-polyatomic molecule

scattering at the static-exchange level which do not use a local exchange potential.

Our calculations in the body-frame include contributions from the $^2A_1$, $^2T_2$, and ^2E symmetries in the trial scattering wave function. All 83 available virtual orbitals from the molecular SCF calculation, i.e., $21a_1$, $18t_{2x}$, $18t_{2y}$, $18t_{2z}$, and 8e, were used in the expansion of $\tilde{\psi}^{(\pm)}$:

$$^2A_1 \; : \; 1a_1^2 2a_1^2 1t_{2x}^2 1t_{2y}^2 1t_{2z}^2 ka_1$$

$$^2T_{2x} \; : \; 1a_1^2 2a_1^2 1t_{2x}^2 1t_{2y}^2 1t_{2z}^2 kt_{2x}$$

$$^2T_{2y} \; : \; 1a_1^2 2a_1^2 1t_{2x}^2 1t_{2y}^2 1t_{2z}^2 kt_{2y}$$

$$^2T_{2z} \; : \; 1a_1^2 2a_1^2 1t_{2x}^2 1t_{2y}^2 1t_{2z}^2 kt_{2z}$$

$$^2E \; : \; 1a_1^2 2a_1^2 1t_{2x}^2 1t_{2y}^2 1t_{2z}^2 ke$$

with ka_1, kt_2, and ke being continuum functions represented by the virtual orbitals. For the partial-wave expansion of the scattering amplitude, we include values of $\ell \leqslant 5$. Further details of these calculations have been discussed elsewhere [31].

In *Figs. 11-16* we present our calculated differential cross-sections for incident energies of 3, 5, 7.5, 10, 15, and 20 eV at the static-exchange level along with the results of some other studies and the measured cross-sections. The calculated differential cross-sections in this range of energies are characterized by forward peaking, a minimum around 120° and a backward peak. The minimum is rather sharp at low energies and becomes broader with increasing energy. In addition, the experimental data show a secondary minimum occurring near 45° at 3 eV. As the incident energy increases to 7.5 eV the minimum moves to 60° and becomes a much less pronounced shoulder-like structure. The calculated static-exchange cross-sections reproduce both the forward and backward peaks as well as the primary minimum near 120°. At low energies the calculated forward peak is much more pronounced than experiment. The secondary minimum is also missing, suggesting that this feature is most probably due to polarization effects. *Figure 17* shows the results of very preliminary calculations on the differential cross-sections at 5 eV in which polarization effects have been included through closed channels. These closed channel effects were obtained using several hundred Slater determinants

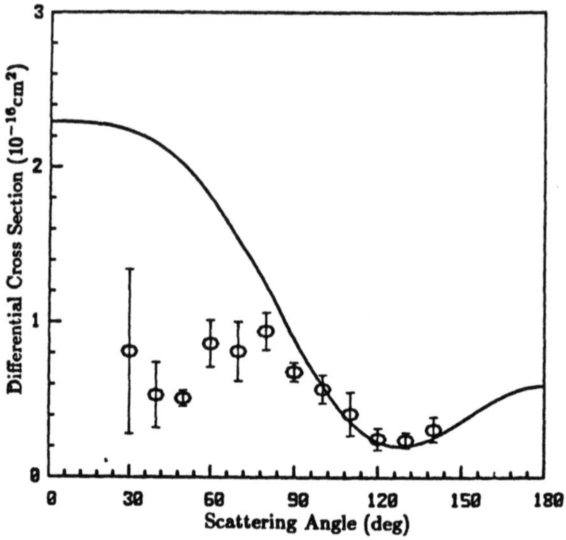

Figure 11. Differential cross-section (DCS) for e-CH_4 collisions at 3 eV: ——, present static-exchange results; Φ, measured values of Ref.32.

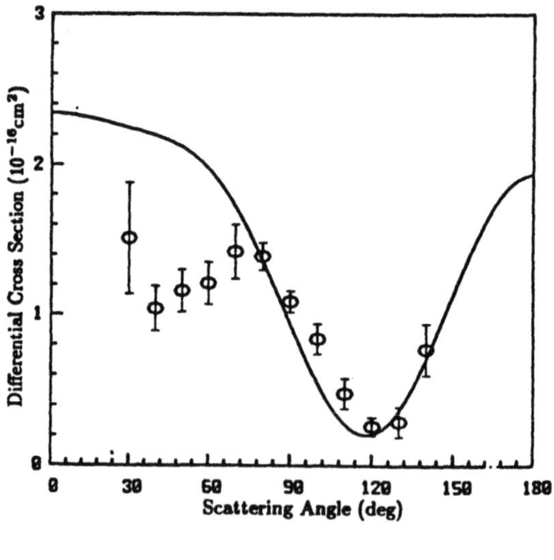

Figure 12. DCS for e-CH_4 scattering at 5 eV: ——, present static-exchange results; Φ, experimental values of Ref.32.

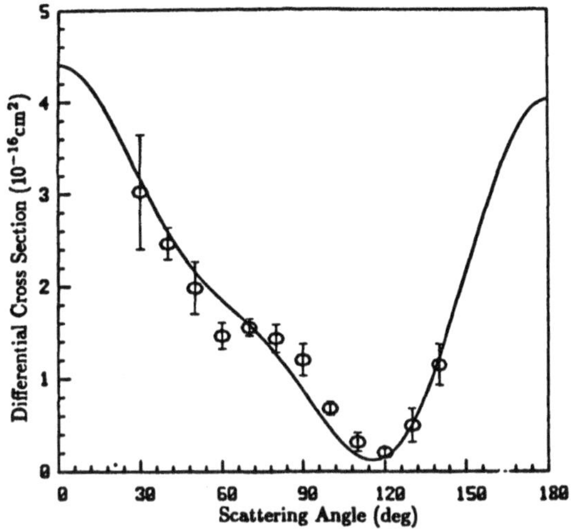

Figure 13. DCS for e-CH_4 scattering at 7.5 eV: ——, present static-exchange results, Φ, experimental values of Ref.32.

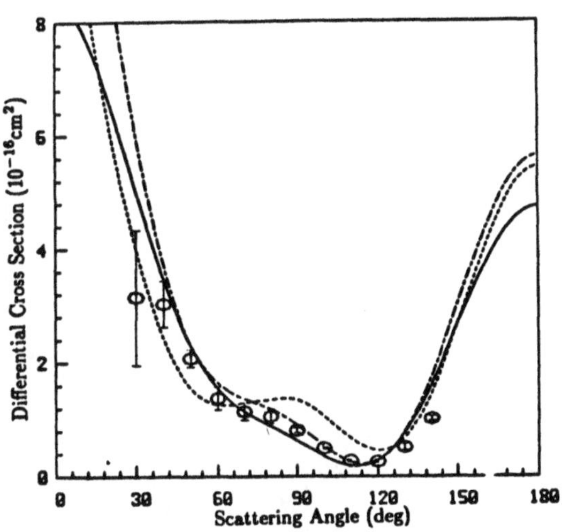

Figure 14. DCS for e-CH_4 collisions at 10 eV: ——, present static-exchange results; ----, static-model-exchange-polarization (SMEP) results using an adiabatic polarization potential (Ref.33); — - — - SMEP results (Ref.33) using a local kinetic energy semiclassical potential; Φ, measured values of Ref.32.

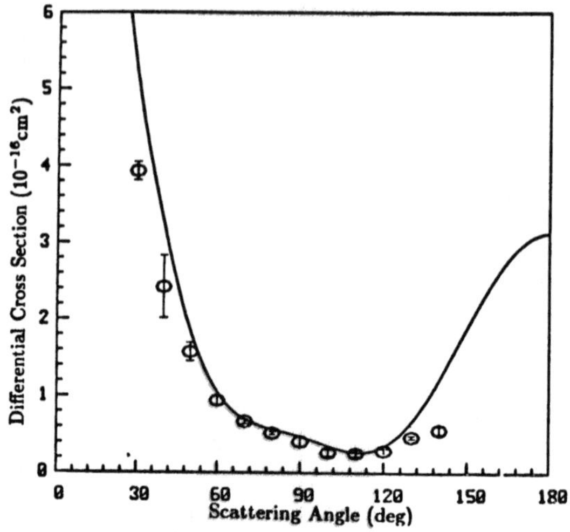

Figure 15. DCS for e-CH_4 collisions at 15 eV: ——, present static-exchange results; Φ, measured values at Ref.32.

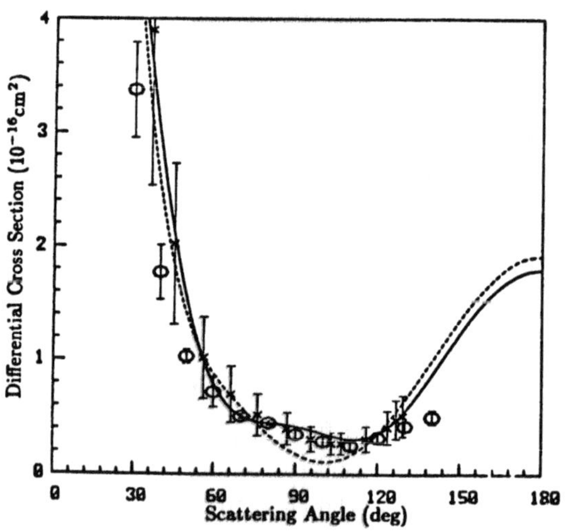

Figure 16. DCS for e-CH_4 collisions at 20 eV: ——, present static-exchange results; ----, SMEP results of Ref.34; Φ, measured values of Ref.32; ×, measured values of Ref.35.

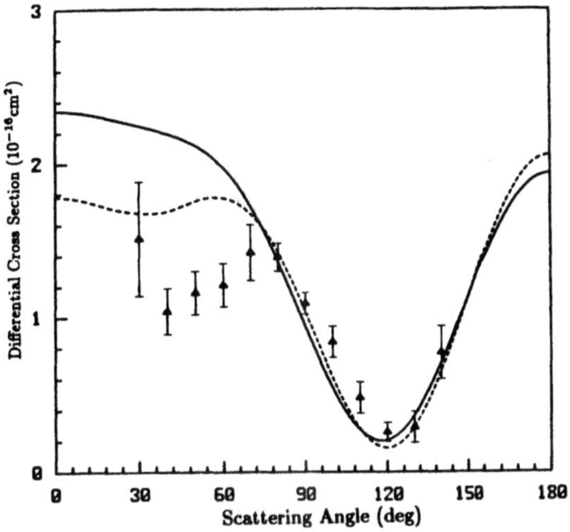

Figure 17. DCS for e-CH_4 collisions at 5 eV: ——, present static-exchange results; ----, present results with limited inclusion of polarization effects; Δ, measured values of Ref.32.

in the variational trial function. The changes in the differential cross-sections between 0° and 60° which arise from this very limited inclusion of polarization effects are encouraging. Further studies of polarization effects on these differential cross-sections are under way.

Figure 18 shows our calculated integral static-exchange elastic cross-sections along with the static-model-exchange-polarization (SMEP) results of Jain and Thompson [36] and Tanaka et al [32]. The maximum in our calculated integral cross-sections is shifted about 2.5 eV to higher energy than is seen experimentally. This difference is most probably due to polarization, which is not included in these calculated cross-sections. Some of the discrepancy between our calculated results and the data of Tanaka et al [32] could be due to uncertainties in extrapolating the measured differential cross-sections to small and large angles. There are clearly significant differences between the measured data of Tanaka et al [32] and Jones [37]. Note that the cross-section data of Jones [37] was obtained from electron transmission experiments.

In *Fig. 19* we show our calculated momentum transfer cross-sections along with those of the SMEP calculations of Jain [38] and the values derived from the data of Tanaka et al [32]. The differences between our calculated momentum

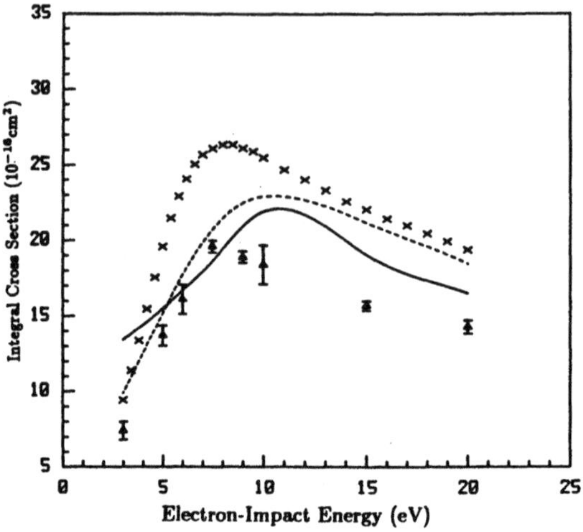

Figure 18. Elastic integral cross-sections for e-CH_4 collisions: ——, present static-exchange results; ----, SMEP calculations of Ref.37; ⫱, data of Ref.32; ×, measurements of Ref.37.

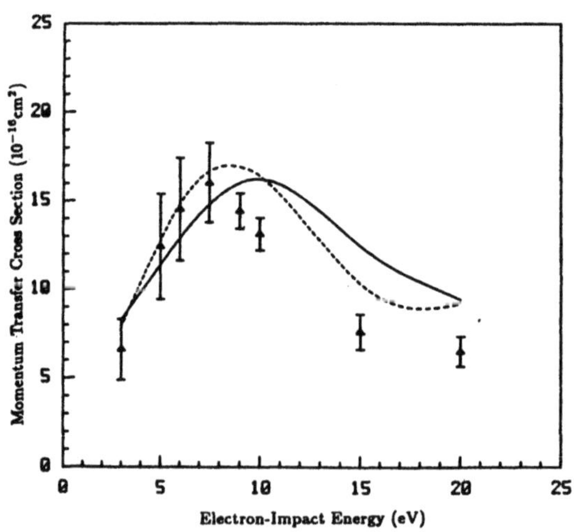

Figure 19. Momentum transfer cross-sections for e-CH_4 collisions: ——, present calculated results; ----, results of Ref.38; ⫱, date of Ref. 32.

transfer cross-sections and the derived values of Tanaka et al [32], particularly at higher energies, are again probably due to uncertainties in extrapolating the measured differential cross-sections out to larger angles. The values of the differential cross-section at higher angles, e.g., beyond 120°, are heavily weighted in the calculation of the momentum transfer cross-sections.

4. *Elastic scattering of electrons by H_2O:* As a final example, we discuss the results of fixed-nuclei calculations of the cross-sections for elastic scattering of electrons by H_2O. These fixed-nuclei cross-sections are well known to diverge in the forward direction for polar targets [39]. In these studies our strategy is to obtain the differential cross-sections at angles away from the forward direction, where a discrete basis can adequately and economically represent the scattering wave function. In the forward direction the contribution of the higher partial waves to the differential cross-section could be conveniently obtained within the Born approximation.

In *Figs. 20-23* we show our calculated elastic differential cross-sections, obtained with a large Cartesian Gaussian basis only, at collision energies of 6, 10, 15, and 20 eV. Details of these calculations are discussed in Ref.42. The calculated cross-sections at 10, 15, and 20 eV are seen to agree well with the measured values of Ref.41. At lower energies, e.g., 6 and 10 eV, the calculated cross-sections show clear signs of

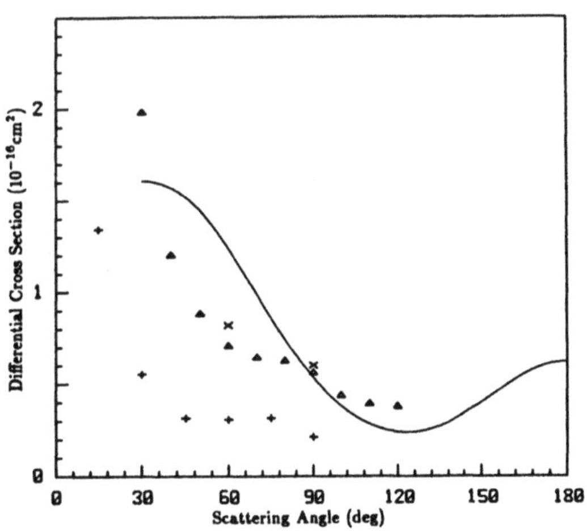

Figure 20. DCS for e-H_2O scattering at 6 eV: ——, present static-exchange results; +, measured values of Ref.40; Δ, measured values of Ref.41; ×, calculated results of Ref.36.

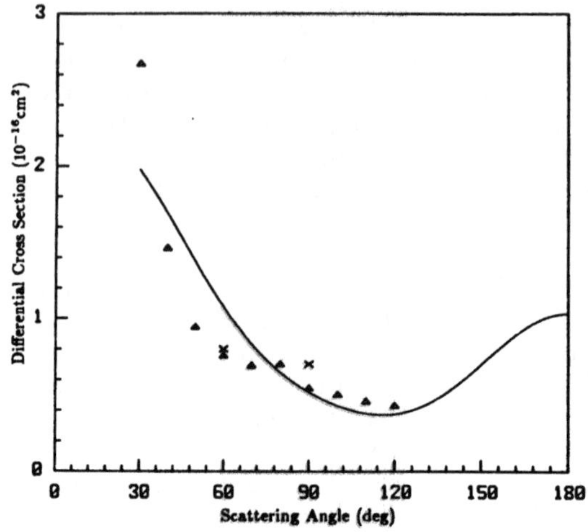

Figure 21. DCS for e-H_2O scattering at 10 eV: ——, present static-exchange results; Δ, measured values of Ref.41; ×, calculated results of Ref.36.

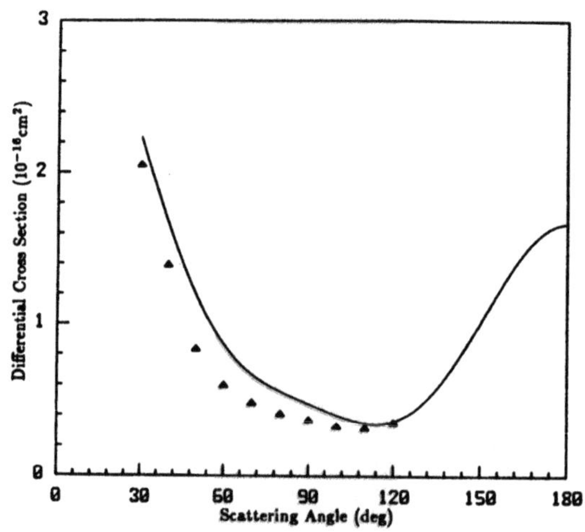

Figure 22. DCS for e-H_2O scattering at 15 eV: ——, present static-exchange results; Δ, measured values of Ref.41.

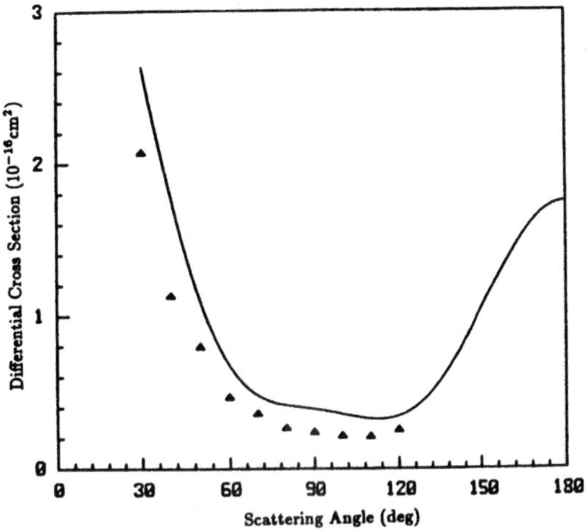

Figure 23. DCS for e-H_2O scattering at 20 eV: ——, present static-exchange results; Δ, measured values of Ref.41.

falling off below 30° as discussed above. An important feature of the calculated cross-sections is their behavior between 120° and 180°. Note that the experimentally accessible region extends only out to 120°. The backward peaking seen in the calculated cross-sections will contribute significantly to the momentum transfer cross-sections derived from these results. The momentum transfer cross-sections derived from the measured cross-sections of Ref.41, and shown in *Fig. 24*,

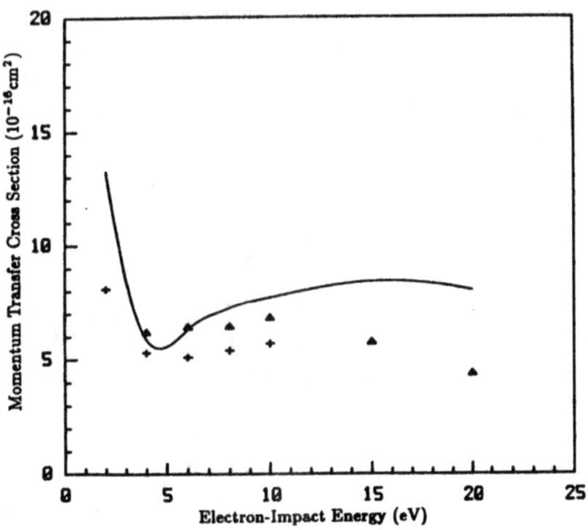

Figure 24. Momentum transfer cross-sections for e-H_2O: ——, present results; Δ, derived from measured data (Ref.41); +, calculated results of Ref.36.

assumed a smooth extrapolation of the data beyond 120°. Such an extrapolation should underestimate the momentum transfer cross-sections. This trend is seen very clearly in the results of *Fig. 24*. We expect that calculated differential elastic and inelastic cross-sections will generally prove very useful in the extrapolation of measured data to both low and high angular regions.

Concluding Remarks

We have discussed some results of initial applications of our Schwinger multichannel formulation to elastic and electronically inelastic scattering of low-energy electrons by molecules. An important feature of this multichannel extension of Schwinger's variational principle is that it is capable of dealing with some important aspects of electron-molecule collisions such as electronically inelastic scattering and nonlinear targets. We believe that the results we have presented for electron impact excitation of the $X^1\Sigma_g^+ \rightarrow b^3\Sigma_u^+$, $a^3\Sigma_g^+$, $c^3\Pi_u$, and $B^1\Sigma_u^+$ transitions in H_2 and elastic scattering by CH_4 and H_2O are very encouraging, and illustrate the potential utility of these calculated cross-sections. The results of such studies clearly complement experimental efforts to determine absolute values of these electron-molecule scattering cross-sections. The calculated and measured integral cross-sections for excitation of the $b^3\Sigma_u^+$ state of H_2 shown in *Fig. 6* are a specific example of this point. Quite generally, however, calculated differential cross-sections are needed to guide the extrapolation of experimental data into the difficult, and often inaccessible, low and high angle regions. Furthermore, such comparisons of calculated and measured cross sections will ultimately provide a realistic assessment of the underlying dynamics of these collisions.

Acknowledgments

This material is based upon research supported by the National Science Foundation under Grant No. PHY-8213992 and by the NASA-Ames Cooperative Agreement No. NCC2-319. One of us (M.A.P.L.) acknowledges financial support from the Comissão Nacional de Energia Nuclear (CNEN), Rio de Janeiro, Brazil. Another (L.M.B.) acknowledges a fellowship from Conselho Nacional de Desenvolvimento Científico e Tecnológico (CNPq), Brazil. Two of us (M.A.P.L. and L.M.B.) acknowledge financial support from the Fundação de Amparo à Pesquisa do Estado de São Paulo (FAPESP), Brazil. The research of W.M.H. is supported by the National Aeronautics and Space Administration NASA-Ames Cooperative Agreement No. NCC2-147.

References

[1] See, for example, Trajmar, S.; Register, D. F.; Chutjian, A. Phys. Repts. 1983, 97, 219.
[2] Chung, S.; Lin, C. C.; Lee, E.T. Phys. Rev. A 1975, 12, 1340.
[3] Cartwright, D. C.; Kuppermann, A. Phys. Rev. 1967, 163, 86.
[4] Hazi, A. U. Phys. Rev. A 1981, 23, 2232.
[5] Rescigno, T. N.; McCurdy, C.W., Jr.; McKoy, V. J. Phys. B. 1974, 7, 2396.
[6] Fliflet, A. W.; McKoy, V. Phys. Rev. A 1980, 21, 1863.
[7] See, for example, Collins, L. A.; Robb, W. D.; Norcross, D. W. Phys. Rev. A 1979, 20, 1838.
[8] Cartwright, D. C.; Chutjian, A.; Trajmar, S.; Williams, W. Phys. Rev. A 1977, 16, 1013.
[9] Lippmann, B. A.; Schwinger, J. Phys. Rev. 1950, 79, 469.
[10] Geltman, S. "Topics in Atomic Collision Theory"; Academic Press: New York, 1969; p. 99.
[11] Maleki, N.; Macek, J. Phys. Rev. 1980, 21, 1403.
[12] Feshbach, H. Ann. Phys. (N. Y.) 1962, 19, 287.
[13] Gibson, T. L.; Lima, M. A. P.; Huo, W. M.; McKoy, V. Phys. Rev. A (to be published).
[14] Lima, M. A. P.; Gibson, T. L.; Takatsuka, K.; McKoy, V. Phys. Rev. A 1984, 30, 1741.
[15] See, for example, Schneider, B. I.; Collins, L. A. Phys. Rev. A 1983, 27, 2847.
[16] Klonover. A.; Kaldor, U. J. Phys. B 1978, 11, 1623.
[17] Gibson, T. L.; Lima, M. A. P.; Takatsuka, K.; McKoy, V. Phys. Rev. A (1984) 30, 3005.
[18] Srivastava, S. K.; Chutjian, A.; Trajmar, S. J. Chem. Phys. 1975, 63, 2659. The results shown here are the renormalized values given in Table 4b of Ref.1.
[19] Gibson, T. L.; Morrison, M. A. Phys. Rev. A 1984, 29, 2497.
[20] Jones, R.; Bonham. R. A. as given in Table 2a of Ref.1.
[21] Khakoo, M. A.; McAdams, R.; Trajmar, S. "Electron Impact Excitation Cross-Sections for the $b^3\Sigma_u^+$ State of H_2" Phys. Rev. A (submitted for publication).
[22] Khakoo, M. A.; Trajmar, S. "Electron Excitation of the $a^3\Sigma_g^+$, $B^1\Sigma_u^+$, $c^3\Pi_u$, $C^1\Pi_u$, and $E(F)^1\Sigma_g^+$ States of H_2" Phys. Rev. A (submitted for publication).
[23] Schneider, B. I.; Collins, L. A. J. Phys. B 1985, 18, L857.
[24] Baluja, K. L.; Noble, C. J.: Tennyson, J. J. Phys. B 1985, 18, L851.
[25] Hall, R. I.; Andrić, L. J. Phys. B 1984, 17, 3815.
[26] Nishimura, H. (private communication).

[27] Lima, M. A. P.; Gibson, T. L.; Huo, W. M.; McKoy, V. J. Phys. B 1985, 18, L865. This paper and those of references 23 and 24 appeared as companion publications.
[28] Chung, S.; Lin, C. C. Phys. Rev. A 1978, 17, 1874.
[29] Lee, M.-T.; Lucchese, R. R.; McKoy, V. Phys. Rev. A 1982, 26, 3240.
[30] Srivastava, S. K.; Jensen, S. J. Phys. B 1977, 10, 3341.
[31] Lima, M. A. P.; Gibson, T. L.; Huo, W. M.; McKoy, V. Phys. Rev. A 1985, 32, 2696.
[32] Tanaka, H.; Okada, T.; Boesten, L.; Suzuki, T.; Yamamoto, T.; Kubo, M. J. Phys. B 1982, 15, 3305.
[33] Abusalbi, N.; Eades, R. A.; Nam, T.; Thirumalai, D.; Dixon, D. A.; Truhlar, D. G. J. Chem. Phys. 1983, 78, 1213.
[34] Jain, A. J. Chem. Phys. 1984, 81, 724.
[35] Vušković, L.; Trajmar, S. J. Chem. Phys. 1983, 78, 4947.
[36] Jain, A.; Thompson, D. G. J. Phys. B 1982, 15, L631.
[37] Jones, R. K. J. Chem. Phys. 1985, 82, 5424.
[38] Private communication (A. Jain).
[39] See, for example, Norcross, D. W.; Collins, L. A. "Adv. Atomic and Molecular Physics"; Academic Press, New York, 1982; 18, 341.
[40] Jung, K.; Antoni, Th.; Müller, R.; Kochem, K. H.; Ehrhardt, H. J. Phys. B 1982, 15, 3535.
[41] Danjo, A.; Nishimura, H. J. Phys. Soc. Japan 1985, 54, 1224.
[42] Brescansin, L. M.; Lima, M. A. P.; Huo, W. M.; Gibson, T. L.; McKoy, V. Phys. Rev. A (to be published).

INELASTIC ELECTRON-MOLECULE SCATTERING USING THE R-MATRIX METHOD

P. G. Burke† and C. J. Noble*

† Department of Applied Mathematics and Theoretical Physics, Queen's University of Belfast, Belfast BT7 1NN, Northern Ireland.

* Theory and Computational Science Division, SERC, Daresbury Laboratory, Warrington WA4 4AD, England.

In this paper we review progress in the development of theoretical and computational methods for studying low-energy inelastic electron-molecule scattering with particular emphasis on the R-matrix method. Recent R-matrix calculations of electronic excitation cross-sections for H_2^+, H_2 and O_2 are discussed and compared with other calculations, and where possible with experiment. The paper concludes with a discussion of a new R-matrix approach which enables many coupled electronic states to be accurately included in the calculation.

Summary of Processes and Methods

The processes of interest in electron molecule scattering are: rotational, vibrational and electronic excitation

$$e^- + AB \rightarrow e^- + AB^*, \tag{1}$$

dissociative attachment

$$e^- + AB \rightarrow A + B^- , \tag{2}$$

and dissociative recombination

$$e^- + AB^+ \rightarrow A + B . \tag{3}$$

Also of interest are the closely related processes of photoionisation

$$h\nu + AB \rightarrow AB^+ + e^- \tag{4}$$

and photodissociation

$$h\nu + AB \rightarrow A + B . \tag{5}$$

These processes can be calculated by a number of different computational methods, as discussed below.

Most recent ab-initio methods are based on the fixed-nuclei approximation discussed by Lane [1] and by Buckley, et al [2]. In this approach the wave function for the scattering electron is calculated in the potential field of the molecule whose nuclei are held fixed in space during the collision, as illustrated in figure 1.

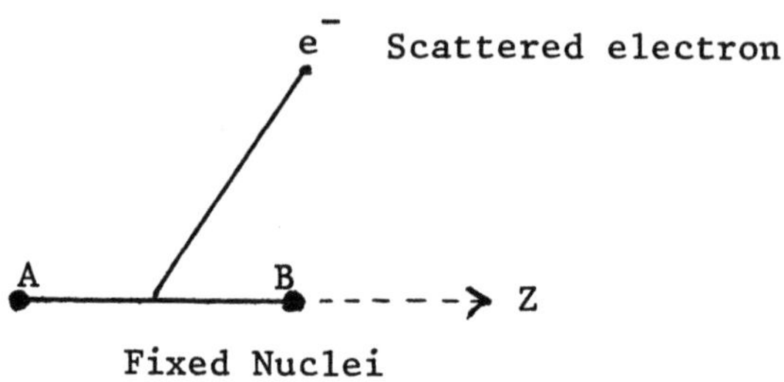

Figure 1. Coordinate frame for the fixed-nuclei approximation.

The inclusion of nuclear motion corresponding to rotational and vibrational excitation and to dissociation is then included in a second step in the calculation.

The computational methods divide into two classes. In the first class the full collision problem is solved yielding the cross-sections directly. This class includes the R-matrix method [3,4], the linear algebraic equations method [5] and the Schwinger variational method [6]. In the second class, which is appropriate when resonances dominate the cross-section, the resonance parameters are first calculated and then the interaction of the resonance with the non-resonant continuum is included in a second step. This class includes the Feshbach projection operator method [7,8], the stabilisation method [9] and the Stieltjes-moment method [10]. In the following section we will discuss the R-matrix method, while some of the other methods will be considered elsewhere in this volume.

The R-Matrix Method

The R-matrix method was first introduced by Wigner [11] and Wigner and Eisenbud [12] in an analysis of resonant nuclear reactions. It was later developed as an ab-initio method for calculating electron-atom and electron-ion collision cross-sections by Burke et al [13, 14, 15] and for calculating electron-molecule cross-sections by Schneider [3] and by Burke et al [4].

The method starts by partitioning configuration space into two regions as illustrated in figure 2.

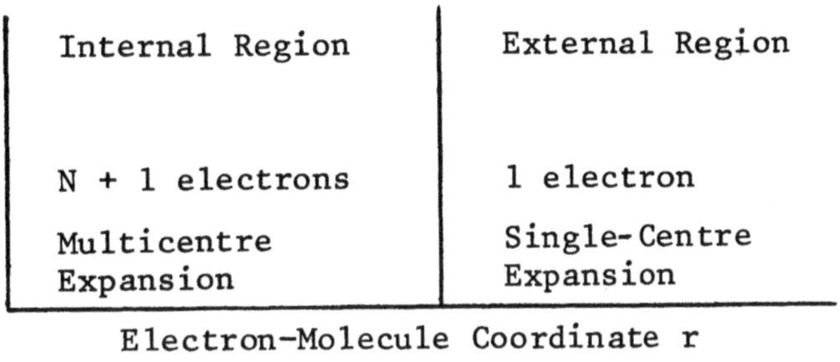

Figure 2. Partition of configuration space in the R-matrix method.

In the internal region exchange and correlation effects between the scattered electron and the N-electron target molecule are important. A multicentre C.I. expansion of the total wave function is used for each fixed internuclear separation. In the external region the scattered electron moves in the long-range multipole potential of the target molecule. A single-centre expansion of the wave function of the scattered electron is adopted and either the fixed-nuclei approximation may be used or the rotational and vibrational motion of the target may be represented by the usual coupled channels formalism [16]. The R-matrix provides the link between the calculations in these two regions.

The expansion of the wave function in the internal region for fixed internuclear separation is given by

$$\Psi_k = A \sum_{ij} \Phi_i(1, \ldots, N)\, \psi_j(N+1)\, a_{ijk} + \sum_i \chi_i(1, \ldots, N+1) b_{ik} \qquad (6)$$

where Φ_i are multicentre target eigenstates or pseudo-states which vanish by the boundary $r = a$, ψ_j are continuum molecular orbitals which are non-zero on the boundary $r = a$, χ_i are multicentre quadratically integrable functions which represent polarisation and correlation effects and A is the usual antisymmetrisation operator. The coefficients a_{ijk} and b_{ik} are determined by diagonalising the operator $H_{e\ell}^{N+1} + L_b^{N+1}$ in the internal region so that

$$\langle\Psi_k|H_{e\ell}^{N+1} + L_b^{N+1}|\Psi_{k'}\rangle = E_k^{N+1}\,\delta_{kk'} \tag{7}$$

where $H_{e\ell}^{N+1}$ is the electronic part of the Hamiltonian for N+1 electrons and L_b^{N+1} is a surface projection operator introduced by Bloch [17] which ensures that the operator $H_{e\ell}^{N+1}+L_b^{N+1}$ is Hermitian in the internal region. It is defined by

$$L_b^{N+1} = \frac{a^2}{2}\sum_{i\ell m}|\Phi_i\, Y_{\ell m}\rangle\, \delta(r_{N+1}-a)\left(\frac{d}{dr_{N+1}} - \frac{b-1}{r_{N+1}}\right)\langle\Phi_i\, Y_{\ell m}| \tag{8}$$

where $Y_{\ell m}$ is a spherical harmonic representing the angular motion of the scattered electron and b is an arbitrary constant.

The Schrödinger equation can be written as

$$(H_{e\ell}^{N+1} + L_b^{N+1} - E)\Psi = L_b^{N+1}\,\Psi \tag{9}$$

which has the formal solution

$$\Psi = (H_{e\ell}^{N+1} + L_b^{N+1} - E)^{-1}\, L_b^{N+1}\,\Psi \tag{10}$$

The inverse operator in this equation can be expanded in terms of the eigenfunctions defined by eq. (7) giving

$$|\Psi\rangle = \sum_k \frac{|\Psi_k\rangle\ \langle\Psi_k|\, L_b^{N+1}|\Psi\rangle}{E_k^{N+1} - E} \tag{11}$$

We now project this equation onto the channel functions $\Phi_i Y_{\ell m}$ and evaluate it on the boundary r = a. We define the reduced radial functions $f_{i\ell m}(r)$ by the equation

$$r_{N+1}^{-1}\, f_{i\ell m}(r_{N+1}) = \langle\, \Phi_i\, Y_{\ell m}|\Psi\,\rangle \tag{12}$$

and the reduced width amplitudes $\gamma_{i\ell mk}$ by the equation

$$\gamma_{i\ell mk} = (a/2)^{\frac{1}{2}}\,[\langle\,\Phi_i\, Y_{\ell m}|\Psi_k\,\rangle]_{r_{N+1}} = a \tag{13}$$

Then eq. (11) gives

$$\underline{f} = \underline{R}\cdot\left(a\,\frac{d\underline{f}}{dr} - b\,\underline{f}\right)_{r=a} \tag{14}$$

The R-matrix is defined by

$$\underline{R} = \sum_k \frac{\underline{\gamma}_k \times \underline{\gamma}_k}{E_k^{N+1}-E} \tag{15}$$

where $\underline{f}$ and $\underline{\gamma}$ are vectors with elements defined by eqs. (12) and (13) respectively. Eqs. (14) and (15) are the basic equations of R-matrix theory. Once the R-matrix has been calculated by diagonalising eq. (7) then the logarithmic derivative of the reduced radial wave function on the boundary r = a is determined at all energies by eq. (14).

In practice care has to be taken in the choice of the expansion basis in eq. (6) to obtain accurate results. The Φ_i are usually taken to be SCF target wave functions while the χ_i are usually 2 particle-1 hole functions where one of the target electrons and the scattered electron occupy excited virtual orbitals. In early work on diatomic molecules the ψ_j were formed from target STO's centred on the nuclei and long-ranged STO's which were non-zero on the boundary and centred on the centre-of-gravity of the molecule [18]. However, it was found that linear dependence problems limited the useful range of this basis to electron impact energies below about 1 Ryd. More recently the STO's on the centre of gravity have been replaced by orthogonal numerical orbitals [19]. In the usual approach these are taken to satisfy a model equation of the form

$$\left(\frac{d^2}{dr^2} - \frac{\ell(\ell+1)}{r^2} + V(r) + k_i^2\right)u_{\ell i}(r) = 0 \tag{16}$$

where

$$u_{\ell i}(0) = 0 \tag{17}$$

and

$$\frac{a}{u_{\ell i}} \left.\frac{du_{\ell i}}{dr}\right|_{r=a} = b \, . \tag{18}$$

The potential V(r) is taken to be the spherically symmetric part of the static potential of the target molecule. Since these orbitals satisfy homogeneous boundary conditions at r=a then the convergence of expansion (15) is slow and a correction first proposed by Buttle [20] has to be included. All of the results reported in the next section used a numerical basis with a Buttle correction.

The final step in the calculation is to solve the scattering problem in the external region. This is carried out using standard asymptotic codes [21,22] and will not be discussed further here.

The extension of the R-matrix method to treat vibrational excitation and dissociative attachment defined by eqs. (1) and (2) has been discussed by Schneider et al [23]. The partition of configuration space in this case is illustrated in figure 3.

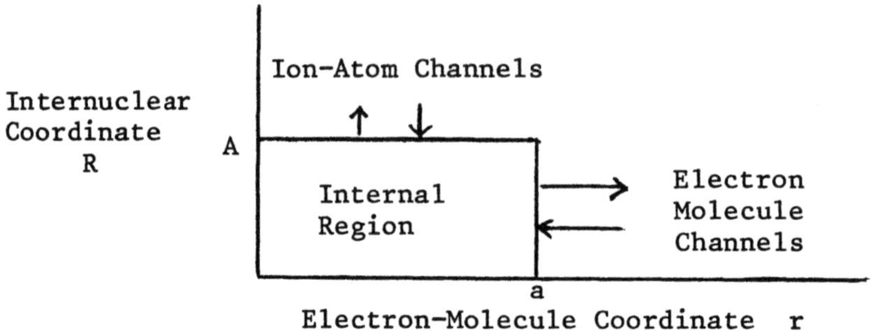

Figure 3. Division of configuration space for vibrational excitation and dissociative attachment.

The internal region is defined by the condition

$$0 \leq r \leq a, \; 0 \leq R \leq A \tag{19}$$

where r is the electron-molecule coordinate and R is the internuclear coordinate. The expansion of the total wave function in this internal region is then given by

$$\psi_k = \sum_{ij} \Psi_i(1,\ldots,N+1)\;\Theta_j(R)\;c_{ijk} \tag{20}$$

where Ψ_i are the R-matrix states defined by eqs. (6) and (7) calculated for each internuclear separation R and Θ_j are functions representing the nuclear motion. The coefficients c_{ijk} are obtained by diagonalising the operator $T_R + H_{e\ell}^{N+1} + \mathcal{L}^{N+1}(b,B)$ in this basis where T_R is the nuclear kinetic energy operator and $\mathcal{L}^{N+1}$ is a generalised surface projection operator. By a similar approach to that which led to the fixed-nuclei R-matrix given by eq. (15) we obtain a generalised R-matrix linking the electron-molecule and the ion-atom channels. This has the form

$$\mathcal{R}(E) = \left[\begin{array}{c|c} \underline{R}_{aa}(E) & \underline{R}_{aA}(E) \\ \hline \underline{R}_{Aa}(E) & \underline{R}_{AA}(E) \end{array}\right] \tag{21}$$

The solution of the coupled equations in the external region defined by

$$r \geq a \text{ or } R \geq A \tag{22}$$

then gives the S-matrix and related cross-sections.

Electronic Excitation Results

Ab-initio calculations have been carried out for the following diatomic molecules using the R-matrix method described in the previous section:

$$H_2^+,\ H_2,\ LiH,\ N_2,\ CO^+,\ CO,\ NO,\ O_2,\ F_2\ . \tag{23}$$

In this section we describe recent electronic excitation calculations which have been carried out for H_2^+, H_2 and O_2.

A. $e^- - H_2^+$ scattering. R-matrix calculations in which two- and four-target electronic states were retained in eq. (6) were carried out by Tennyson et al [24,25]. The processes of interest are

$$e^- + H_2^+ (X\ {}^2\Sigma_g^+) \rightarrow \begin{cases} e^- + H_2^+ (X\ {}^2\Sigma_g^+) \\ e^- + H_2^+ (A\ {}^2\Sigma_u^+) \\ H + H \end{cases} \tag{24}$$

where in the energy region below the ${}^2\Sigma_u^+$ threshold Feshbach resonances converging to this threshold were calculated. A potential energy diagram of this system is shown in figure 4. The two-state calculations included the $X^2\Sigma_g^+$ and $A^2\Sigma_u^+$ LCAO-MO-SCF target states of Cohen and Bardsley [26]. In the four-state calculations the two $B^2\Pi_u$ states were also included. For the two-state calculations an extra $p\pi$ STO on each nucleus was added to provide π virtual MO's to represent polarisation effects by the second expansion in eq. (6). In the four-state calculations, π_g virtual MO's were obtained from the complement of the π_u basis of Cohen and Bardsley [26].

The positions and widths of the Feshbach resonances were obtained by calculating the eigenphase shift in the resonance region and then fitting to a Breit-Wigner formula [27]. These calculations were carried out as a function of internuclear separation, and as an example we show in figure 5 the results obtained for the lowest ${}^1\Sigma_g^+$ resonances compared with other calculations.

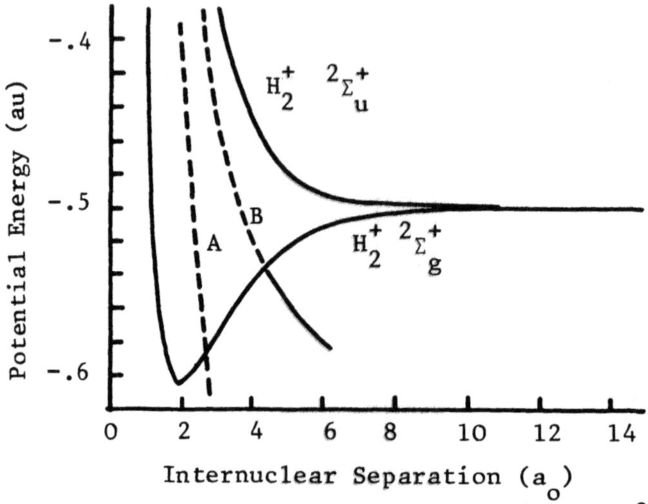

Figure 4. Potential energy curves for the lowest $^2\Sigma_g^+$ and $^2\Sigma_u^+$ states of H_2^+. The two lowest $^1\Sigma_g^+$ Feshbach resonance curves A and B are also shown.

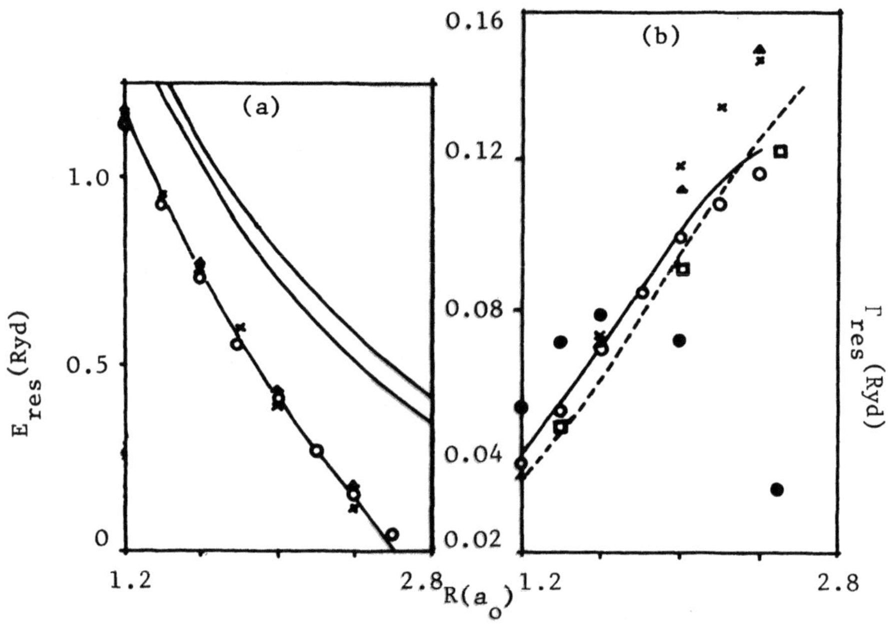

Figure 5. (a) Positions of the three lowest resonances and (b) width of the lowest resonance of $^1\Sigma_g^+$ symmetry as a function of H_2^+ separation. Full curve R-matrix (two-state close coupling plus polarisation); broken curve R-matrix(two-state close coupling); X, Takagi and Nakamura [28]; O, Sato and Hara [29]; ▲, Collins and Schneider [5] (four-state); ●, Bottcher and Docken [30]; □, Hazi [10].

For the lowest $^1\Sigma_g^+$ resonance position there is a good measure of agreement between theoretical calculations, so for clarity not all have been included in figure 5(a). The situation for the resonance width shown in figure 5(b) is less clear. If we ignore the early work of Bottcher and Docken [30], which is probably in error, then the R-matrix results agree best with the Stieltjes-moment method results of Hazi [10] and the Feshbach projection operator method results of Sato and Hara [29]. On the other hand, agreement with the Kohn variational method results of Takagi and Nakamura [28] and the four-state close coupling results of Collins and Schneider [5] at larger internuclear separation is less satisfactory. Further work needs to be done to clarify these discrepancies, since they will be significant in the dissociative recombination process defined by eqs. (3) and (24).

We show in figure 6 the $^2\Sigma_g^+ \rightarrow {}^2\Sigma_u^+$ electronic excitation cross-section calculated at the equilibrium internuclear separation.

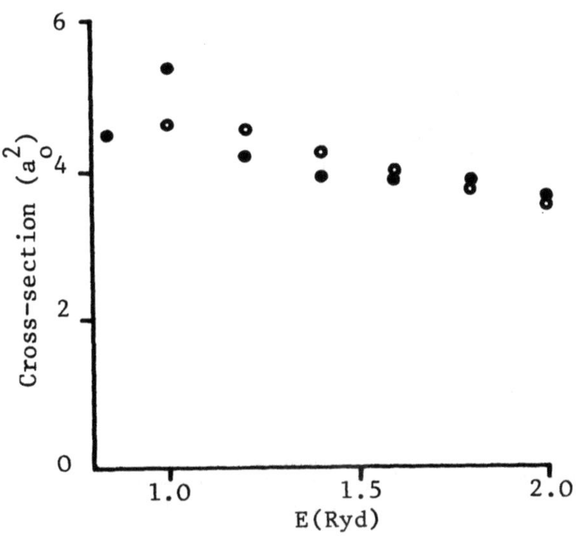

Figure 6. $e^- - H_2^+$ $^2\Sigma_g^+ \rightarrow {}^2\Sigma_u^+$ excitation cross-section. ● R-matrix (two-state close coupling plus polarisation). O, Collins and Schneider [5] (two-state).

There is reasonably good agreement between the two calculations illustrated in this figure, the small differences probably being accounted for by the inclusion of some allowance for polarisation in the R-matrix calculation.

B. $e^- - H_2$ scattering. R-matrix calculations in which the two lowest target electronic states were retained in eq. (6) were carried out by Baluja et al [31]. The processes of interest are:

$$e^- + H_2(X^1\Sigma_g^+) \begin{array}{l} \nearrow e^- + H_2(X^1\Sigma_g^+) \\ \searrow e^- + H_2(b^3\Sigma_u^+) \end{array} \qquad (25)$$

The $X^1\Sigma_g^+$ and $b^3\Sigma_u^+$ states of H_2 were represented by the LCAO-MO SCF wave function of Fraga and Ransil [32]. Their basis was augmented by a $2p\pi$ STO on each nucleus to provide π virtual MO's to represent polarisation effects by the second expansion in eq. (6).

We show in figure 7 the total elastic cross-section calculated at the equilibrium internuclear separation compared with other calculations and with experiment.

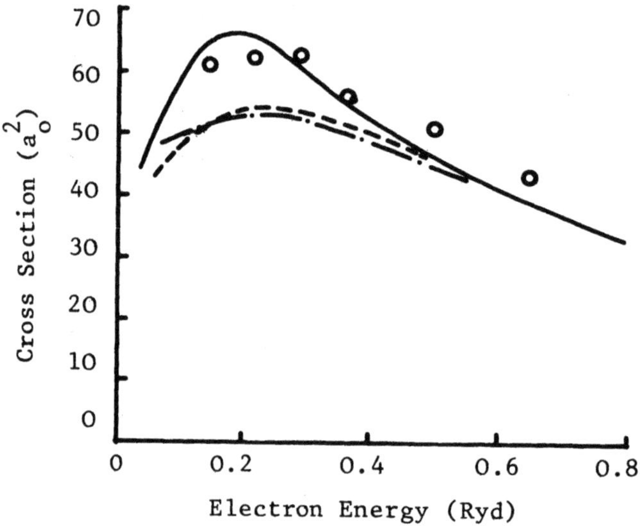

Figure 7. $e^- - H_2$ total elastic cross-section ——— , R-matrix; - - -, Lima et al [33]; -· - ·- , Schneider and Collins [34]; ○, experiment of Hoffman et al [35].

There is good agreement between the R-matrix results and experiment [35] both exhibiting a low energy peak due to the presence of the broad $^2\Sigma_u^+$ H_2^- shape resonance. The Schwinger variational method results [33] and the linear algebraic equations method results[34] lie lower in the resonance region which may be due to a different representation of the polarisation potential. However, before any firm conclusion can be drawn, it will be necessary to average the R-matrix calculation over the internuclear separations appropriate to the ground vibrational state.

In figure 8 we compare the $X^1\Sigma_g^+ \rightarrow b^3\Sigma_u^+$ measured and calculated excitation cross-sections.

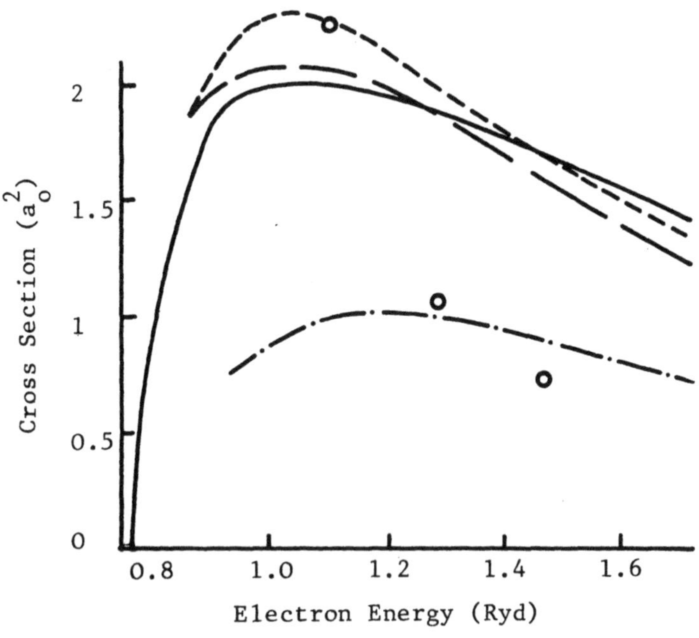

Figure 8. $e^- - H_2$ $X^1\Sigma_g^+ \rightarrow b^3\Sigma_u^+$ excitation cross-section. ——, R-matrix; - - - , Lima et al [36], —— ——, Schneider and Collins [37]; — . — , Chung and Lin [38]; o , experiment of Khakoo et al [39].

The R-matrix calculation is in good agreement with the most recent Schwinger variational method results [36] (but not with earlier calculations [33]) and the linear algebraic equations method results [37]. The close coupling results of Chung and Lin [38] are in error since they omit important orthogonality terms [37]. However, the recent calculations are almost a factor of two higher than the recent experiments of Khakoo [39]. This may be because the calculations include only a limited number of target electronic states and no pseudo-states, and hence do not properly take into account the loss of flux into the omitted channels. This is known to lead to a similar overestimate in the $e^- - H\ 1s - 2s$ excitation cross-section [40]. In the next section we discuss a new approach, which will enable more electronic states to be included in the calculation and hence will avoid errors from this source.

C. $e^- - O_2$ scattering. Recently, R-matrix calculations have been carried out by Noble and Burke [41] in which the three electronic terms of the ground state configuration were retained in eq. (6). The processes of interest are:

$$e^- + O_2(X^3\Sigma_g^-) \rightarrow e^- + O_2(X^3\Sigma_g^-,\ a^1\Delta_g,\ b^1\Sigma_g^+) \quad (26)$$

where the $X^3\Sigma_g^-$, $a^1\Delta_g$ and $b^1\Sigma_g^+$ terms all belong to the configuration

$$1\sigma_g^2\ 2\sigma_g^2\ 3\sigma_g^2\ 1\sigma_u^2\ 2\sigma_u^2\ 1\pi_u^4\ 1\pi_g^2 \ . \quad (27)$$

Approximate potential energy curves for these three terms as well as for the $X^2\Pi_g$ ground state of O_2^- are illustrated in figure 9.

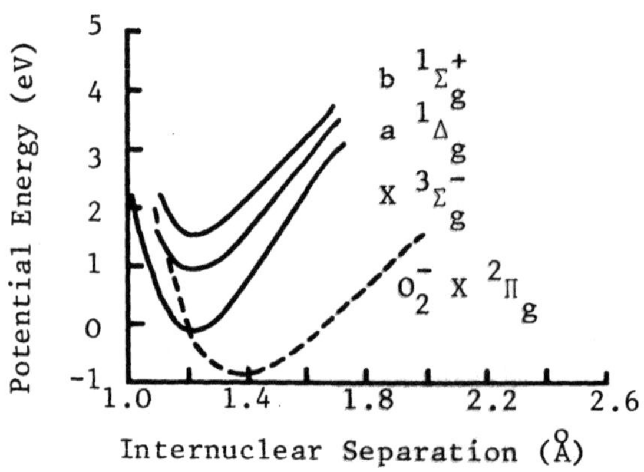

Figure 9. Approximate potential energy curves for the $X^3\Sigma_g^-$, $a^1\Delta_g$ and $b^1\Sigma_g^+$ states of O_2 and for the $X^2\Pi_g$ state of O_2^-.

At the equilibrium internuclear separation of the $X^3\Sigma_g^-$ ground state of O_2 the $X^2\Pi_g$ state of O_2^- gives rise to a shape resonance, while at larger internuclear separations it becomes bound, giving rise to the observed negative ion state of O_2^- [42].

The three electronic states were represented by the LCAO-MO-SCF wave function of Saxon and Liu [43]. In addition, the resonant configuration $1\sigma_g^2\ 2\sigma_g^2\ 3\sigma_g^2\ 1\sigma_u^2\ 2\sigma_u^2\ 1\pi_u^4\ 1\pi_g^3$ was included in the second expansion in eq. (6). The resonant $^2\Pi_g$ symmetry of O_2^- was found to dominate the low-energy electronic excitation cross-sections.

We show in figure 10 the $X^3\Sigma_g^- \rightarrow a^1\Delta_g$ and the $X^3\Sigma_g^- \rightarrow b^1\Sigma_g^+$ excitation cross-sections calculated at the equilibrium internuclear separation compared with the experimental measurements of Trajmar et al [44].

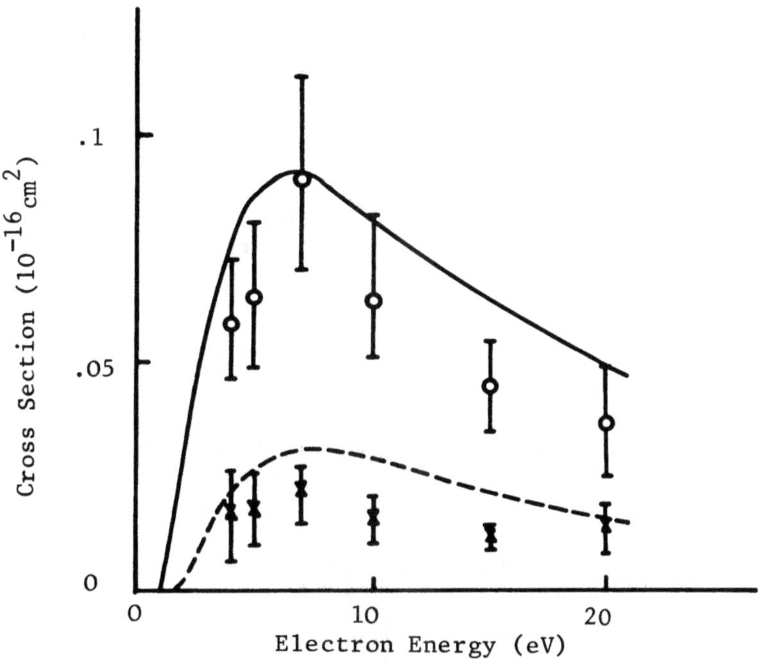

Figure 10. $e^- - O_2$ excitation cross-sections. $X^3\Sigma_g^- \rightarrow a^1\Delta_g$; ——— theory, ○ experiment [44]. $X^3\Sigma_g^- \rightarrow b^1\Sigma_g^+$; - - - theory, X experiment [44].

The measurements of Linder and Schmidt [45] at lower energies and of Konishi et al [46] at higher energies are consistent with these results. The main features of the observed cross sections, such as the position of the maximum and the overall magnitude, are well represented by the theory. It is now necessary to extend the calculation to other internuclear separations, in order to estimate the vibrational excitation and dissociative attachment cross-sections.

A New R-Matrix Approach

We have pointed out in our discussion of $e^- - H_2$ scattering in the previous section that many excited target electronic states may need to be included in eq. (6) in order to obtain reliable electronic excitation cross-sections close to threshold. This raises serious computational difficulties, particularly in defining an orthonormal orbital basis which gives an accurate representation of these target states. In this section we describe a new approach which eliminates this difficulty.

We consider the excitation of an (N+1)-electron target molecule

$$e^- + AB_i \rightarrow e^- + AB_f \ . \tag{28}$$

In our new approach we represent both the initial and final electronic states of the molecule in this equation as bound states of an electron plus an N-electron target ion

$$AB_i, \ AB_f \equiv e^- + AB_k^+ \tag{29}$$

We thus have to carry out the calculation in two steps. In the first step we calculate the bound states of the electron + N electron target ion. In the second step we solve the collision problem for an electron plus (N+1)-electron target using the target wave functions from the first step. In both steps we adopt an R-matrix expansion analogous to eq.(6).

The partition of configuration space corresponding to this new approach is illustrated in figure 11.

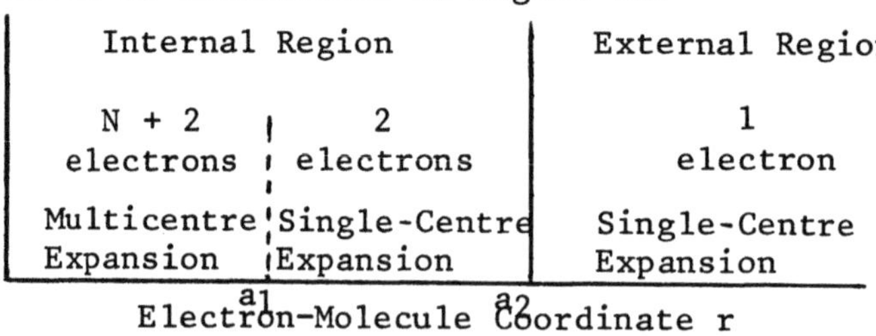

Figure 11. Partition of configuration space in the new R-matrix approach.

In comparing this partitioning with that illustrated in figure 2 for the usual R-matrix method we note that the internal region is now divided into two sub-regions. The radius a_1 of the inner sub-region is chosen to just envelope the electronic states of the N-electron molecular ion AB^+ which are retained in a multicentre CI expansion of the (N+1)-electron wave function. The radius a_2 of the outer sub-region is chosen to just envelope the excited electronic states which are retained in a multicentre CI expansion of the (N+2)-electron wave function. In the region $a_1 \leq r \leq a_2$ only the valence electron of the excited target AB and the scattered electron are present, and both are represented by numerical orbitals centred on the centre-of-gravity of the molecule. In practice a_2 is much larger than a_1 since the core orbitals of AB^+ are usually quite compact compared with the valence orbitals of AB. Finally the external regions in both figures 2 and 11 are treated in the same way. A single-centre expansion of the wave function of the scattered electron is adopted in both cases.

The expansion of the total wave function for the (N+1)-electron system defined by eq. (29) is

$$\Psi_k = \mathcal{A} \sum_{ij} \Phi_i(1,\ldots,N)\psi_j(N+1)\, a_{ijk} + \sum_i \chi_i(1,\ldots,N+1) b_{ik} \tag{30}$$

where Φ_i are target eigenstates of AB^+ which are constructed from nuclear centred STO's which vanish by the boundary $r=a_1$, ψ_i are continuum orbitals which are constructed from the STO's used in defining AB^+ and numerical orbitals obtained by solving eqs. (16), (17) and (18) which extend out to $r = a_2$ and finally χ_i are multicentre quadratically integrable functions which are again constructed from the STO's used in defining AB^+. The coefficients a_{ijk} and b_{ik} are determined as in eq. (7) by diagonalising

$$\langle \Psi_k | H^{N+1}_{e\ell} + L^{N+1}_b | \Psi_{k'} \rangle = E^{N+1}_k \delta_{kk'} \tag{31}$$

over the internal region. The bound states of the (N+1)-electron system are determined as in the R-matrix theory of photoionisation [47] by solving the asymptotic equations in the external region with bound state boundary conditions and iterating until the logarithmic derivative on the boundary $r = a_2$ is consistent with that given by the R-matrix defined by eq. (15). The eigenstates of AB are then given by eq. (11).

The expansion of the total wave function for the (N+2)-electron system defined by eq. (28) is then

$$\Theta_k = \mathcal{A}\sum_{ij}\Psi_i(1,\ldots,N+1)\psi_j(N+2)\alpha_{ijk} + \sum_i \chi_i(1,\ldots,N+2)\beta_{ik} \quad (32)$$

where the Ψ_i are obtained by solving eqs. (30) and (31) and the remaining quantities in this equation have the same definitions as in eq. (30). The coefficients α_{ijk} and β_{ik} are determined by diagonalising

$$\langle\Theta_k|H_{e\ell}^{N+2} + L_b^{N+2}|\Theta_{k'}\rangle = E_k^{N+2}\,\delta_{kk'}, \quad (33)$$

over the internal region. We can then obtain the R-matrix corresponding to eq. (28) by rewriting expansion (32) in terms of the bound states of the (N+1)-electron system.

As an example, we have carried out the first step of the calculation for $e^- - H_2$ scattering to obtain bound states of the $e^- - H_2^+$ system. Our approximation was the same as that adopted in the two-state calculation for $e^- - H_2^+$ scattering discussed in the previous section. That is, we included in eq. (30) the $X^2\Sigma_g^+$ and $A^2\Sigma_u^+$ target states of H_2^+ as well as appropriate quadratically integrable functions formed from the virtual H_2^+ orbitals. We present in table 1 the energies which we obtained for some of the bound states of H_2.

Table 1. H_2 bound state energies calculated using the new R-matrix approach

State	R(a.u.)	R-matrix energy (a.u.)	Accurate energy (a.u.)
$X^1\Sigma_g^+$	1.5	−1.1681	−1.1729
	1.7	−1.1579	−1.1625
EF $^1\Sigma_g^+$	1.5	−0.7005	−0.7030
	1.7	−0.7124	−0.7149
GH $^1\Sigma_g^+$	1.5	−0.6377	−0.6389
HH $^1\Sigma_g^+$	1.5	−0.6348	−0.6362
B $^1\Sigma_u^+$	1.4	−0.7030	−0.7057
B' $^1\Sigma_u^+$	1.4	−0.6272	
a $^3\Sigma_g^+$	1.4	−0.7115	−0.7136
b $^3\Sigma_u^+$	1.4	−0.7827	−0.7841
c $^3\Pi_u$	1.4	−0.7040	
C $^1\Pi_u$	1.5	−0.6971	−0.7005
I $^1\Pi_g$	1.5	−0.6374	

With this quite simple basis, we obtain over 90% of the correlation energy for the $X^1\Sigma_g^+$ ground state, while the excited state energies are even more accurate. More importantly, this basis provides a uniform way of representing the ground and excited electronic states of H_2 which it is hoped will enable reliable cross-sections to be obtained when substituted into the second step of the calculation defined by eqs. (32) and (33). This step of the calculation is underway at present.

Finally, we remark that this new approach may also be used to obtain ionisation cross-sections. This is because the valence electron and the scattered electron are both represented by continuum orbitals defined by eqs. (16), (17) and (18) over the internal region. This enables a generalised R-matrix corresponding to the escape of two electrons through the boundary $r = a_2$ to be defined. This generalised R-matrix can then be matched to an asymptotic solution of the wave function for two electrons in the external region given by Peterkop [48] (see also Fano and Inokuti [49]).

Acknowledgments

We are grateful to our colleagues at Queen's University, Royal Holloway College and the Daresbury Laboratory for permission to quote recent R-matrix results on H_2^+, H_2 and O_2 and for discussions on the theory.

References

[1] Lane, N. F.; Rev. Mod. Phys. 1980, 52, 29.
[2] Buckley, B. H.; Burke, P. G.; Noble, C. J. in "Electron Molecule Collisions" Shimamura, I. and Takayanagi, K. Eds Plenum Press: New York and London, 1984; Chapter 7.
[3] Schneider, B. I.; Chem. Phys. Lett. 1975, 31, 237; Phys. Rev. 1975, A11, 1957.
[4] Burke, P. G.; Mackey, I.; Shimamura, I. J. Phys. B. (Atom. Molec. Phys) 1977, 10, 2497.
[5] Collins, L. A.; Schneider, B. I. Phys. Rev. 1983, A27, 101.
[6] Lucchese, R. R.; Watson, D. K.; McKoy, V. Phys. Rev. 1980, A22, 421.
[7] Feshbach, H.; Ann. Phys. (N.Y.) 1958, 5, 357; ibid 1962 19, 287.
[8] Berman, M.; Domcke, W. Phys. Rev. 1984, A29, 2485.
[9] Hazi, A.U.; Taylor, H.S. Phys. Rev. 1970, A1, 1109.

[10] Hazi, A. U.; in "Electron-Atom and Electron-Molecule Collisions" Hinze, J. Ed. Plenum Press: New York and London, 1983; Chapter 7.
[11] Wigner, E. P.; Phys. Rev. 1946, 70, 15; ibid 1946, 70, 606.
[12] Wigner, E. P.; Eisenbud, L. Phys. Rev. 1947, 72, 29.
[13] Burke, P. G.; Hibbert, A. in "Proceedings of the VI ICPEAC" Massachusetts Institute of Technology, 1969, p. 367.
[14] Burke, P. G.; Hibbert, A.; Robb, W. D. 1971, J. Phys. B. (Atom. Molec. Phys.) 1971, 4, 153.
[15] Burke, P. G.; Robb, W. D. Adv. Atom. Molec. Phys. 1975, 11, 143.
[16] Arthurs, A. M.; Dalgarno, A. Proc. Roy. Soc. London, 1960, A256, 540.
[17] Bloch, C.; Nucl. Phys. 1957, 4, 503.
[18] Noble, C. J.; Burke, P. G.; Salvini, S. J. Phys. B. (Atom. Molec. Phys.) 1982, 15, 3779.
[19] Burke, P. G.; Noble, C. J.; Salvini, S. J. Phys B. (Atom. Molec. Phys.) 1983, 16, L113.
[20] Buttle, P.J.A.; Phys. Rev. 1967, 160, 719.
[21] Baluja, K. L.; Burke, P. G.; Morgan, L. A. Comput. Phys. Commun. 1982, 27, 299.
[22] Noble, C. J. and Nesbet, R. K. Comput. Phys. Commun. 1984, 33, 399.
[23] Schneider, B. I.; LeDourneuf, M.; Burke, P.G. J. Phys. B. (Atom. Molec. Phys.) 1979, 12, L365.
[24] Tennyson, J.; Noble, C. J.; Salvini, S. J. Phys. B. (Atom. Molec. Phys.) 1984, 17, 905.
[25] Tennyson, J.; Noble, C. J. J. Phys. B. (Atom. Molec. Phys.) 1985, 18, 155.
[26] Cohen, J. S.; Bardsley, J. N. 1980, private communication to Robb and Collins.
[27] Tennyson, J.; Noble, C. J. Comp. Phys. Commun. 1984, 33, 421.
[28] Takagi H.; Nakamura, H. Phys. Rev. 1983, A27, 691.
[29] Sato, H.; Hara, S. 1983 quoted by Takagi and Nakamura [28].
[30] Bottcher, C.; Docken, K. J. Phys. B. (Atom. Molec. Phys.) 1974, 7, L5.
[31] Baluja, K. L.; Noble, C. J.; Tennyson, J. J. Phys. B. (Atom. Molec. Phys.) 1985, to be published.
[32] Fraga, S.; Ransil, B. J. Chem. Phys. 1961, 35, 1967.
[33] Lima, M.A.P.; Gibson, T. L.; Takatsuka, K. ; McKoy, V. Phys. Rev. 1984, A30, 1741.
[34] Schneider, B. I.; Collins, L. A. Phys. Rev. 1983, A27, 2847.

[35] Hoffman, K. R.; Dababneh, M. S.; Hsieh, Y. F.; Kauppila, W. E.; Pol, V.; Smart, J. H.; Stein, T.S. Phys. Rev. 1982, A25, 1393.
[36] Lima, M.A.P.; Gibson, T. L.; Takatsuka, K.; McKoy, V. 1985, to be published.
[37] Schneider, B. I.; Collins, J. A. J. Phys. B. (Atom. Molec. Phys.) 1985, to be published.
[38] Chung, S.; Lin, C. C. Phys. Rev. 1978, A17, 1874.
[39] Khakoo, H. A.; McAdams, R.; Trajmar, S. Phys. Rev. 1985, to be published.
[40] Burke, P. G.; Webb, T. G. J. Phys. B. (Atom. Molec. Phys.) 1970, 3, L131.
[41] Noble, C. J.; Burke, P. G. J. Phys. B. (Atom. Molec. Phys.) 1985, to be published.
[42] Schulz, G. J.; Rev. Mod. Phys. 1973, 45, 423.
[43] Saxon, R. P.; Liu, B. J. Chem. Phys. 1977, 67, 5432.
[44] Trajmar, S.; Cartwright, D. C.; Williams, W. Phys. Rev. 1971, A4, 1482.
[45] Linder, F.; Schmidt, H. Z. Naturforsch. 1971, 26a, 1617.
[46] Konishi, A.; Wakiya, K.; Yamamoto, M.; Suzuki, H. J. Phys. Soc. Japan. 1970, 29, 526.
[47] Burke, P. G.; Taylor, K. T. J. Phys. B. (Atom. Molec. Phys.) 1975, 8, 2620.
[48] Peterkop, R. K. in "Theory of Ionization of Atoms by Electron Impact"; Colorado Associated University Press, 1977, Chapter 2.
[49] Fano, U.; Inokuti, M. "On the Theory of Ionisation by Electron Collisions"; Argonne National Laboratory Report ANL-76-80 1976.

PART VI

SINGLE AND MULTICOLLISION PHENOMENA: DISSOCIATION, INTERNALLY EXCITED TARGETS, AND CLUSTERS

FRAGMENTATION DYNAMICS AND ENERGY PARTITIONING IN DISSOCIATIVE ATTACHMENT ON TRIATOMIC MOLECULES

Michel Tronc

Laboratoire de Chimie Physique
Université Pierre et Marie Curie, 11 rue Pierre et Marie Curie 75231 Paris, Cedex 05, France

Introduction

The dissociative attachment process on a triatomic molecule can be considered as a two step process: a vertical electron attachment to the molecule in its ground electronic and vibrational state, followed by dissociation of the short lived negative ion (resonance state) on its potential energy surface. The dissociation step, that can be seen as a half-collision process, has some specific characteristics: i) it is always competing with the autodetaching process of the extra electron, yielding rotational-vibrational or (and) electronic excitation of the neutral molecule; ii) the resonant ABC^- state can be formed in a well-defined internal energy level fixed by the incident electron energy E_i. The internal energy defines the excess energy E_{av} above the thermodynamical threshold for the fragments formation:

$$e(E_i) + ABC \rightarrow ABC^- \rightarrow A^- \ (E_K) + BC \ (E_{v,j})$$

where E_K (A^-) is the kinetic energy of the A^- anion, $E_{v,j}(BC)$ is the internal energy of BC which is most often reduced to the rotational-vibrational energy of BC at low incident energy. The kinetic energy of the BC fragment will be $E_K(BC)$.

The available energy E_{av} can be calculated from energy conservation, and from thermodynamical data:

$$E_{av} = E_i - [D(A\text{-}BC) - E.A\ (A)]$$

with D(A-BC) being the dissociation energy of the A-BC bond, and E.A. (A) the electron affinity of the A atom.

The excess energy has to be shared between translational and internal degrees of freedom of the recoiling fragments:

$$E_{av} = E_K\ (A^-) + E_K\ (BC) + E_{v,j}\ (BC).$$

From the kinetic energy measurement of the A^- ion, one can deduce the kinetic energy of BC and its total rotational-vibrational energy:

$$E_K\,(A^-) = (1 - \frac{m_{A^-}}{M_{ABC}})\;[E_i - [D(A\text{-}BC) - E.A.(A)] - E_{v,j}(BC)]$$

$$E_K(BC) = \frac{m_A}{m_{BC}}\,E_K(A^-)$$

$$E_{v,j}(BC) = E_{av} - E_K\,(A^-)\;[1 + \frac{m_A}{m_{BC}}]$$

where m_A, m_{BC} and M_{ABC} are the mass, respectively of the A atom, the BC fragment and the ABC molecule.

It appears that the measurement of translational energy of the ionic fragment as a function of the total excess energy over the resonance energy band yields information on the distribution of energy among the various degrees of freedom of the fragment products (T - VR coupling), and consequently on the dynamics of the dissociation process, and on the topology of the potential surface which mainly determines the energy partitioning. In some cases, these measurements can give accurate thermodynamical parameters of the dissociation products.

Quasi-Classical Trajectory Calculations

While in large molecules one can expect energy partitioning to be essentially random, which gives an average kinetic energy equal to the total excess energy divided by the number of degrees of freedom [1] or the active number of vibrational modes [2], because the transfer of energy among oscillators is fast compared to the dissociation time; statistical theories are questionable for unimolecular dissociation of triatomic molecules.

Consequently, a classical trajectory procedure for model potential energy surfaces has been developed by Goursaud et al. [3] and Sizun and Goursaud [4] for dissociative electronic attachment on triatomic molecules.

Briefly, the dissociation is considered as a half-collision, so that it can be studied by a quasi-classical trajectory method, as reactive collisions for which a classical approximation has been shown to be capable of mean value calculations [5]. Moreover this classical calculation provides the mechanism of the dissociation process. In this method the initial conditions on the resonance state potential surface are represented by a Wigner distribution, which simulates the quantum distribution of momenta when integrated over positions, and the quantum distribution of positions when integrated over momenta. The potential surface of the ABC^- resonance is represented by: i) a LEPS function [6] or a Wall and Porter analytic surface [7] for a repulsive surface having a saddle point between the two valleys of dissociation, ii) by a VFM function (vanishing function method) [3] for an attractive surface or iii) by a REC surface (rotated exponential curve) for a purely repulsive surface, i.e. having one valley of dissociation. These surfaces have free parameters and parameters fixed by the dissociation limit, and by the shape and the width of the dissociative attachment cross-section in the Franck-Condon region. The sampling in the initial conditions is made by a quasi-random method, and trajectories are calculated by step-by-step integration of Hamilton's equations of motion. The competition with the autodetachment of the extra electron is taken into account with a unimolecular decay constant rate $\Gamma/\hbar$ (where Γ is the autodetaching width). Each trajectory is weighted by multiplying the probability density given by the Wigner function by the survival probability, $\exp(-\frac{\Gamma t}{\hbar})$, where t is for each trajectory the critical time beyond which auto-detachment is energetically not possible. From all the trajectories we obtain the mean kinetic energy, the mean vibrational energy, the mean dissociation time, and the quantified mean energy.

The main results [3,4,8] show that, besides the role of the autodetaching width, which suppresses the slow trajectories, it is the character of the surface that plays a prominent role in the energy distribution: i) for a saddle point surface, the translational-vibrational coupling is small and most of the excess energy is released as kinetic energy. ii) for a potential well surface, irrespective of the details of the surface, there is always a strong coupling, giving strong vibrational excitation of the neutral fragment with population inversion. iii) for a purely repulsive surface (three-body fragmentation) all the trajectories emerge out of the central three-body region, nearly parallel to the diagonal of the surface and without oscillation.

Experimental

Since the early work of Chantry and Schulz [9] and Changry [10], using a Wien filter for the ion velocity analysis, very few experimental setups have been specially devoted to kinetic energy distribution measurements of negative ions. The techniques include i) a time-of-flight arrangement for measuring translational energies of the negative ions formed by interaction with a pulsed electron beam [2,11,12]; ii) interaction of electrons selected by a trochiodal monochromator, with a static gas in a target cell, with the negative ions being extracted at 90° to the electron beam, and energy analyzed by a 90° cylindrical condenser [13]; iii) crossed beam experiments using electrostatic filters: the electron beam is energy selected by a 127° cylindrical electrostatic filter. A crossed electron

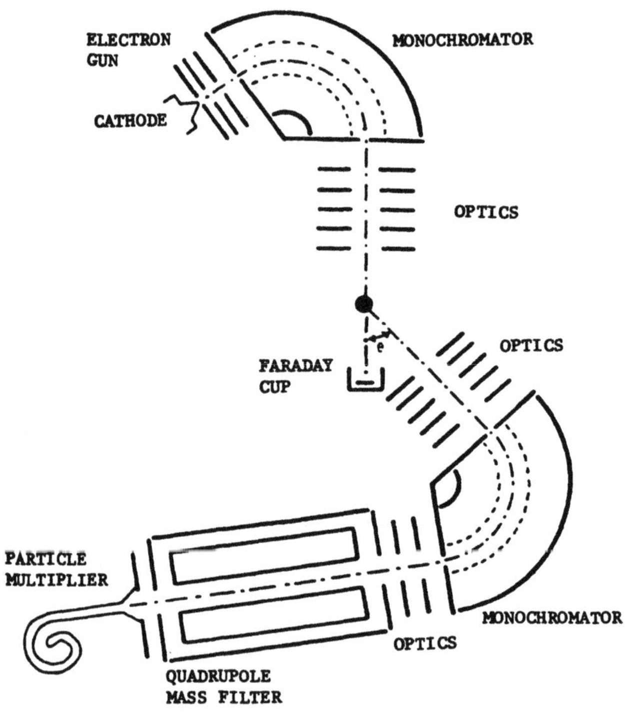

Fig. 1. Electron impact spectrometer for the study of kinetic energy and angular distributions of negative ions.

beam-molecular beam geometry is used and both scattered electrons and negative-ion products of the dissociation are energy-analyzed at a selected angle. Ions and electrons are then separated by a quadrupole mass spectrometer (Fig. 1) for which the mass resolution can be adjusted to a specific ionic mass (high resolution) or to separate only the very light electrons from the negative ions of same translational energy (low resolution) [14]. This apparatus is an evolution of a crossed-beam electron impact spectrometer using a very simple magnetic momentum filter developed by Hall et al. [15] and Tronc et al. [16], and described by Scherman et al. [17].

The apparatus is used in two modes of operation: i) Energy Loss Spectra, obtained by maintaining the incident electron energy fixed and sweeping the analyzer energy, give ion kinetic energy spectra; ii) Constant ion energy spectra, obtained by setting the analyzer energy and sweeping the incident energy, give the yield of negative ions produced with a specific kinetic energy. If the ion energy is fixed at zero, threshold ion spectra are obtained. For these two modes, the scattering angle can be varied from 0 to 120 degrees, and the angular distirbution obtained for the A^- ions give the symmetry of the resonance state [18]. The resolution of the apparatus is limited by thermal broadening which is small at low kinetic energy, but increases rapidly at higher ion energy. The thermal motion of the target molecule in the plane of observation can be reduced from an equivalent temperature of 300 K in a cell to 50 K in a hypodermic needle and to 20 K in a multicapillary array. The overall resolution is 130-150 meV for most of the spectra presented in the following sections

Results

I. Dissociation on a Saddle Point Surface.

H^-/H_2O

Figure 2 shows the kinetic energy distribution of H^- ions produced in H_2O through the $^2B_1$ resonance state around 6.5 eV as observed by Belic et al. [19], with a multicapillary array beam to reduce the thermal notion of the target molecules to an equivalent temperature of 20 K.

At 6 eV all the excess energy goes into kinetic energy of the H^- ions, with the OH fragments left in the v = o level. When the incident electron energy is increased, vibrational levels of OH are populated, but some H^- ions are still produced with the highest allowed kinetic energy, which is roughly the excess energy because of the very small H^- ion mass.

Calculations on a LEPS repulsive surface [3] show that at 6 eV, the faster trajectories (i.e. having no excursion in the symmetric stretch motion) can only survive to an autodetachment width of 0.15 eV and lead to dissociation. But when the incident electron energy is increased, the excess energy is increased, and trajectories of the representative point on the surface making an excursion in the symmetric stretch motion are fast enough to lead to dissociation, so that part of this excess energy is released as vibrational excitation of the OH fragment.

A similar situation was observed for the two higher energy processes giving H^- in H_2O at 8.6 and 11.8 eV [19] and for H^- ions formation in H_2S [20].

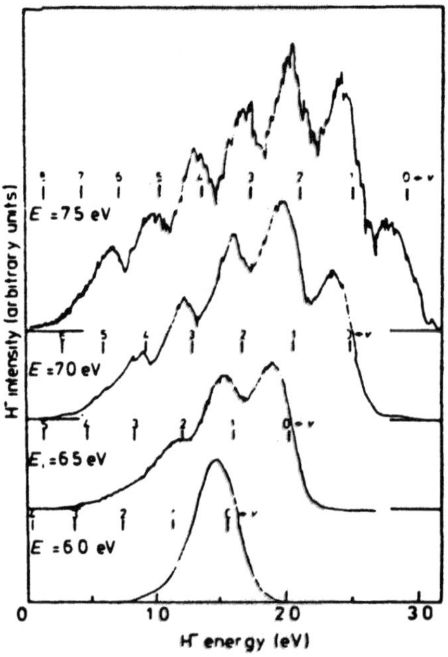

Fig. 2. Kinetic energy distribution of H^- ions produced in H_2O at 90° scattering angle and at the indicated incident electron energies Ei. The vertical bars correspond to excitation of the vibrational levels v of OH. (From Ref. 19).

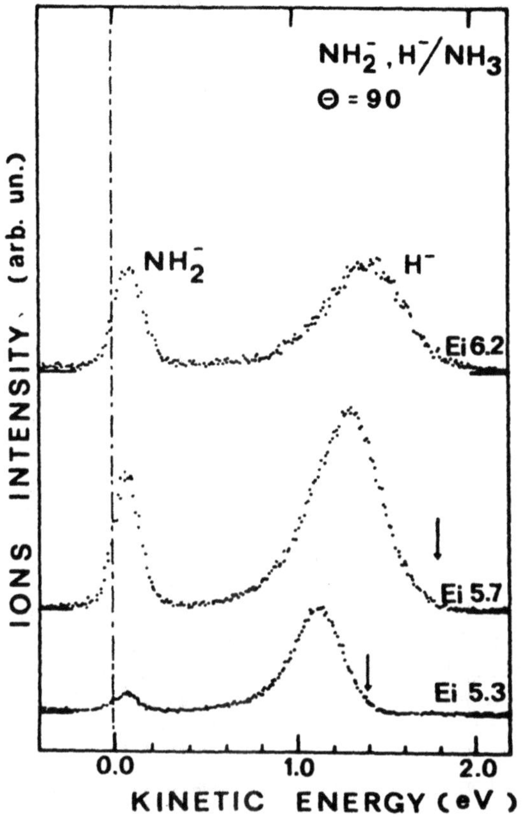

Fig. 3. Kinetic energy distribution of NH_2^- and H^- ions in NH_3 at a 90° scattering angle and at the indicated incident electron energies Ei.

H^-/NH_2

H^- and NH_2^- ions appear in the same energy band in NH_3 [21]. Figure 3 shows the kinetic energy distribution of NH_2^- ions formed with very low energy because of their mass, and of H^- ions flying away with most of the excess energy. The H^- ion distribution is broad and shifted to higher energy when the incident electron energy increases, but the vibrational structure of NH_2 is smeared out because the thermal broadening was important in this experiment, and the NH_2 fragments are probably highly rotationally excited [21].

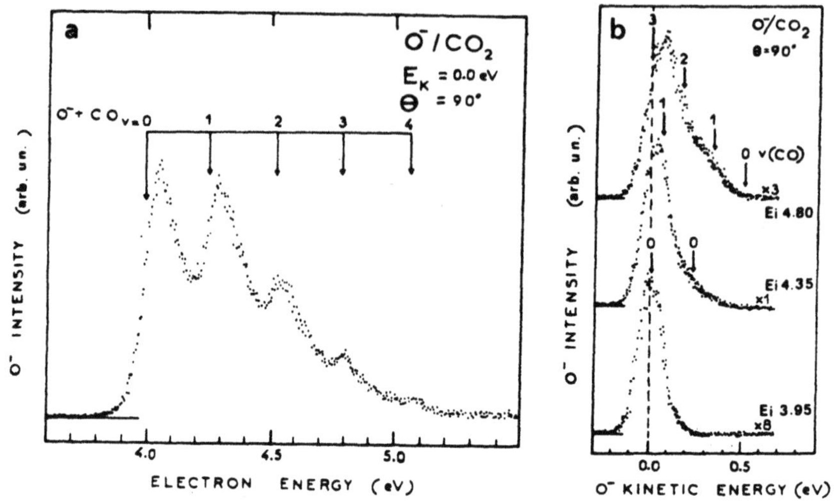

Fig. 4. a) Yield of zero energy O^-/CO_2 ions at 90° scattering angle through the first Π_u shape resonance; and b) kinetic energy distribution of O^- ions at the indicated electron energies Ei. The arrows show the vibrational levels of CO.

II. Dissociation on a Potential Well Surface

O^-/CO_2

Figure 4 shows the kinetic energy distribution and the yield of zero energy O^- ions produced in CO_2 through the first Π_u shape resonance.

These angular differential results [24] are similar to the previous angular integrated observations of Chantry [10] and Stamatovic and Schulz [23] and the recent results of Dressler and Allan [13]. O^- ions are produced with zero kinetic energy whatever the incident electron energy. As zero energy ions can appear only at thresholds of $O^- + CO_v$ because of energy conservation, excitation up to the $v = 4$ level of CO is observed.

The O^- cross-section shows additional structures attributed to interference effects in the nuclear motion of CO_2 [25,13].

Trajectory calculations on a VFM surface [3] reveal that when the excess energy is increased, the average dissociation time decreases, and the high-energy tail of the kinetic energy distribution is increased, but the percentage of excess energy transferred to translation of the fragments is kept nearly constant and small (10 - 13% for O^-/CO_2 between 4.3 and 5.3 eV).

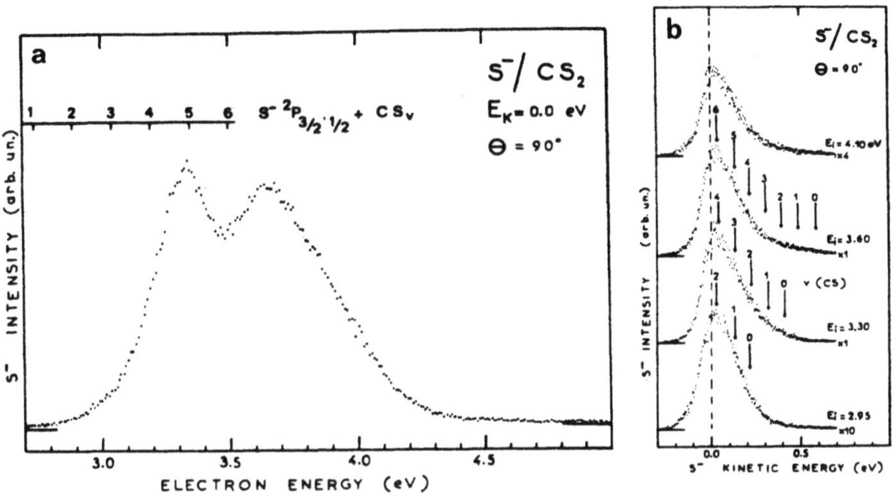

Fig. 5. a) Yield of zero-energy S^-/CS_2 ions at 90° scattering angle and b) kinetic energy distribution of S^- ions. The vibrational levels v of CS are shown.

S^-/CS_2

In the first S^- band at 3.3 eV in CS_2, the kinetic energy distribution is peaked at zero and is nearly constant as in CO_2 (Fig. 5). The smearing of the expected vibrational structure of the zero-energy ion yield (Fig. 5) results from the high degree of rotational excitation and the small vibrational spacing (159 meV) of the CS fragment, together with the existence of the two spin-orbit states 2 $P_{3/2}$, 2 $P_{1/2}$ of S^- lying 60 meV apart. Considering the close similarity between the shape and separation of the two S^- peaks with the electron energy loss spectrum [26], the structure [27], has been attributed to two polarization-bound Feshbach resonances having the two lowest $(\Pi)^3$ (Π^*) excited valence triplet states as parents [26].

O^-/SO_2

Figure 6 shows that O^- ions in SO_2 are produced with zero kinetic energy. The zero-energy ion yield shows two series of structures corresponding to the vibrational spacing of the two lowest $^3\Sigma^-$ and $^1\Delta$ states of the SO fragments. These structures are similar to the structures observed by Abouaf et al. [25] in an angular integrated cross-section experiment under high discrimination conditions. The S^- ions yield shows likewise two series of structures originating from the vibrational excitation of O_2 in its ground and first excited state.

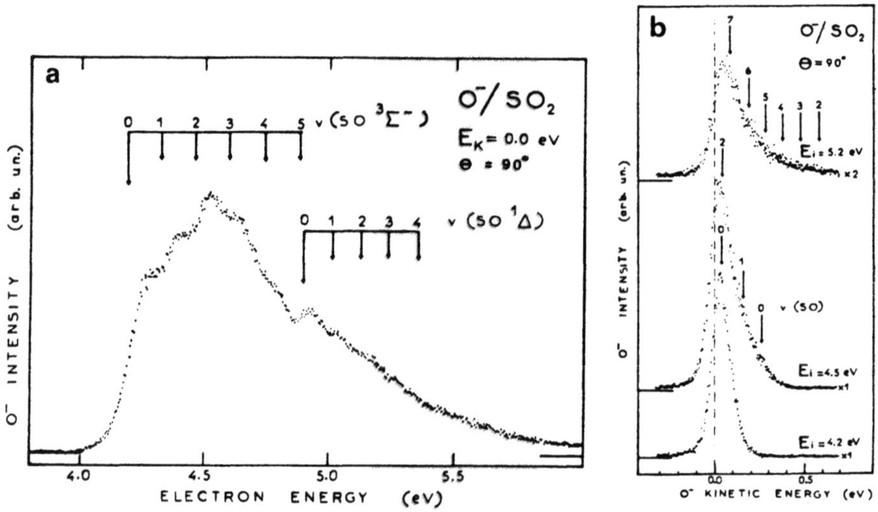

Fig. 6. a) Yield of zero-energy O^-/SO_2 ions at 90° scattering angle, the vibrational levels of the ground and the first excited states of SO are shown and b) kinetic energy distribution of O^- ions at the indicated incident electron energies.

III. Dissociation on a Purely Repulsive Surface

Cl^-/HG cl_2, B_r^-/Hg Br_2

In the collinear three-body fragmentation of a ABC^- resonance state, $ABC \rightarrow A^- + B + C$, the kinetic energy of the A^- and C fragments is:

$$E_K\ (A^-) = p^2{}_A/2\ m_A$$

$$E_K\ (C) = p^2{}_C/2\ m_c$$

where m_A and m_C are the masses of the A and C atoms, and p_A and p_C are the conjugate momenta of AB and BC respectively.

Since at the end of the three-body dissociation the potential vanishes, when the two bonds are broken the kinetic energy of the central atom B is $E_K(B) = E_{av}(A) - E_K(C) = (p_A - p_C)^2/2m_B$.

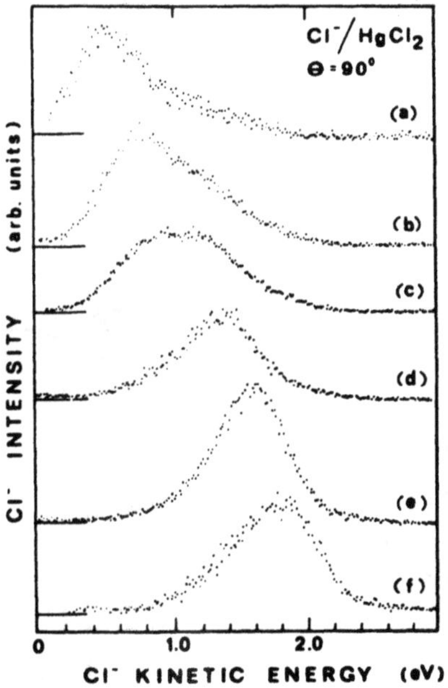

Fig. 7. Kinetic energy distribution of Cl^- ions from $HgCl_2$ at 90° scattering angle and for an electron energy of (a) 3.0 eV, (b) 3.4 eV, (c) 3.6 eV, (d) 4.0 eV, (e) 4.5 eV and (f) 5.0 eV. The accuracy of the kinetic energy scale is ± 0.15 eV.

The sharing of the available energy will be very different in a simultaneous mechanism in which the two bonds break at the same time, and in a sequential mechanism which is the result of two successive two-body dissociations: i) if the dissociation is simultaneous $p_A = p_C$ and $E_K(b) = 0$. Moreover, in the case of a symmetric molecule ABA, $E_K(A) = 1/2\ E_{av}$; ii) if the dissociation is sequential and the bond A-BC is broken first, the first step is as described in the introduction. At the end of dissociation the kinetic energy of A will be maximum if $E_{v,j}$ (BC) = 0, so that:

$$E_K\ (A^-) \leqslant E_K^{max}\ (A^-) = \frac{m_B+m_C}{M_{ABC}}\ E_{av}$$

$$E_K\ (B) \leqslant E_K^{max}\ (B) = \frac{m_A m_B}{M_{ABC}(m_C+m_B)}$$

$$E_K\ (C) \geqslant E_K^{min}\ (C) = \frac{m_A m_C}{M_{ABC}(m_B+m_C)}\ E_{av}$$

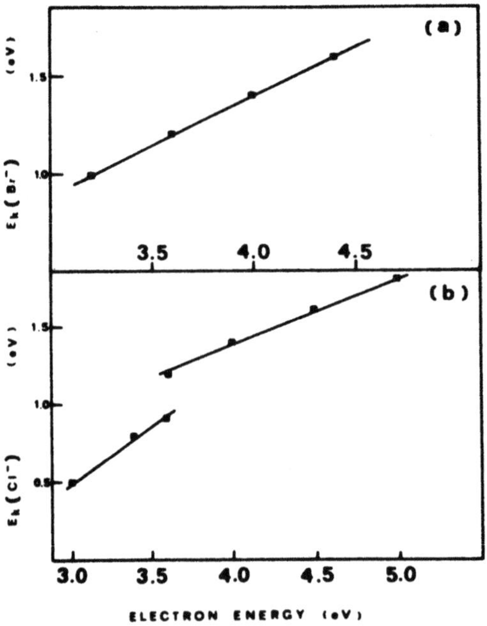

Fig. 8. Position of the maximum of the kinetic energy distribution against the electron energy (a) for Br^- ions in $HgBr_2$, (b) for Cl^- ions in $HgCl_2$.

If the first step of the dissociation is the breaking of the AB-C bond, the kinetic energies of A and C have to be interchanged.

Figure 7 shows the kinetic energy of Cl^- ions produced in Hg Cl_2 between 3 and 5 eV through the $^2\Pi_{3/2,1/2}$ resonance states with the final state products Cl^- (^1S_o) + Hg (^1S_o) + Cl ($^2P_{3/2}$, $^2P_{1/2}$) (Fig. 9). The position of the maximum of the kinetic energy distribution versus the electron energy is on a straight line with a slope of 0.5 (Fig. 8), which means that 0.5 E_{av} is taken by the Cl^- ion, the Cl atom carrying the remaining 0.5 E_{av} and the Hg atom being left with no kinetic energy. This observation is in favor of a simultaneous breaking of the two Cl-Hg bonds, while for a sequential mechanism we expect a broad distribution ranging from 0.5 E_{av} to 0.87 E_{av} if the Cl^- - Hg bond was broken first and from 0.02 E_{av} to 0.5 E_{av} if the Hg-Cl bond was broken first.

Very similar results are obtained for the Br^- formation in Hg Br_2 around 4 eV [29].

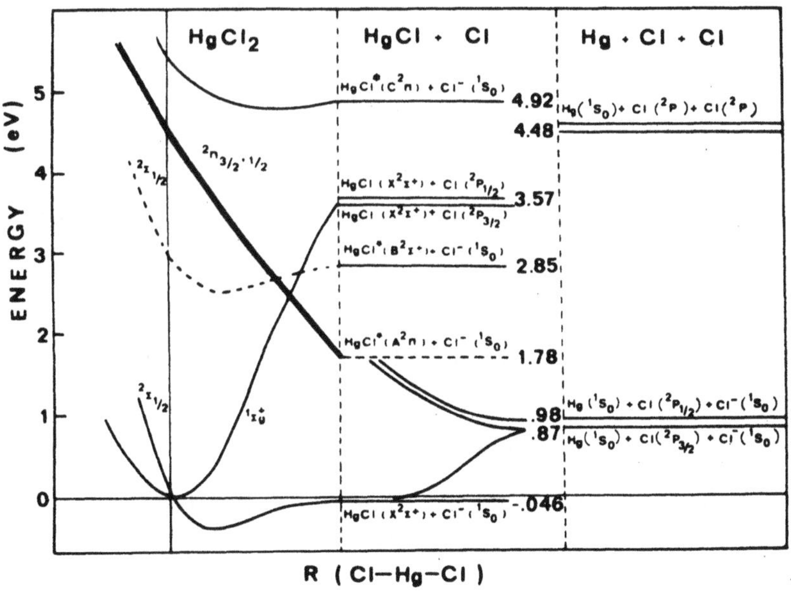

Fig. 9. Schematic potential energy curves for dissociative attachment processes in $HgCl_2$.

Classical trajectory calculations on a REC surface (Fig. 10) can reproduce qualitatively the experimental results and show that all the trajectories are nearly parallel to the diagonal of the surface, without any oscillation, which appears as a typical characteristic of a simultaneous dissociation on a purely repulsive surface.

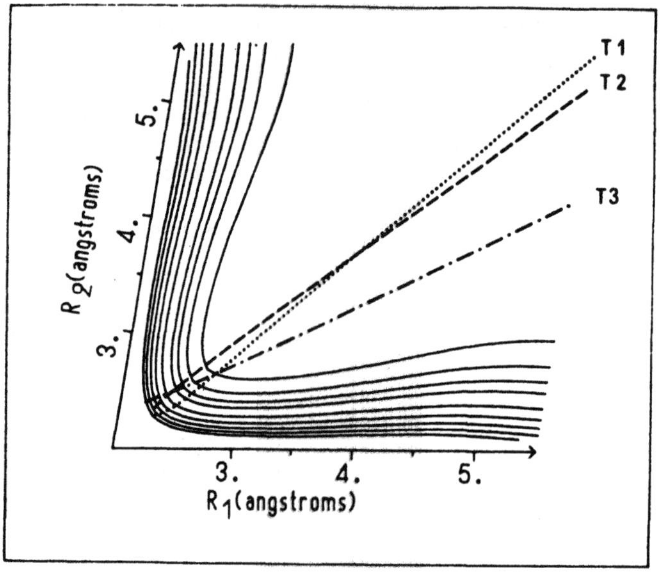

Fig. 10. REC (Rotated exponential curve) surface for $HgCl_2^-$ in skewed coordinates with typical trajectories: $T_1(t_d = 11\ 10^{-14}s,\ E_k = 1.755\ eV)$; $T_2(t_d = 10.3\ 10^{-14}s,\ E_k = 1.991\ eV)$; $T_3(t_d = 9\ 10^{-14}s,\ E_k = 2.472\ eV)$. The equipotential lines are separated by 0.25 eV between 0 and 2 3V and by 0.5 eV beyond 1 eV.

Conclusion

Ion kinetic energy analysis and angular distributions give detailed information on the dynamics of triatomic negative ion fragmentation. A precise analysis of the internal energy of the neutral fragment by laser-induced fluorescence would be of great interest to have a full description of the final-state energy distribution.

References

[1] Decorpo, J. J.; Bafus, D.A.; Franklin, J. L. J. Chem. Phys., 1971, 54, ±592.
[2] Harland, P.W.; Franklin, J. L.; J. Chem. Phys., 1974, 61, 1621.
[3] Goursaud, G.; Sizun, M.; Fiquet-Fayard, F. J. Chem. Phys., 1976, 65, 5453.
Goursaud, S.; Sizun, M.; Fiquet-Fayard, F. J. Chem. Phys., 1978, 68, 4310.
[4] Sizun, M.; Goursaud, S. J. Chem. Phys., 1979, 71, 4042.
[5] Le Forestier, C. Thèse doctorate état ORSAY, 1980.
[6] Kuntz, P. J.; Nemeth, E. H.; Polanyi, J. C.; Rosner, S. D.; Young, C. F. J. Chem. Phys., 1966, 44, 1168.
[7] Wall, F. T.; Porter, R. N. J. Chem. Phys., 1962, 36, 3256.
[8] Sizun, M.; Goursaud, S.; Ziesel, J. P.; Azring, R., Tronc, M. Chem. Phys., 1985, 95, 189.
[9] Chantry, P. F.; Schulz, G. J. Phys. Rev., 1967, 156, 134.
[10] Chantry, P. J. J. Chem. Phys., 1972, 57, 3180.
[11] Illenberger, E.; Scheuneman, H. U.; Baungartel, H. Chem. Phys., 1979, 37, 21.
[12] Illenberger, F. Chem. Phys Let., 1981, 80, 153.
[13] Dressler, R.; Allan, M. Chem. Phys., 1985, 92, 449.
[14] Le Coat, Y.; Azria, R.; Tronc, M. J. Phys. B., 1982, 15, 1569.
[15] Hall, R. I.; Cadez, I.; Scherman, C.; Tronc. M. Phys. Rev. A, 1977, 15. 599.
[16] Tronc, M.; Fiquet-Fayard, F.; Scherman, C.; Hall, R. I. J. Phys. B, 1977, 10, 305.
[17] Scherman, C.; Cadez, I., Delon, P.; Tronc, M.; Hall, R. I. J. Phys. E.: Sci. Inst., 1978, 11, 746.
[18] O'Malley, T. F.; Taylor, H. S. Phys. Rev., 1968, 176, 207.
[19] Belic, D. S.; Landau, M.; Hall, R. I. J. Phys. B., 1981, 14, 175.
[20] Azria, R.; Le Coat, Y.; Lefevre, G.; Simon, D. J. Phys. B., 1979, 12, 679.
[21] Sharp, T. E.; Dowell, J. T. J. Chem. Phys., 1969, 50, 3024.
[22] Tronc, M. to be published
[23] Stamatovic, A.; Schulz, G. J. Phys. Rev. A, 1973, 7, 589.
[24] Tronc, M.1 Malegat, L.; Azria, R. Chem. Phys. Lett., 1982, 5, 551.

[25] Abouaf, R.; Paineau, R.; Fiquet-Fayard, F. J. Phys. B, 1976, 9, 303.
[26] Dressler, R.; Allan, M. to be published.
[27] Ziesel, J. P.; Schulz, F. J. J. Chem. Phys., 1975, 62, 1936.
[28] Abouaf, R.; Fiquet-Fayard, F. J. Phys. B, 1976, 9, L 323.
[29] Tronc, M.; Ziesel, J. P.; Azria, R.; Sizun, M. Goursand, S. Chem. Phys., 1985, 95, 179.

EFFECTS OF TEMPERATURE ON DISSOCIATIVE AND NONDISSOCIATIVE ELECTRON ATTACHMENT

L. G. Christophorou,* S. R. Hunter, J. G. Carter, and S. M. Spyrou†

Atomic, Molecular and High-Voltage Physics Group
Health and Safety Research Division
Oak Ridge National Laboratory
Oak Ridge, Tennessee 37831

Results of recent studies on the effects of temperature, T, on the dissociative and nondissociative electron attachment to molecules are presented and discussed. These show the delicate and large effects of T on the cross-section $\sigma_{da} = \sigma_c\, p$ of resonance dissociative $[e + AX \underset{\tau_a^{-1}}{\overset{\sigma_c}{\rightleftharpoons}} AX^{-*} \xrightarrow{p} A + X^-]$ and $\sigma_{nd} = \sigma_c' p'$ of resonance nondissociative $[e + AX \underset{\tau_a'^{-1}}{\overset{\sigma_c'}{\rightleftharpoons}} AX^{-*} \xrightarrow{p'} AX^-]$ electron attachment to a molecule AX. For AX molecules where only dissociative attachment processes occur, the effect of T on σ_{da} is an increase in σ_{da} resulting from an increase in p principally because of a decrease in the separation time of A and X^-; the energy integrated σ_{da} increases with increasing average internal energy of AX. For AX molecules with pure nondissociative attachment, the effect of T is a decrease of σ_{nd} with increasing T resulting from a decrease in p' (i.e., an increase with T of $\tau_a'^{-1}$). For AX molecules with both dissociative and nondissociative processes the total rate constant (or cross-section) increases or

---

*Also, Department of Physics, The University of Tennessee, Knoxville, Tennessee 37996.

†Present address: Theoretical and Physical Chemistry Institute, The National Hellenic Research Foundation, 48, Vassileos Constantinou Avenue, Athens 501/1, Greece.

decreases with T depending on the relative contribution of the dissociative and nondissociative processes. It appears that for both dissociative and nondissociative attachment the effect of T on σ_c or σ_c' is small except in those cases where electron capture by the hot molecule is accompanied by geometrical changes. Besides their intrinsic value, these results are of applied significance in many areas where the operating temperatures are higher than ambient and where the number density of electrons and negative ions crucially affects the performance of the device.

Introduction

Resonance electron attachment processes occur at low ($\lesssim$20 eV) energies and are generally discussed within the formalism of the resonance scattering theory and the formation of transient negative ions. Thus, resonance dissociative and nondissociative electron attachment to a molecule AX is viewed as occurring in two steps: (a) capture of the electron by AX to form the transient anion AX^{-*} and (b) the subsequent decay or stabilization of AX^{-*}; viz.,

$$e + AX \underset{\tau_a^{-1}}{\overset{\sigma_c}{\rightleftharpoons}} AX^{-*} \xrightarrow{p} A + X^- \text{ (dissociative)}, \quad (1)$$

$$e + AX \underset{\tau_a'^{-1}}{\overset{\sigma_c'}{\rightleftharpoons}} AX^{-*} \xrightarrow{p'} AX^- \text{ (nondissociative)}. \quad (2)$$

In reactions (1) and (2), σ_c and σ_c' are the respective electron capture cross-sections, p and p′ are the probabilities for AX^{-*} to decay by stable fragment [Eq. (1)] or parent [Eq. (2)] anion formation, and τ_a^{-1} and $\tau_a'^{-1}$ are the respective constants for AX^{-*} to decay by autodetachment. While many negative ion states (NISs) are usually involved in process (1), only one NIS (the lowest) is usually involved in process (2) [process (2) also requires that the electron affinity of AX is positive (>0 eV)]. In certain cases reactions (1) and (2) can proceed concomitantly and be in competition.

Processes (1) and (2) can be classified [1,2] according to the internal state of excitation of AX, viz.

$$e + AX(G;\ v \simeq 0) \xrightarrow{\sigma_c} AX^{-*}(G \text{ or } E) \rightrightarrows \text{decay}\ , \quad (3)$$

$$e + AX^*(G;\ v > 0;\ j > 0) \xrightarrow{\sigma_{c,v,j}} AX^{-*}(G \text{ or } E) \rightrightarrows \text{decay}\ , \quad (4)$$

$$e + AX^*(E) \xrightarrow{\sigma_{c,E}} AX^{-*}(E) \rightrightarrows \text{decay}\ . \quad (5)$$

In reaction (3), AX(G = 0 v = 0) is a molecule in its ground electronic state G and predominantly in its lowest (v = 0) vibrational state of excitation, and AX^{-*}(G or E) is the transient anion formed in either the field of the ground (G) or the field of an excited (E) electronic state with a capture cross-section σ_c. In reaction (4), AX*(G = 0, v > 0, j > 0) is a molecule in its ground electronic state, but in higher vibrational (v)/rotational (j) states, and AX^{-*} (G or E) is the respective transient anion formed with a cross-section $\sigma_{c,v,j}$. In reaction (5) the target molecule AX*(E) is electronically excited, and the electron is captured in the field of an excited electronic state producing AX^{-*}(E) with a cross section $\sigma_{c,E}$. Most studies to date concerned themselves with reaction (3). Swarm studies on reaction (5) are in progress at our laboratory. Electron attachment to "hot" molecules [reaction (4)] (the vibrationally/rotationally excited molecules can be formed by either laser excitation or by gas heating) have been reviewed [2].

In this paper we discuss reaction (4) with reference to published data and with reference to new results obtained at our laboratory on polyatomic halogenated compounds. Studies of the effects of temperature on the various electron attachment processes are of both intrinsic and of applied significance. With regard to the latter, in many applied areas the operating temperatures are higher than ambient, and the performance of the various devices is crucially affected by the number density of electrons and negative ions (such is the case, for example, in diffuse discharge switches) and thus by T. Our discussion of resonance electron attachment to hot molecules will be separated into three parts: (a) electron attachment to molecules where only dissociative attachment processes occur, (b) electron attachment to molecules where only nondissociative electron attachment takes place, and (c) electron attachment to molecules where both dissociative and nondissociative electron attachment processes occur over an energy range.

Effects of Temperature on Electron Attachment to Molecules Where Only Dissociative Attachment Occurs

Diatomic Molecules

The cross-section, σ_{da}, for (1) can be expressed as

$$\sigma_{da} = \sigma_c \, p \; . \tag{6}$$

In Eq. (6) the capture cross-section σ_c depends [2-4] on the autodetachment width Γ_a and the dissociation width Γ_d and varies inversely with the resonance energy ε_{max}; the probability p is usually expressed as [2-4]

$$p = e^{-\tau_s/\tau_a} \; , \tag{7}$$

where τ_s is the average separation time of A and X^-, and τ_a is the average lifetime of AX^{-*}. As T increases, higher-lying vibrational levels of AX are populated, for which the internuclear distances increase significantly (and hence the Franck-Condon region is broadened), and the magnitude of σ_{da} for molecules, AX^*_{vibr}, in such excited nuclear motion states increases significantly; also, the threshold energy is lowered, and the Γ_d is increased (e.g., see Refs. 2, 5). Such an enhancement in σ_{da}, however, can be small in cases where the dissociative attachment process is exoergic and the potential energy curve for the transient negative ion AX^{-*} crosses that of the neutral molecule close to the equilibrium separation.

The increase in σ_{da} with T results from an increase in both σ_c [as higher vibrational levels of AX are populated, progressively lower energy electrons, for which σ_c is larger [2,6], are captured (also the Franck-Condon factors change)] and p. However, the increase in σ_c is usually small [except perhaps in those cases (e.g., N_2O [7]) where geometrical changes concomitant with electron capture occur] compared with that in p; the latter dominates the T dependence of σ_{da} and results from a shortening of τ_s associated with the spatially more extended wavefunctions of AX^*_{vibr}. Theoretical calculations [2,5,8,9] have shown that the effect of rotational excitation on σ_{da} is usually small and that the effect of vibrational excitation substantially accounts for the observed increases in σ_{da} with T.

The aforementioned conclusions are based on experimental and theoretical results on diatomic molecules (O_2, H_2, D_2, HCℓ, DCℓ) [2,5,8-11]. Examples of these findings are shown in

Figs. 1 and 2. In Fig. 1 are plotted the calculated [8] values of σ_{da} close to the vertical onset for H^-/H_2 and D^-/D_2 for H_2 and D_2 in various vibrational levels v. In Fig. 2 the experimental (see figure caption) σ_{da} for $C\ell^-$ from $HC\ell$ and $DC\ell$ are shown for $HC\ell/DC\ell$ in the v = 0, 1, and 2 levels. The σ_{da} increases dramatically as the vibrational quantum number increases. For a given pair of isotopic molecules, the lower the vibrational energy $h\nu_x$ of a given mode x is, the larger is the effect of T on σ_{da} since at a fixed T higher levels v_x are populated for which τ_s is shorter (p larger). It should be realized, however, that unless T is very large or ε_{max} small, the increase of the *measured* σ_{da} with T is much smaller than indicated in Figs. 1 and 2 because σ_{da} is the Boltzmann-factor-weighted σ_{da} for all levels v_x, and only a small fraction (itself a function of the size of $h\nu_x$) of molecules are in higher vibrational levels.

It has recently been pointed out [10] that the data in Figs. 1 and 2 show that the isotope effects observed [2,6] in the σ_{da} for H_2/D_2 and $HC\ell/DC\ell$ (and for other molecules [2,6]) depend on T. As T increases, τ_s decreases and hence the isotope effects become less pronounced; for a given T, higher vibrational levels (for which p is larger) are populated in the heavier molecule ($h\nu_D < h\nu_H$) and thus the increase in σ_{da} with T is larger for the heavier than for the lighter analog (see insets in Figs. 1 and 2). Actually (see inset in Fig. 2), when $HC\ell/DC\ell$ have vibrational energy $\gtrsim 0.1$ eV this increase in σ_{da} overtakes the opposite effect (decrease) introduced by the larger reduced mass of $D\text{-}C\ell^-$ compared with $H\text{-}C\ell^-$, so that the ratio $[\sigma_{da}(\varepsilon_{max})]_{DC\ell}/[\sigma_{da}(\varepsilon_{max})]_{HC\ell}$ which for $HC\ell$ and $DC\ell$ in the v = 0 level is equal to 0.71 (Ref. 12) becomes >1.

The cross-section data for the various v levels of H_2 and D_2 in Fig. 1 have been used [10] to determine the contributions to the total $\sigma_{da}(T)$ from the various vibrational levels at a number of T; at each value of T the cross-section for a particular vibrational level v (see Fig. 1) was multiplied by the fractional population of that level. The resultant cross sections $\sigma_{da}(v)$ are shown in Fig. 3 along with the total $\sigma_{da}(T)$ [the sum of $\sigma_{da}(v)$ over all contributing v levels]. Although these results are approximate (the cross-sections in Fig. 1 for the various v are threshold values [8]), it is clear that as T increases the isotope effect decreases (see inset in Fig. 3).

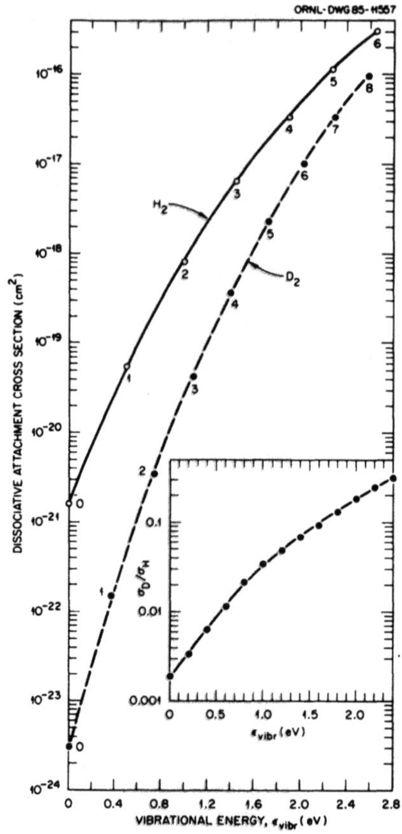

Figure 1. Calculated [8] σ_{da} for H^-/H_2 and D^-/D_2 with H_2/D_2 in various vibrational levels. The data are for energies close to the vertical onset. Inset: Ratio σ_D/σ_H of the σ_{da} for D^-/D_2 and H^-/H_2 as a function of the neutral molecule's vibrational energy (from Ref. 10).

Similar calculations [10] for HCℓ and DCℓ are more limited since for these molecules only cross-section data for the v = 0, 1, and 2 levels are available (see Fig. 2) and the effect of rotational excitation of $\sigma_{da}(T)$ may not be insignificant [9,11] as was the case for H_2 and D_2 [8,14]. Nevertheless, the contributions to the total dissociative attachment cross section $\sigma_{da}(T)$ from the v = 0, 1, and 2 vibrational levels in Fig. 4 show that the total $\sigma_{da}(T)$ of DCℓ exceeds that of HCℓ at $T \gtrsim 650$ K (see inset in Fig. 4), while it is only 70% that of HCℓ at $T \simeq 300$ K.

It is thus apparent [10] that the isotope effects observed in dissociative attachment depend on gas temperature; they are

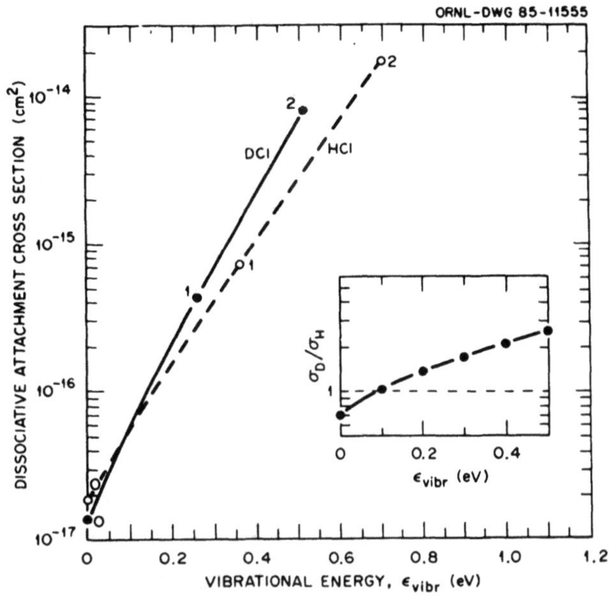

Figure 2. Experimental [11] σ_{da} for $C\ell^-/HC\ell$ and $C\ell^-/DC\ell$ for $HC\ell$ and $DC\ell$ in the v = 0, 1, and 2 levels. These cross sections were obtained from the values reported in Ref. 11 for the ratios $\sigma_{da}(v = 1,2)/\sigma_{da}(v = 0)$ for $HC\ell$ and $DC\ell$ [$\sigma_{da}(v = 1)/\sigma_{da}(v = 0)$ and $\sigma_{da}(v = 2)/\sigma_{da}(v = 0)$ were reported [11] to be, respectively, 38 and 880 for $HC\ell$ and 32 and 580 for $DC\ell$] and by normalizing the $\sigma_{da}(v = 0)$ relative cross-section of Ref. 11 to the cross-section measured [12] at T ≃ 300 K (the peak value of σ_{da} for $HC\ell$ and $DC\ell$ is, respectively, equal to 1.95 and 1.4 $\times 10^{-17}$ cm^2 at ∿0.8 eV [12]) (from Ref. 10).

the largest when the isotopic molecules are in their v = 0 levels. It is also apparent that while for diatomic molecules the ratio $\sigma_{da}(v > 0)/\sigma_{da}(v = 0)$ increases with increasing vibrational energy, for a given T the internal energy is a function of the magnitude of $h\nu_x$ and for polyatomic molecules also of the number of vibrational degrees of freedom N.

Polyatomic Molecules

Earlier work on the effect of T on σ_{da} of polyatomic molecules has been reviewed [2]. Recent work on freons, which are of interest as additives in multicomponent gas mixtures for use as gaseous dielectrics or in diffuse discharge switches, has been undertaken at our laboratory, and some of the results we obtained are presented and discussed in this and the following sections.

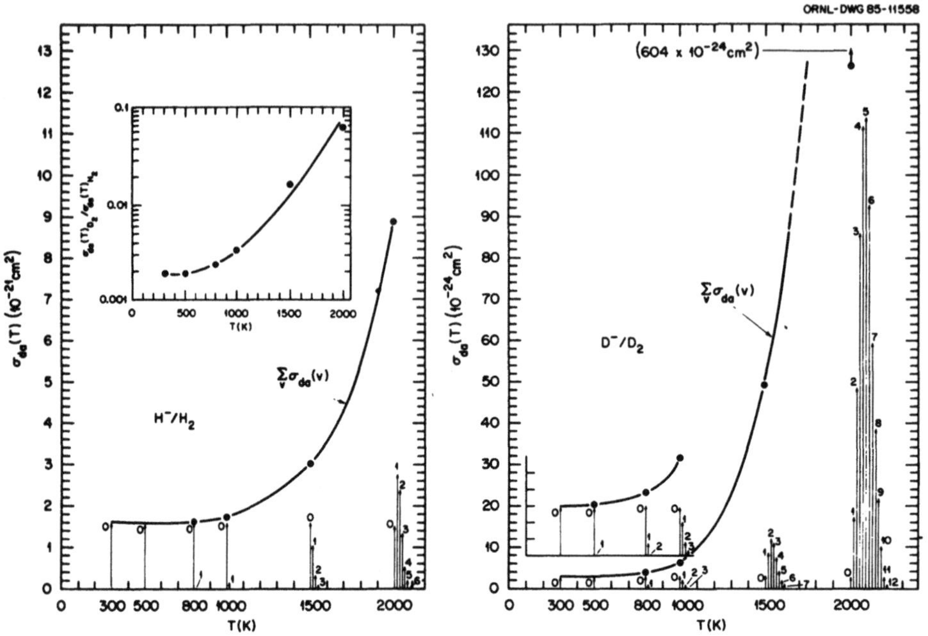

Figure 3. Dissociative attachment cross-section $\sigma_{da}(T)$ for H^-/H_2 and D^-/D_2 at various T determined as described in the text. For each T the length of the vertical arrows designated by 0,1,2...gives the contribution $\sigma_{da}(v)$ to the total $\sigma_{da}(T) = \sum_v \sigma_{da}(v)$ of molecules, respectively, in the v = 0,1,2...levels. The energy, E_v, of the v = 0,1,2...vibrational levels was determined using the formula $E_v = hc\omega_e(v + \frac{1}{2}) - hc\omega_e\chi_e(v + \frac{1}{2})^2$, where h is the Planck constant, c is the speed of light, and ω_e and $\omega_e\chi_e$ are the vibrational constants given in Ref. 13. As T increases, progressively larger contributions to $\sigma_{da}(T)$ come from molecules in higher vibrational quantum states. Inset: Ratio $\sigma_{da}(T)_{D_2}/\sigma_{da}(T)_{H_2}$ at various T (from Ref. 10).

$CC\ell F_3$. In Fig. 5 are given the measured [15] total electron attachment rate constants $k_a(\langle\varepsilon\rangle)$ as a function of the mean electron energy $\langle\varepsilon\rangle$ for $300 \leq T \leq 700$ K. As T increases, k_a increases, especially at low $\langle\varepsilon\rangle$. In Fig. 6 are shown the total electron attachment cross-sections $\sigma_a(\varepsilon)$ obtained [15] at each value of T from the respective $k_a(\langle\varepsilon\rangle,T)$ in Fig. 5 via the swarm unfolding technique [16]. The peak at ~1.5 eV is especially sensitive to changes in T. The peak value of $\sigma_a(\varepsilon)$ is increased by a factor of 3, and the energy position, ε_{max}, of the peak and the energy onset, AO, shift progressively to lower energy as T increases. Electron beam studies (inset,

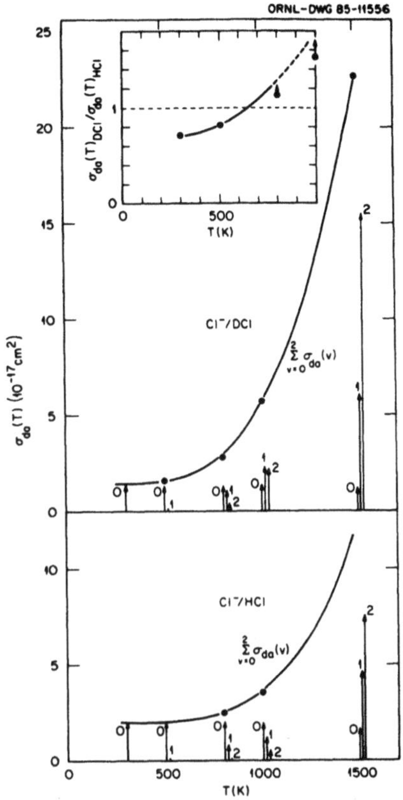

Figure 4. Dissociative attachment cross-section $\sigma_{da}(T)$ for $C\ell^-/HC\ell$ and $C\ell^-/DC\ell$ at various T determined as described in the text. For each T the length of the vertical arrows designated by 0, 1, and 2 gives the contribution $\sigma_{da}(v)$ to the total $\sigma_{da}(T)$ of molecules, respectively, in the v = 0, 1, and 2 vibrational levels (the energy of each vibrational level was determined as described in the caption of Fig. 3). The sum, $\sum_{v=0}^{2} \sigma_{da}(v)$, of the $\sigma_{da}(v)$ for the v = 0, 1, and 2 levels is also given in the figure. Since for T $\gtrsim$ 1000 K the contributions to $\sigma_{da}(T)$ of molecules in v > $\tilde{2}$ is substantial, the values of $\sigma_{da}(T)$ [$\equiv \sum_{v=0}^{2} \sigma_{da}(v)$] in the figure for 1000 and 1500 K are grossly underestimated. As a consequence of this, the values of the ratio $[\sigma_{da}(T)]_{DC\ell}/[\sigma_{da}(T)]_{HC\ell}$ for 1000 and 1500 K (see inset) are lower than their true values (this is indicated in the inset by the data points ↑) (from Ref. 10).

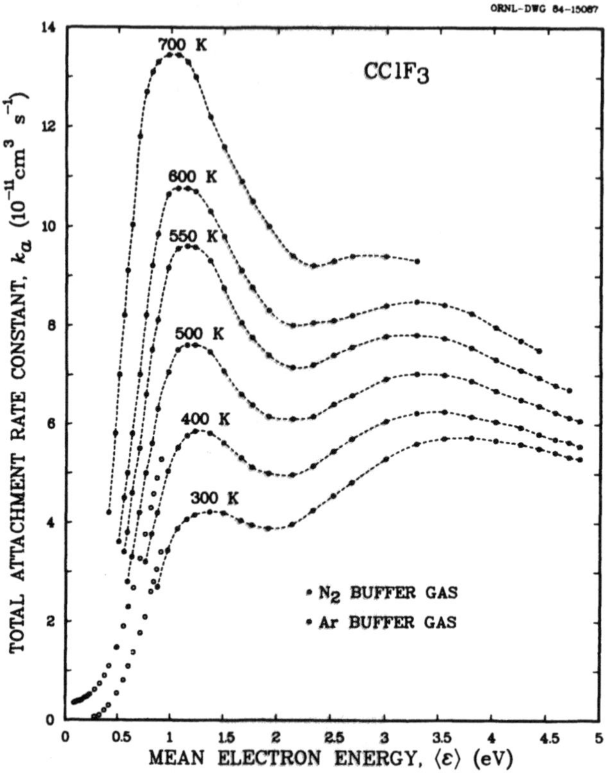

Figure 5. Total electron attachment cross-section versus electron energy for $CC\ell F_3$ measured [15] in the buffer gases N_2 or Ar at various temperatures; the $k_a(\langle\varepsilon\rangle)$ were independent of gas number density.

Fig. 6) have shown [15] that at low gas pressures $CC\ell F_3$ captures electrons exclusively via dissociative attachment and that the peaks at ∿1.5 and ∿4.7 eV are the former due to $C\ell^-$ and the latter due to $C\ell^-$, F^-, $CC\ell F_2^-$, and $C\ell F^-$ ions.

C_2F_6. In Fig. 7 the $k_a(\langle\varepsilon\rangle,T)$ are given along with the relative abundance of the fragment negative ions observed (inset, Fig. 7) in a beam study [17]. No parent negative ions were observed in the low-pressure beam study, and this is consistent with the absence of any pressure dependence of $k_a(\langle\varepsilon\rangle,T)$ in the swarm study. In Fig. 8 are plotted the swarm unfolded cross-sections which show a single peak due to F^- and CF_3^- (see inset of Fig. 7). The decrease in ε_{max} and AO and the increase in FWHM (full width at half maximum) of $\sigma_a(\varepsilon)$ with T are shown in the inset of Fig. 8.

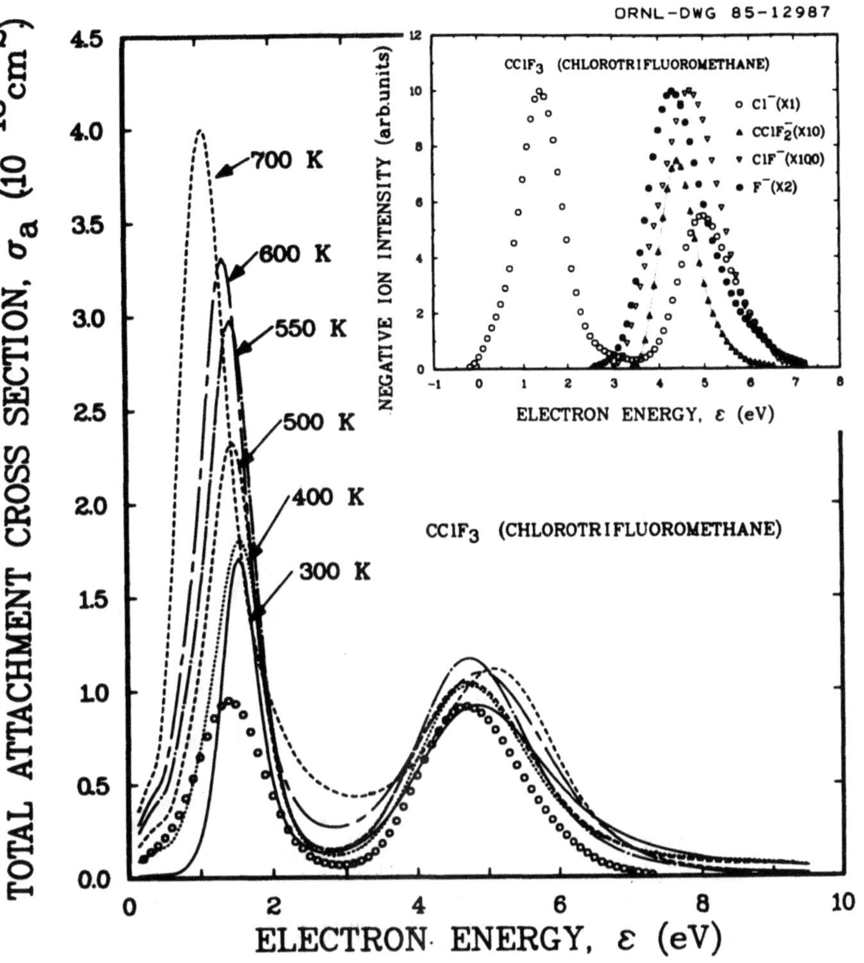

Figure 6. Total electron attachment cross-section versus electron energy for $CC\ell F_3$ unfolded from the $k_a(\langle\varepsilon\rangle,T)$ data in Fig. 5 at various T. The curve designated by the open circles (o) is the electron beam total attachment cross-section normalized to the high energy peak of the swarm unfolded cross section for 300 K. Inset: Relative intensity of the dissociative attachment negative ions produced by low-energy electron impact on $CC\ell F_3$ as a function of electron energy measured in a beam study (these spectra were corrected for the finite width of the electron pulse) [15].

In addition to the $k_a(\langle\varepsilon\rangle),T)$ we measured in mixtures with Ar, we also measured the electron attachment, $\eta/N_a(E/N)$, and ionization, $\alpha/N_a(E/N)$, coefficients in pure C_2F_6 at 300 and

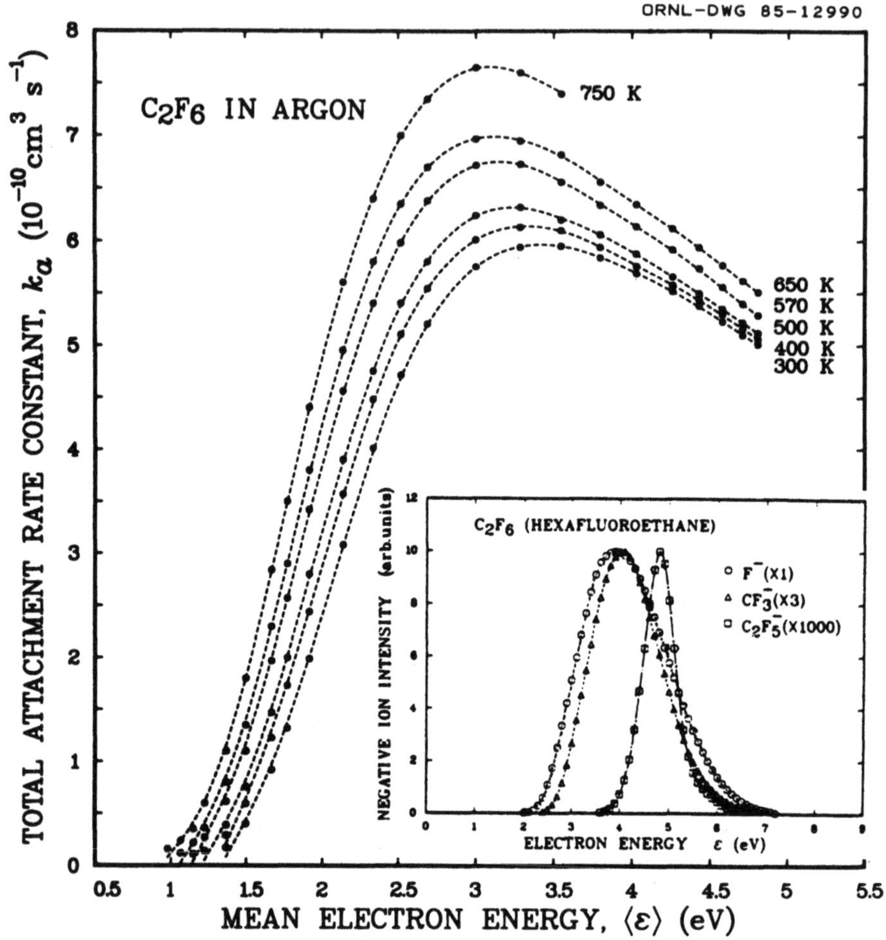

Figure 7. Total electron attachment rate constant as a function of mean electron energy for C_2F_6 in Ar buffer gas at various temperatures. The $k_a(\langle\varepsilon\rangle,T)$ were independent of gas number density [15]. Inset: Relative intensity of the dissociative attachment fragment anions measured [17] in a low-pressure beam study.

500 K. These measurements are shown in Fig. 9. The η/N_a data are consistent with those obtained in mixtures of C_2F_6 with Ar (Fig. 7).

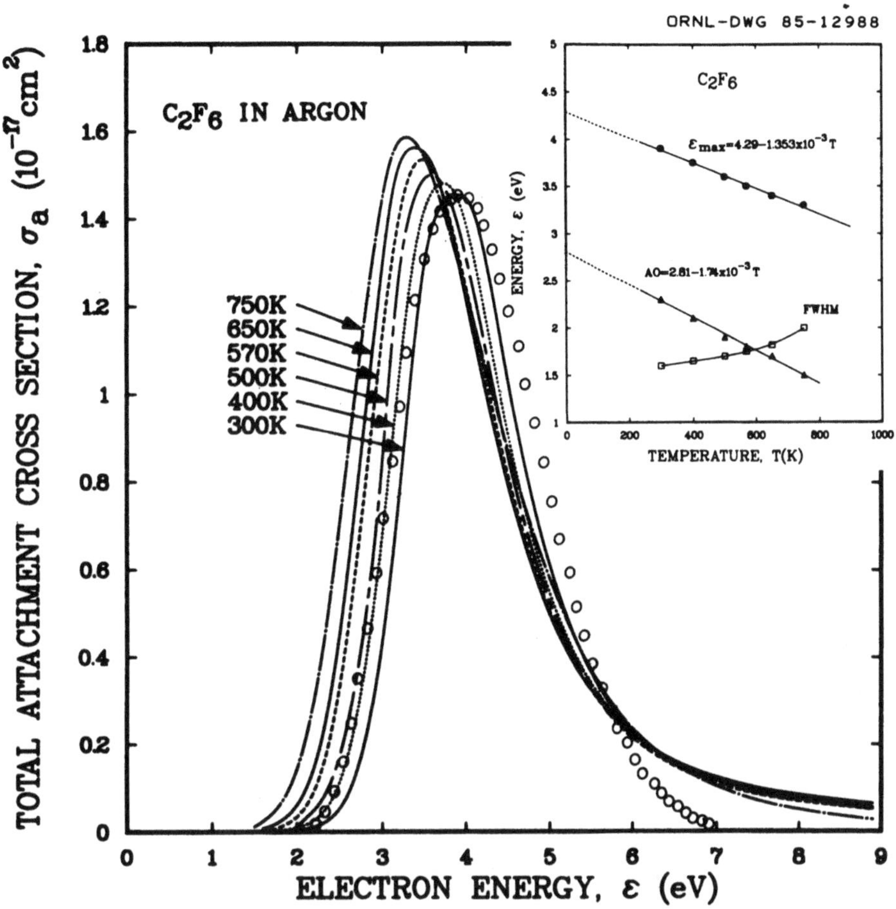

Figure 8. Total electron attachment cross-section versus electron energy unfolded from the $k_a(\langle\varepsilon\rangle,T)$ data in Fig. 7 at the indicated temperatures. The open circles (o) are the total electron beam cross-section normalized to the peak of the swarm unfolded cross-section for 300 K. Inset: Variation of the cross-section peak position (ε_{max}), cross-section onset energy (AO), and cross-section full width at half maximum (FWHM) with temperature (from Ref. 15).

Variation of the Energy-Integrated Attachment Cross-Section with Temperature and with the Molecule's Internal Energy

We have determined the energy-integrated attachment cross section

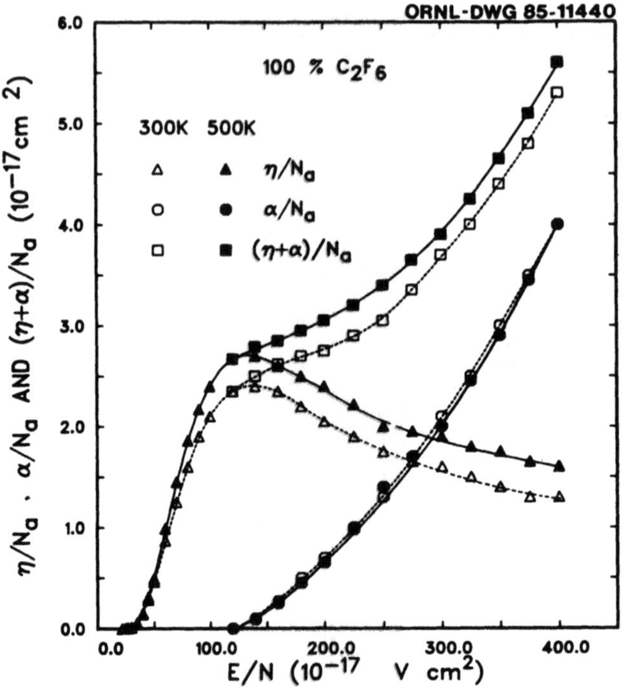

Figure 9. Electron attachment, η/N_a, and ionization, α/N_a, coefficients and their sum, $(\eta + \alpha)/N_a$, as a function of E/N at 300 and 500 K for pure C_2F_6.

$$\int_{\varepsilon_{min}}^{\varepsilon_{max}} \sigma_{da}(\varepsilon,T)\, d\varepsilon \equiv \sigma_{EIA}(T) \tag{8}$$

from the respective $\sigma_{da}(\varepsilon,T)$ measured for O^-/O_2 [18], $C\ell^-/HC\ell$ [11,19], $C\ell^-/CC\ell F_3$ [15], O^-/N_2O [7,20], SF_5^-/SF_6 [22,23], C_2F_6 [15] (all ions), and C_3F_8 [26] (all ions). These are plotted in Fig. 10a. The σ_{EIA} increases with T; this increase varies from molecule to molecule but not, however, in the simple fashion (i.e., the lower the σ_{EIA} at T = 300 K the faster its increase with T) stated earlier [27].

The fast increase of σ_{EIA} with T for SF_5^-/SF_6 and O^-/N_2O is most interesting. For SF_6 this may be due to the larger increase in p with increasing T probably because almost all 15 vibrational frequencies of SF_6 are small [774 cm^{-1} (singly degenerate); 642 cm^{-1} (doubly degenerate); 948, 616, 525, and 347 cm^{-1} (all triply degenerate)] [28] and hence high-lying levels of each mode are populated at relatively low T; also, it should be noted that the ε_{max} for SF_5^-/SF_6 is low (~0.37 eV

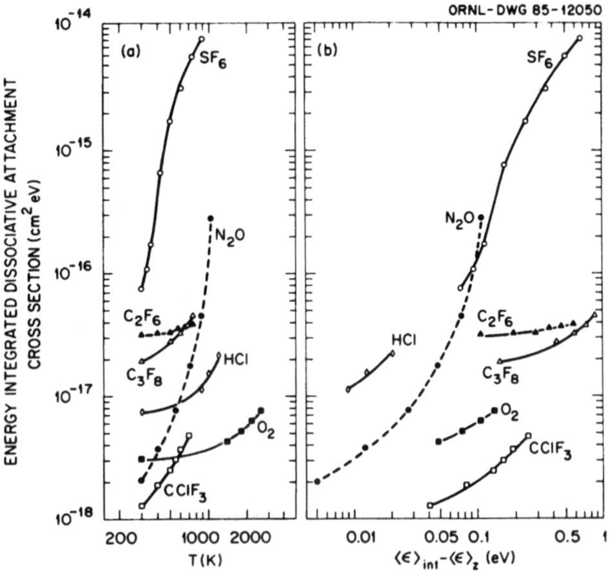

Figure 10. Energy-integrated dissociative attachment cross-section as a function of temperature (Fig. 10a) and excess internal energy $\langle\varepsilon\rangle_{int}-\langle\varepsilon\rangle_z$ (Fig. 10b) for O_2, HCℓ, N_2O, SF_6, $CC\ell F_3$, C_2F_6, and C_3F_8 (see the text).

[24]) and that all six SF_5-F coordinates lead to SF_5^-. For N_2O the large increase of σ_{EIA} with T may result from the fact that as T increases the hot N_2O^* molecule (in the bending mode) [7,29] better facilitates upon electron collision the geometrical changes (from a straight N_2O^* to a bend N_2O^{-*}) which are known to occur concomitantly with electron capture. It has been suggested [7] that the increased excitation in the bending mode of N_2O results in a lowering of the position of the NIS which leads to O^- formation; this would increase greatly the magnitude of σ_c and, thus, σ_{da}.

At any T there is a Boltzmann distribution B_v of the population of the vibrational levels v of each vibrational mode x. For a molecule with N normal modes, the vibrational energies of the normal mode x in the v = 0,1,2,....levels (if we neglect anharmonicity) are $\varepsilon_{v,x} = (v + \frac{1}{2})h\nu_x$ and for each x

$$B_v = e^{-\varepsilon_v/kT} / \sum_{v=0}^{\infty} e^{-\varepsilon_v/kT} \qquad (9)$$

If we neglect the effect of rotational excitation and consider only the effect of vibrational excitation to be significant, then for a diatomic molecule the cross-section $\sigma_{da}(\varepsilon,T)$ can be expressed as

$$\sigma_{da}(\varepsilon,T) = \sum_{v=0}^{\infty} B_v \sigma_{da}^{v} (\varepsilon,T) \ . \tag{10}$$

For a polyatomic molecule the summation in (10) must be carried out for all x. However, even if this were possible the x are not independent.

Let us then assume that as T increases, each vibrational mode x of a polyatomic molecule is excited by an equal probability, and that the total average internal energy $\langle\varepsilon\rangle_{int}$ of the molecule is principally the sum of the energy in the various normal modes, x, viz.

$$\langle\varepsilon\rangle_{int} = \sum_{x=1}^{N} \sum_{v=0}^{\infty} D_x B_v \varepsilon_{v,x} \ , \tag{11}$$

where D_x is the degeneracy of the mode x. If we take $\langle\varepsilon\rangle_{int}$ to be the molecule's *total* internal energy, then the molecule's excess energy would be $\langle\varepsilon\rangle_{int}-\langle\varepsilon\rangle_z$, where $\langle\varepsilon\rangle_z$ $(\equiv \sum_{x=1}^{N} \frac{1}{2} h\nu_x)$ is the zero-point energy. If now, $\langle\varepsilon\rangle_{int}-\langle\varepsilon\rangle_z$ is distributed quickly among the molecule's N vibrational degrees of freedom and can thus become available for the dissociative attachment reaction, one might expect a relationship between σ_{EIA} and $\langle\varepsilon\rangle_{int}-\langle\varepsilon\rangle_z$. Indeed, σ_{EIA} increases with $\langle\varepsilon\rangle_{int}-\langle\varepsilon\rangle_z$ (see Fig. 10b), although this increase differs--as expected--from one molecule to another. Actually, a better comparison might have been a plot of σ_{EIA} versus $(\langle\varepsilon\rangle_{int}-\langle\varepsilon\rangle_z)/\varepsilon_{max}$. This would shift the C_2F_6, C_3F_8, O_2, and $CC\ell F_3$ curves to lower energies compared with SF_6 for which ε_{max} = 0.37 eV. The fact that the curves in Fig. 10b for SF_6 and N_2O mesh reasonably well, although ε_{max} for SF_6 is 0.37 eV and for N_2O it is 2.25 eV is consistent with the arguments presented earlier in this section that the ε_{max} for electron attachment to N_2O^* (bending mode) is lower than the ε_{max} for electron attachment to unexcited N_2O.

While further experimental and theoretical work is necessary (especially on polyatomic molecules) to fully understand the effect of temperature on $\sigma_{da}(\varepsilon)$ and $\sigma_{EIA}(T)$, it is clear that for both diatomic and polyatomic molecules the changes in $k_{da}(\langle\varepsilon\rangle)$, $\sigma_{da}(\varepsilon)$, and $\sigma_{EIA}(T)$ with T result principally from an increase with T of the internal energy ($\simeq$ vibrational) of the molecule.

Effect of Temperature on Nondissociative Electron Attachment

The cross-section, σ_{nd}, for nondissociative electron attachment--as that, σ_{da}, for dissociative varies profoundly with the gas temperature. Based on the data outlined in this section, this dependence arises from an effect of T on both σ_c' and p'. However, while for dissociative attachment σ_{da} generally *increases* with T, for nondissociative attachment σ_{nd} generally *decreases* with T. Furthermore, while in dissociative attachment the increase in σ_{da} with T is predominantly due to a decrease in τ_s (and thus increase in p), the decrease in σ_{nd} for nondissociative attachment is due to a decrease in τ_a (and thus decrease in p) and σ_c'. These conclusions are based on the following results.

SF_6

The cross-section for the formation of SF_5^- from SF_6 at ~0.0 eV has been found to increase dramatically with increasing T (Fig. 10; Refs. 2, 22). However, a number of studies [2] have shown that the total attachment cross-section or rate constant for SF_6 is independent of T to ~1200 K. This implies that the formation of SF_6^- (whose σ_{nd} peaks at ~0.0 eV [2]) decreases with increasing T. Direct evidence for this is provided by the early work of Hickam and Berg [30]. It is presently not possible to which quantity, σ_c' or p', to ascribe this decrease in σ_{nd} with T, although p is expected to decrease with increasing T because τ_a is expected to decrease as the internal energy of SF_6^{-*} increases [2,6].

$1\text{-}C_3F_6$

A large decrease of the attachment rate constant for nondissociative electron attachment to perfluoropropylene ($1\text{-}C_3F_6$) with increasing T has been observed (Fig. 11; Refs. 2, 31, 32). This has been attributed [31,32] to a decrease in the τ_a of $1\text{-}C_3F_6^{-*}$ with increasing T.

C_6F_6

A profound decrease in the rate constant for electron attachment to perfluorobenzene (C_6F_6) with T has been reported for C_6F_6 (Fig. 12; Ref. 33). At T = 300 K, C_6F_6 forms parent $C_6F_6^-$ ions by capturing near-zero energy electrons [2]; the τ_a of $C_6F_6^{-*}$ was found to be ~10 μs [2]. Spyrou and Christophorou [33] concluded that the decrease in $k_a(\langle\varepsilon\rangle)$ with T (Fig. 12) cannot be attributed to a decrease in τ_a (decrease in p) with T [the $k_a(\langle\varepsilon\rangle)$ did not depend on the gas number

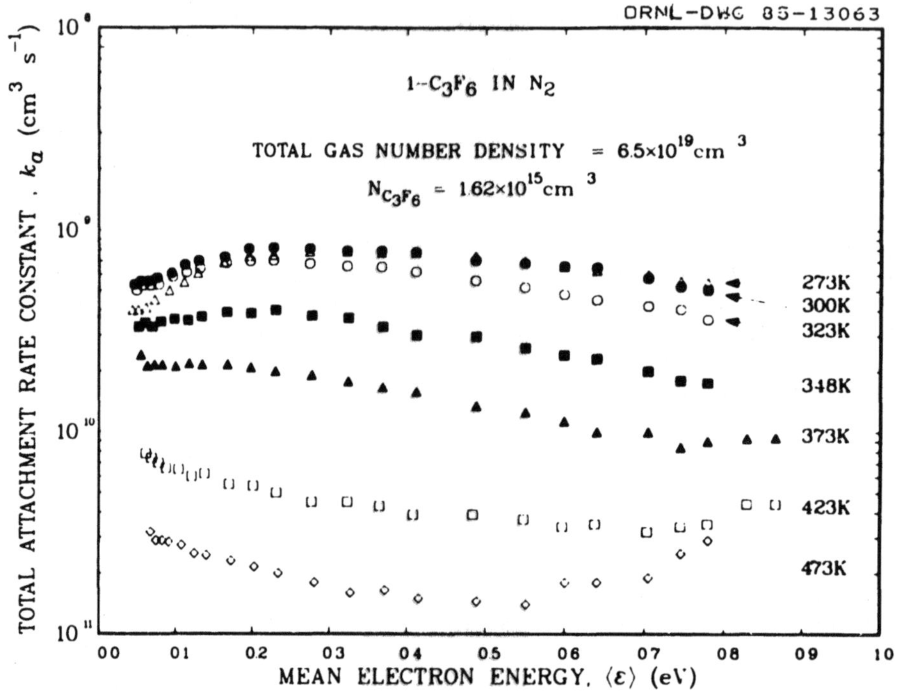

Figure 11. Total attachment rate constant versus mean electron energy for 1-C_3F_6 in N_2 at 273 ≤ T ≤ 473 K (from Ref. 32).

density at any T] or other by-products resulting from gas heating. They attributed it to a decrease in σ_c' and suggested that the increase in the internal energy of C_6F_6 affects rather profoundly the rate for the capture transition (i.e., to differences in the magnitude of σ_c' for the reactions e + C_6F_6 → $C_6F_6^{-*}$ and e + $C_6F_6^*$ → $C_6F_6^{-*}$).

While much improvement in our understanding of the effects of the internal energy of a molecule on its electron attachment properties is still desirable, it is clear that as a rule σ_{da} increases and σ_{nd} decreases with increasing internal energy, that is, increasing T. It is also apparent that for both dissociative and nondissociative electron attachment p is the determining factor unless geometrical changes concomitant with electron capture effect changes in σ_c or σ_c'.

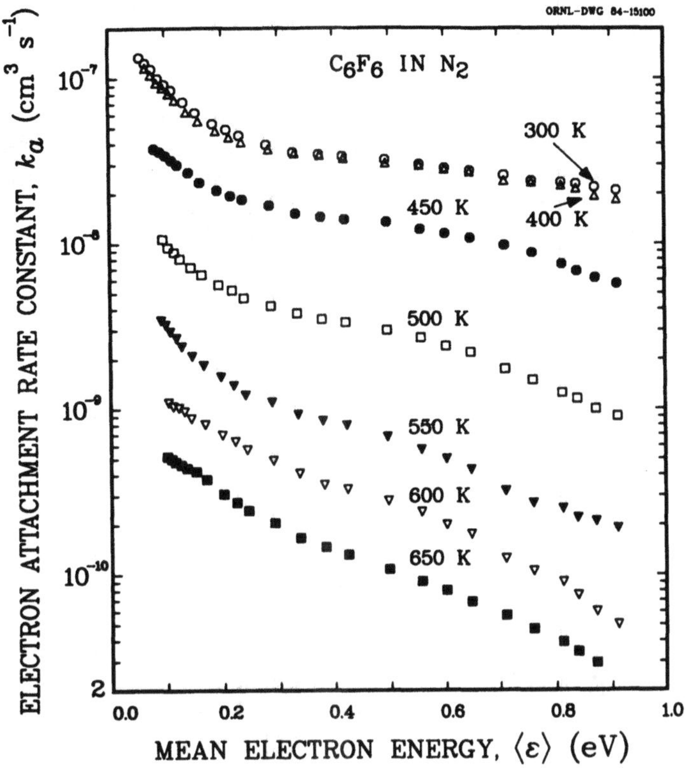

Figure 12. $k_a(<\varepsilon>,T)$ for C_6F_6 measured in a buffer gas of N_2. The C_6F_6 gas number density varied from 0.41 to 46.3 × 10^{13} cm^{-3} and that of N_2 from 2.25 to 6.44 × 10^{19} cm^{-3} (from Ref. 33).

Effect of Temperature on the Measured Attachment Rate Constant and Cross-Section for Molecules for Which Both Dissociative and Nondissociative Electron Attachment Occur Over an Energy Range

Recently, we measured [26] the total electron attachment rate constant $k_a(<\varepsilon>)$ for C_3F_8 in Ar in the temperature range from 300 to 750 K. At T $\lesssim$ 425 K the $k_a(<\varepsilon>)$ were found to increase with increasing total gas number density N_t over the entire $<\varepsilon>$ range (∿0.5 to ∿5 eV) covered in these experiments. At 450 K, the $k_a(<\varepsilon>)$ increased with N_t only for $<\varepsilon> \lesssim 1.2$ eV and at T > 450 K the $k_a(<\varepsilon>)$ were independent of N_t. The $k_a(<\varepsilon>)$ also showed a weak dependence on the attaching gas number density N_a due to the effect of the presence of the attaching gas on the distribution functions of pure Ar used in the analysis; this effect was taken into account by measuring, for a fixed N_t, the $k_a(<\varepsilon>)$ as a function of N_a and extrapolating at each $<\varepsilon>$ the $k_a(N_a)$ to $N_a \rightarrow 0$.

In Fig. 13 are plotted the values, $k_1(<\varepsilon>)$, of $k_a(<\varepsilon>)$ for $N_a \rightarrow 0$ and $N_t \rightarrow 0$ for all values of T that data were taken, and in Fig. 14 k_1 is plotted as a function of T for two values of $<\varepsilon>$. It is evident from these data that $k_1(<\varepsilon>)$ decreases to a minimum around 450 to 500 K and that it then increases as T increases.

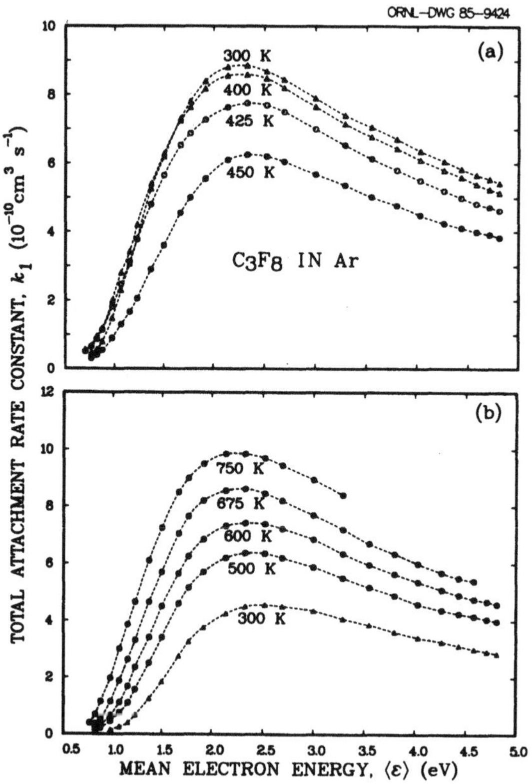

Figure 13. Electron attachment rate constant k_1 ($N_a \rightarrow 0$; $N_t \rightarrow \infty$) for C_3F_8 measured as a function of mean electron energy $<\varepsilon>$ in a buffer gas of Ar at (a) 300, 400, 425, and 450 K and (b) 500, 600, 675, and 750 K. The 300 K curve in Fig. 13b is the dissociative attachment contribution to the measured $k_a(<\varepsilon>)$ at this temperature (see the text and Ref. 26).

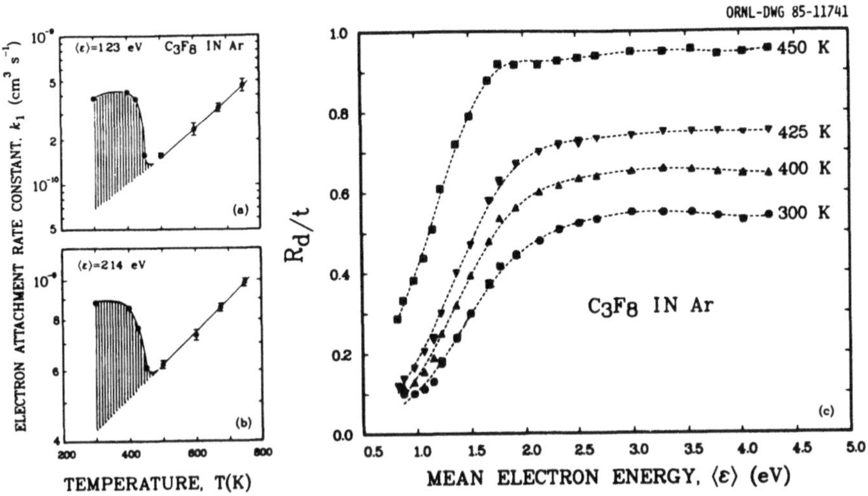

Figure 14. (a) and (b) Electron attachment rate constant k_1 ($N_a \to 0$; $N_t \to \infty$) for C_3F_8 versus temperature T at the mean electron energies <ε> = 1.23 and 2.14 eV. (c) Ratio $R_{d/t}$ of the attachment rate constant due to dissociative attachment to the total attachment rate constant (see the text) versus the mean electron energy at 300, 400, 425, and 450 K obtained from extrapolation of the k_a(<ε>) measured at T ≥ 500 K to lower T (see the text).

The delicate dependence of k_a(<ε>) on T can be understood by considering the results of electron beam and electron swarm studies. Single collision beam experiments on C_3F_8 indicated the presence of only dissociative attachment anions and established their identity and energy dependence; they also showed the existence of a number of NISs which lead to dissociative attachment [17]. On the other hand, the results of high-pressure swarm experiments on C_3F_8 determined the magnitude of the total attachment rate constant and cross section as a function of electron energy and their total pressure dependence [26,34]; they indicated that in addition to the NISs which lead to dissociative attachment (observed in single collision beam experiments) there exists another, lower-lying NIS which is attractive and which leads to the formation of parent negative ions with $\tau_a < 10^{-6}$ s [26,34]. These findings and the observed effects of T on k_a(<ε>) (Figs. 13 and 14) and $\sigma_{da}(\varepsilon)$ (Fig. 15) have been ascribed to electron attachment via an attractive NIS (with a positive electron affinity and a steep repulsive part) leading to parent anions and to one (or more) repulsive NISs leading to fragment anions.

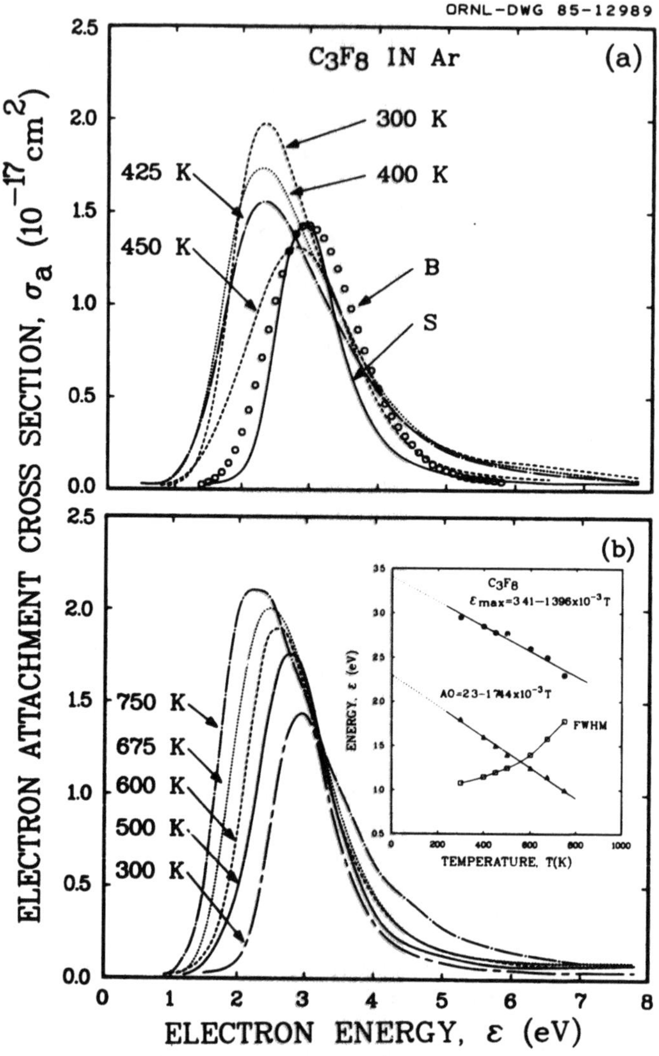

Figure 15. Swarm unfolded *total* electron attachment cross section $\sigma_a(\varepsilon)$ for C_3F_8 obtained [26] from the $k_1(\langle\varepsilon\rangle)$ in Ar shown in Fig. 13 at (a) 300, 400, 425, and 450 K and (b) 300, 500, 600, 675, and 750 K. In Fig. 15a is plotted also the *dissociative* attachment cross-section $\sigma_{da}(\varepsilon)$ obtained from the swarm data at 300 K (curve S) and the dissociative cross section measured in an electron beam experiment (curve B) which has been normalized to the peak of $\sigma_{da}(\varepsilon)$ (see the text). In Fig. 15b the curve for 300 K is the curve S of Fig. 15a. Inset: Variation of peak energy (ε_{max}), appearance onset (AO), and full width at half maximum (FWHM) of the total dissociative attachment cross – section of C_3F_8 with temperature.

The delicate dependence of $k_a(\langle\varepsilon\rangle)$ on T (Figs. 13 and 14) can thus be considered as the result of two opposite effects of T: one on the rate constant for nondissociative and the other on the rate constant for dissociative electron attachment. As it has been shown in the previous section, as a rule, the rate constant for pure nondissociative attachment processes decreases and that for pure dissociative attachment processes increases with increasing T. At each $\langle\varepsilon\rangle$, the magnitude of k_a is determined by the relative magnitudes of the rate constants for nondissociative and dissociative electron attachment, both of which depend on T. From the data in Fig. 14, it is apparent that for $T \gtrsim 500$ K the principal contribution to the measured k_a originates from dissociative attachment; this is supported by the lack of any dependence of k_a on N_t at high T and from the observed increases in k_a with T^a (500 to 750 K) which are characteristic of molecules which attach electrons dissociatively. We then assumed [26] that for $T \gtrsim 500$ K the measured k_a is due entirely to dissociative attachment and extrapolated (at various values of $\langle\varepsilon\rangle$) the measured k_a at $T \gtrsim 500$ K to lower T (see Figs. 14a,b) in an effort to estimate the dissociative attachment contribution to the measured k_a at $T \lesssim 500$ K, where nondissociative attachment takes place and becomes progressively more significant with decreasing T. From plots such as those in Figs. 14a,b we estimated [26] the ratio $R_{d/t}(\langle\varepsilon\rangle)$ of the dissociative to the total attachment rate constant as a function of $\langle\varepsilon\rangle$ for 300, 400, 425, and 450 K. These estimates are given in Fig. 14c and show that the contribution of dissociative attachment processes to the measured rate constant is both a function of $\langle\varepsilon\rangle$ and T.

The total electron attachment rate constants $k_1(\langle\varepsilon\rangle)$ (Fig. 13) were unfolded [26] and the total attachment cross sections $\sigma_a(\varepsilon,T)$ obtained are shown in Fig. 15. They decrease in magnitude with increasing T from 300 to ∿450 K (Fig. 15a) because in this T range the total cross-section contains a large contribution (which decreases as T increases) from *nondissociative* attachment. An increase in T beyond ∿450 K (Fig. 15b) results in an overall increase in the magnitude and full width at half maximum--and a shift to lower energy of the onset and energy of the peak (see inset of Fig. 15b)--resulting from the increasingly larger contribution of the *dissociative* attachment component to the total cross-section.

In Fig. 15a are also compared the cross-sections due to only the dissociative attachment contribution to the total cross-section at 300 K (curve S) and the total (for all fragment anions) dissociative attachment cross-section for C_3F_8 measured in a single-collision electron beam study [17];

the latter was normalized to the peak value of the former. It is seen that the peak positions of the two cross-section functions agree well, and that both lie at a higher energy than the total unfolded cross-section $\sigma_a(\varepsilon, 300\ K)$.

Conclusions

While much improvement in our understanding of the effects of internal energy of a molecule on its electron attaching properties is still desirable, it is clear that as a rule σ_{da} increases and σ_{nd} decreases with increasing T. It is also apparent from the data obtained to date that for both dissociative and nondissociative electron attachment the survival probability is the determining factor (shortening of τ_s in dissociative and shortening of τ_a in nondissociative electron attachment with increasing T) unless geometrical changes concomitant with electron capture effect changes in $\sigma_c(\sigma_c')$.

From the practical point of view, both the increases and the decreases in $k_a(\langle\varepsilon\rangle)$ with T are significant because they affect the conductivity/dielectric strength properties of the gaseous medium. The sensitivity of $k_a(\langle\varepsilon\rangle)$ to changes in T requires that proper attention be given to the operating temperature range of a given device. Interestingly, the sensitivity of $k_a(\langle\varepsilon\rangle)$ to T (e.g., C_6F_6; see Fig. 12) can perhaps be employed to change the conducting/insulating properties of a gaseous medium by varying T.

Acknowledgments

Research sponsored in part by the Office of Health and Environmental Research, U.S. Department of Energy, under contract DE-AC05-84OR21400 and in part by the Office of Naval Research under interagency agreement 43 01 24 60 2 with Martin Marietta Energy Systems, Inc.

References

[1] Christophorou, L. G. *Environ. Health Perspect.* 1980, 36, 3.

[2] Christophorou, L. G.; McCorkle, D. L.; Christodoulides, A. A. In "Electron-Molecule Interactions and Their Applications"; Christophorou, L. G., Ed.; Academic Press: New York, 1984; Volume 1, Chapter 6.

[3] O'Malley, T. F. Phys. Rev. 1966, 150, 14.
[4] Bardsley, J. N.; Herzenberg, A.; Mandl, F. Proc. Phys. Soc. (London) 1966, 89, 321.
[5] O'Malley, T. F. Phys. Rev. 1967, 155, 59.
[6] Christophorou, L. G. "Atomic and Molecular Radiation Physics"; Wiley-Interscience: New York, 1971; Chapter 6.
[7] Chantry, P. J. J. Chem. Phys. 1969, 51, 3369.
[8] Bardsley, J. N.; Wadehra, J. M. Phys. Rev. 1979, 20, 1398.
[9] Bardsley, J. N.; Wadehra, J. M. J. Chem. Phys. 1983, 78, 7227.
[10] Christophorou, L. G. J. Chem. Phys. (submitted).
[11] Allan, M; Wong, S. F. J. Chem. Phys. 1981, 74, 1687.
[12] Christophorou, L. G.; Compton, R. N.; Dickson, H. W. J. Chem. Phys. 1968, 48, 1949.
[13] Huber, K. P.; Herzberg, G. "Molecular Spectra and Molecular Structure. IV. Constants of Diatomic Molecules", Van Nostrand Reinhold Company: New York, 1979.
[14] Allan, M.; Wong, S. F. Phys. Rev. Lett. 1978, 41, 1791.
[15] Spyrou, S. M.; Christophorou, L. G. J. Chem. Phys. 1985, 82, 2620.
[16] Christophorou, L. G.; McCorkle, D. L.; Anderson, V. E. J. Phys. B 1971, 4, 1163.
[17] Spyrou, S. M.; Sauers, I.; Christophorou, L. G. J. Chem. Phys. 1983, 78, 7200.
[18] For O_2 we used the calculated values of O'Malley [5] which fitted well the experimental results.
[19] These cross-sections were obtained using the relative cross-section data of Ref. 11 at various T and by normalizing these to the room temperature cross-section values of Ref. 12.
[20] These cross-sections were obtained using the relative cross-section data of Ref. 7 at various T and normalizing the room-temperature intensity at 2.3 eV to 8.3×10^{-18} cm^2 [21].
[21] Chaney, E. L.; Christophorou, L. G. J. Chem. Phys. 1969, 51, 883.
[22] Chen, C. L.; Chantry, P. J. J. Chem. Phys. 1979, 71, 3897.
[23] The cross-sections at various T were obtained using the relative cross-section of Ref. 22 and taking for the cross-section maximum for SF_5^-/SF_6 at 0.37 eV [24] the value [25] of 9.8×10^{-16} cm^2. Since in the time-of-flight studies of Ref. 24 the temperature was higher (~350 K) than ambient (due to heating of the collision chamber by the filament), it was assumed that the 9.8×10^{-16} cm^2 value corresponds to $T \simeq 355$ K.

[24] Christophorou, L. G.; McCorkle, D. L.; Carter, J. G. J. Chem. Phys. 1971, 54, 253.
[25] Christophorou, L. G.; McCorkle, D. L.; Carter, J. G. J. Chem. Phys. 1972, 57, 2228.
[26] Spyrou, S. M.; Christophorou, L. G. J. Chem. Phys. (in press).
[27] Spence, D; Schulz, G. J. J. Chem. Phys. 1973, 58, 1800.
[28] Shimanouchi, T. NSRDS-NBS 39, June 1972.
[29] N_2O has three normal modes: NN stretch (2224 cm^{-1}), bend (589 cm^{-1}), and NO stretch (1285 cm^{-1}) [28].
[30] Hickam, W. M.; Berg, D. J. Chem. Phys. 1958, 29, 517.
[31] Hunter, S. R.; Christophorou, L. G.; McCorkle, D. L.; Sauers, I; Ellis, H. W.; James, D. R. J. Phys. D 1980, 16, 573.
[32] McCorkle, D. L.; Christophorou, L. G.; Hunter, S. R. In "Proceedings of the Third International Swarm Seminar"; Lindinger, W.; Villinger, H.; Federer, W., Eds.; Innsbruck, Austria, 1983; p. 37.
[33] Spyrou, S. M.; Christophorou, L. G. J. Chem. Phys. 1985, 82, 1048.
[34] Hunter, S. R.; Christophorou, L. G. J. Chem. Phys. 1984, 80, 6150.

ELECTRON IMPACT IONIZATION CROSS-SECTIONS FOR ATOMS, RADICALS, AND METASTABLES

Robert S. Freund

AT&T Bell Laboratories,
Murray Hill, New Jersey 07974

Many areas of science require accurate values of electron impact ionization cross-sections. These areas include modeling of electrical discharges under various conditions of pressure, power, and frequency, such as for semiconductor processing, gas lasers, fluorescent lighting, the upper atmosphere, the aurora, and the interstellar medium. Quantitative mass spectrometry also needs ionization cross sections and accurate ratios of parent and fragment ion formation for qualitative analysis of gas mixtures and for measurements of thermochemical quantities such as heats of formation.

Most cross-section measurements in the past have been made on two classes of atoms and molecules which are simple to handle experimentally, those which are chemically stable and therefore can be handled conveniently in the gas phase [1], and ions which can be prepared and handled as beams [2]. Discharge models and mass spectrometry require, in addition, cross-section values for unstable and highly reactive species: free radicals, metastables, and many atoms.

The reliable experimental data for these three classes of unstable and reactive species is sparse. This statement is true by any standard for free radicals and metastables; absolute cross-section measurements are available for only five metastables and five free radicals. It is surprising that even for atoms there are very few well-known ionization cross-sections. Data for *single* ionization from threshold to several hundred eV exist for only 22 atoms, and of those, only about a half dozen have been measured by two or more different laboratories with agreement to better than 15%. For another 11, only *total* ionization has been measured. For the rest there are no published measurements.

This paper begins by describing the principal experimental methods used for absolute cross-section measurements, emphasizing the fast neutral beam method which has proven to be most useful for measurements on highly reactive species. It continues with a review of the available measurements. Finally, because reliable data are so scarce, it concludes with a qualitative discussion of trends along series and scaling based on orbital energies and occupation numbers.

Experimental Methods

One of the earliest and most widely used methods for measuring ionization cross-sections is the condenser plate method of Tate and Smith [3]. This is the method later used by Rapp and Englander-Golden [4] to produce *total* ionization cross-sections for the 5 rare gases and 9 small molecules. A magnetically collimated electron beam is directed through a gas of accurately measured pressure. All ions which form in a well-defined region are collected. The major limitation of the method is that it requires an accurate pressure measurement, a difficult or impossible task for many species. Also, it works only for pure species, because all ions are collected. Thus it is unsuitable for metastables, radicals, and many atoms.

Excellent data on *partial* ionization cross sections have been obtained recently by Märk and co-workers [1], using a modified mass spectrometer. The classic problem of mass spectrometers for quantitative cross-section measurements has been discrimination in the source between ions of different charge, or in the case of molecules, discrimination against fragment ions with translational kinetic energy. Märk solved the discrimination problem by modifying the ion source to assure complete ion collection, and by exhaustively characterizing the behavior of the entire mass spectrometer. The remaining limitation of Märk's apparatus is that it is not capable of measuring absolute cross-sections. A calibration must be made against either an independent measurement of the total ionization cross-section for the same species, (such as the measurements of Rapp and Englander-Golden), or by normalizing against an accurately known cross-section such as that of argon, relying on the theory of free molecular flow between the gas reservoir and the ion source.

Measurements for a number of atoms have been made by thermal beam methods. The advantage of a beam method is that condensable species such as metal atoms can be handled with relative ease. The difficulty is that direct pressure measurement is not possible; rather, the absolute beam flux must be measured. Several different approaches have been used. Alkali atom fluxes have been measured by surface ionization which can be 100% efficient [5]. The fluxes of other atoms have been determined by measuring the mass deposited on a surface with a quartz crystal microbalance [6], by quantitative chemical analysis [7,8], or by direct weighing of the deposit [9,10]. Magnesium [11] and barium [12] atom densities have been determined by measuring the absolute intensity of light emitted from an excited state with a known excitation cross-section.

Recent experimental results suggest that the most general method is to cross an electron beam with a *fast neutral beam*, prepared by charge transfer neutralization of a mass-selected ion beam. The first use of this approach for atomic ionization cross-section measurements was by Peterson [13]; it has since been used by Ziegler et al [14] and has been highly refined, by Harrison and coworkers [15-19]. Extensions at AT&T Bell Laboratories have made it a powerful method for studying molecular ionization and dissociative ionization [20-22]. The biggest advantage of this method is that it permits preparation of a pure beam, even of such hard-to-handle species as metal atoms, metastables, and free radicals. The high velocity, typically 1 to 5 keV, permits accurate flux measurements by the power deposited in any one of several different thermal detectors. The high collimation of a focused ion beam is preserved in the charge transfer process, simplifying beam handling and permitting complete collection of ions, even fragment ions from molecular dissociation. The biggest difficulty with the method is that charge transfer can produce neutral atoms or molecules in various excited states, so care must be taken to characterize the species under study.

In the fast neutral beam apparatus at Bell Labs [23] (Fig. 1), atomic or molecular ions are extracted from a dc discharge, accelerated to 3 keV, and mass separated with a crossed electric and magnetic field velocity (Wien) filter. The single-mass focussed ion beam is then charge transfer neutralized by a gas selected to have an ionization potential energy resonant with that of the fast ions. The pressure is adjusted to neutralize several percent of the ions (10^{-5}-10^{-6} Torr), with the remainder being deflected to a collector.

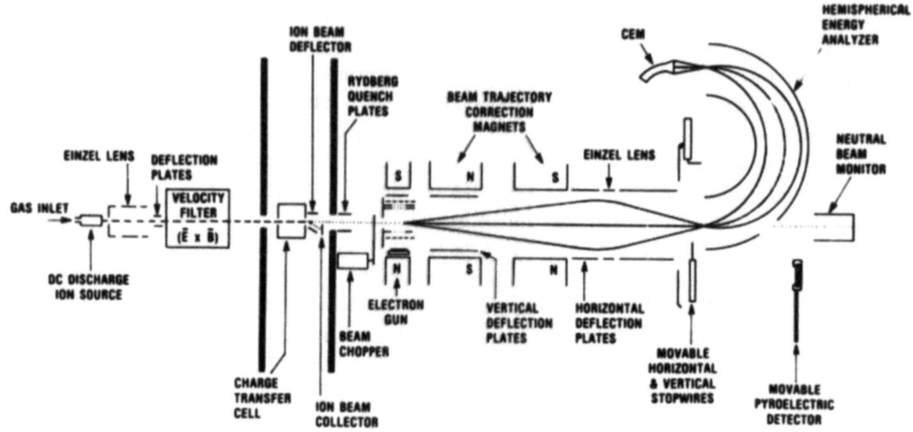

Figure 1. The fast beam apparatus at AT&T Bell Labs.

The resulting neutral beam generally has a flux of about 10^{10}/sec. Its relative intensity is measured by kinetic secondary ejection of electrons from a metal surface. For accurate flux measurements, a pyroelectric crystal is used to calibrate the secondary electron ejection coefficient against an ion beam current.

Ionization is produced by crossing the fast neutral beam with a well-characterized electron beam. The resulting ions are steered and focused with magnetic and electrostatic fields to a hemispherical energy analyzer. This analyzer separates ions of different charge or mass, since all ions retain essentially the same velocity as the 3 keV parent neutral beam. For molecular species, 100% collection of fragments is simplified, since in the laboratory frame, all fragments remain bunched together in the forward direction. Ions are finally collected and counted by a channel electron multiplier.

Cross-sections are given (except for a <1% correction due to the velocity of the neutral beam) by

$$\sigma = I_i(E)/[I_e(E)RF] \qquad (1)$$

where $I_i(E)$ and $I_e(E)$ are the ion and electron currents, respectively, R is the neutral beam flux, and F is a measure of the overlap between the neutral and electron beams. At present, the largest uncertainties are in correcting for reflection of electrons from the electron gun anode and in calibrating the efficiency of the CEM. The overall uncertainty of measurements with the present apparatus is calculated to be ±15%.

Experimental Data

This section lists the atoms, metastables, and radicals for which absolute cross-sections have been measured, and presents selected single ionization data for the region from threshold to 200 eV. Several recent critical reviews [1,24,25] elaborate on the information collected here.

Kieffer and Dunn [26], in their definitive 1966 review of electron impact ionization cross-sections, concluded that the experiments they reviewed were all subject to many systematic errors, and that none of them were reliable to within 10%. The situation has improved over the intervening 20 years, but not as much as one might hope. This appears to be due to the inherent difficulty of absolute measurements and a lack of research attention.

Helium is the atom for which the most accurate absolute measurements have been reported (Fig. 2A), and therefore is useful as a benchmark for future work. Rapp and Englander-Golden [4] reported a total cross-section with estimated ±7% uncertainty. Their total cross-section is essentially equal to the single ionization cross-section, since for helium the double ionization cross-section is 200 times smaller [27]. Montague, Harrison, and Smith [19] recently measured the single ionization cross-section by the fast beam method and reported a value with only ±4% uncertainty. The agreement between these two measurements is excellent, within ±5% (Fig. 2B). Two other measurements of the helium cross-section are included in Fig. 2. The shape of the measurement by Stephan, Helm, and Märk [27] agrees very well with the first two, but it is not absolute; rather it is normalized to the measurement of Rapp and Englander-Golden. The measurement from AT&T Bell Laboratories[23] by the fast beam method agrees to better than its stated ±15% uncertainty. The percentage differences between each pair of these measurements, given in Figs. 2B-2G, are seen to be well under 10%, except near threshold where a small shift of the energy scales gives a larger percentage difference. The near 0% difference shown in Fig. 2C is to be expected, since the SHM data was normalized to the REG measurements. The scatter apparent in Figs. 2E-2G results from scatter in the Bell Labs data.

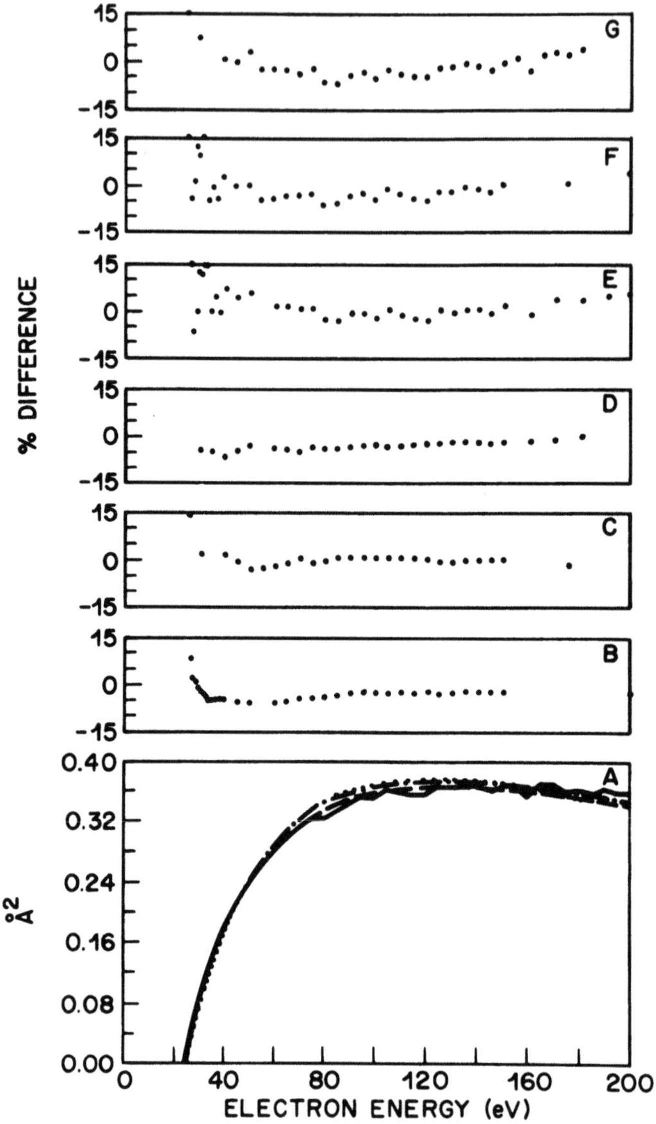

Figure 2. A) Measured single ionization cross-sections for helium (-.-.-REG [4],----MHS [19],....SHM [27],——ATT [23]), and comparisons showing percentage differences between them: B) MHS - REG, C) SHM - REG, D) MHS - SHM, E) ATT - MHS, F) ATT - REG, G) ATT - SHM.

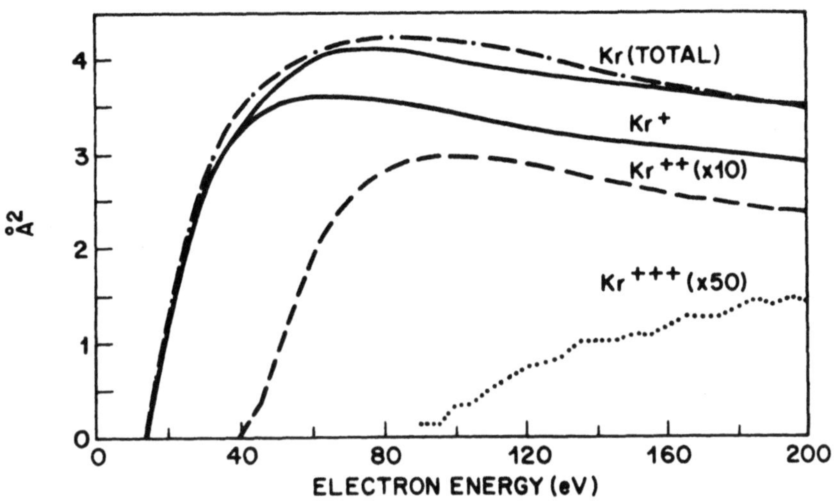

Figure 3. Partial and total ionization cross-sections of krypton from ATT measurements, and the total cross-section of REG -.-.-.

As a further example of rare gas data, Fig 3 shows krypton ionization to three charge states. The total cross section of the AT&T Bell Labs data [23], given by the charge weighted sum, agrees well with the data of Rapp and Englander-Golden [4]. Cross-sections for the individual charge states agree well with those of Stephan, Helm, and Märk [27] (not shown).

Single ionization cross-sections for all five rare gases show similar agreement; the Bell Labs data (Fig. 4) agrees with that of Rapp and Englander-Golden and of Stephan, Helm, and Märk to better than 10% over the entire energy range.

All other atoms for which absolute *single* ionization cross section measurements exist are given in Table 1, along with the original authors' estimated uncertainty (combined statistical and systematic). Only half of these uncertainties are even claimed to be smaller than 10%. Unfortunately, direct intercomparisons of these quoted uncertainties may not be meaningful, because some are given as one standard deviation, some as 90% confidence limits, and others as the "maximum possible error". Fig. 5 shows many of the measured cross-sections (as digitized from the original figures when tabular data were not given).

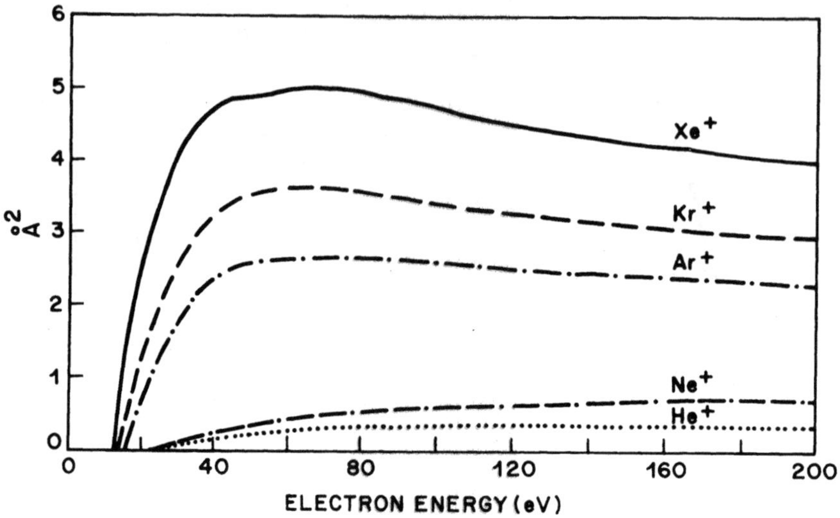

Figure 4. Single ionization cross-sections of the rare gases.

Table 1. Absolute cross-sections measurements for single ionization in order of increasing atomic number.

Atom	±%	Reference	Atom	±%	Reference
H	7	[25]	S	20	[14]
H*	12	[17]	Cl	15	[23]
	10	[37]	Ar	8	[4], [27]
He	4	[19]	Ar*	50	[15]
	8	[4], [27]	Br	15	[23]
He*	6	[16]	Kr	8	[4], [27]
Li	15	[6], [25]	Ag	22	[7]
C	5	[18]	I	15	[23]
N	5	[18]	Xe	8	[4], [27]
O	5	[18]	Cs	7	[6], [31]
F	15	[23]	Ba	20	[12]
Ne	8	[4], [27]	Hg	?	[29]
Ne*	30	[15]	Pb	15	[9]
Mg	10	[11]	U	17	[30]

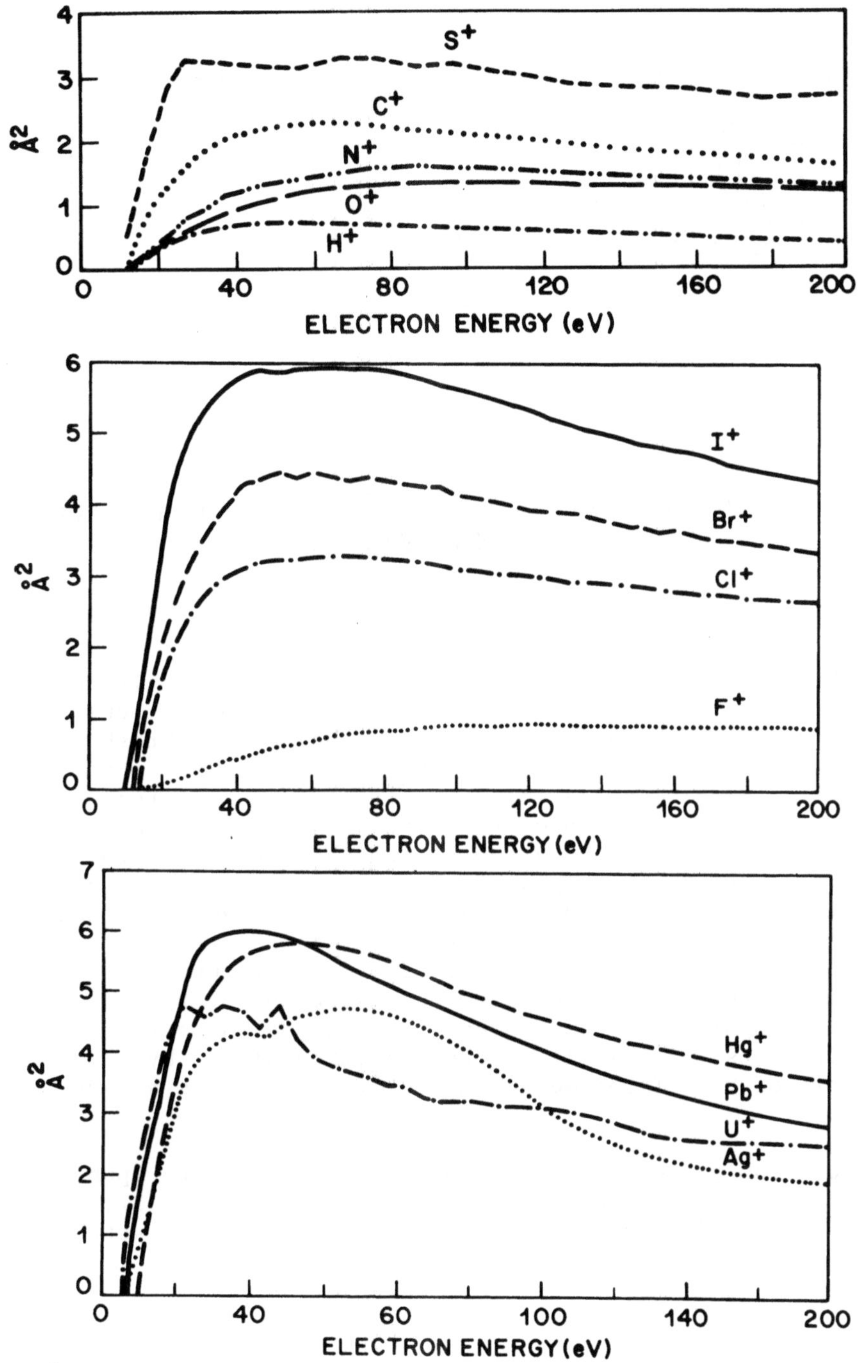

Figure 5. Measured single ionization cross-sections for atoms (see Table 1 for references).

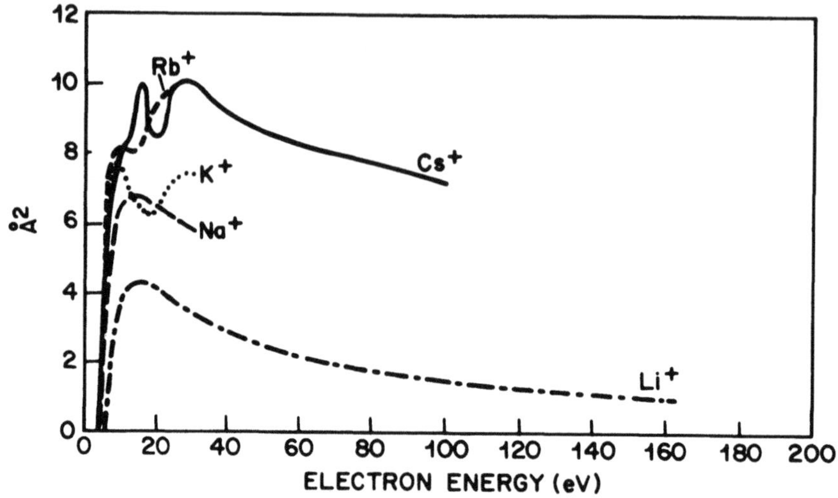

Figure 5 (continued).

The list of single ionization cross-sections in Table 1 is supplemented in Table 2 by the measured total ionization cross sections [32-36]. For energies below the threshold for double ionization, of course, the total cross-section equals that for single ionization. At least one measurement has been reported for each of the alkalis, the alkaline earths (except beryllium), the group IIIA metals (except boron), the halogens, and the rare gases. The most striking lack of data is for the transition metals. The agreement between different measurements, for several of these atoms where more than one measurement has been made, is quite unsatisfactory. For magnesium, three measurements differ by over a factor of 2 in magnitude, and even the shapes disagree. Two measurements of the silver cross-section differ by almost a factor of 2. Three measurements of barium span a range of 50%.

For ionization from excited states, there are published measurements for the four metastable atoms H* [17,37], He* [16], Ne* [15], and Ar* [15] (Fig 6), and the metastable N_2 molecule [20] in its lowest triplet state (Fig. 7).

Absolute measurements of free radical cross-sections have been published for CD_2 and CD_3 (Fig 8) [21], and will soon be published [22] for SiF, SiF_2, and Si_3; the preliminary data shown in Fig 9 are probably accurate to better than 30%.

Table 2. Absolute cross-sections for total ionization.

Atom	±%	Reference	Atom	±%	Reference
Mg	23	[8]	Rb	12	[5]
	17	[10]	Sr	19	[34]
Na	15	[6]		17	[10]
Al	20	[32]	Ag	15	[35]
K	15	[6]	In	20	[32]
	12	[33]	Ba	17	[10]
Ca	19	[34]		19	[34]
	17	[10]	Yb	30	[36]
Cu	15	[35]	Tl	20	[32]
Ga	20	[32]			

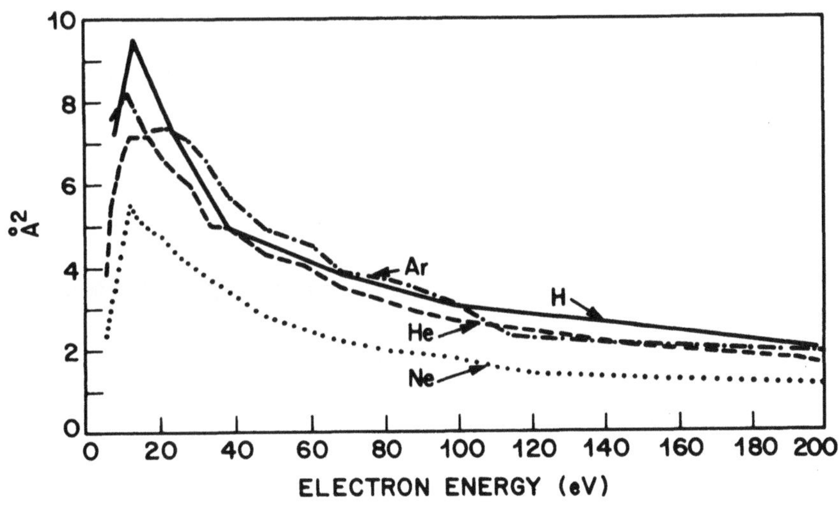

Figure 6. Cross-sections for ionization of metastable atoms.

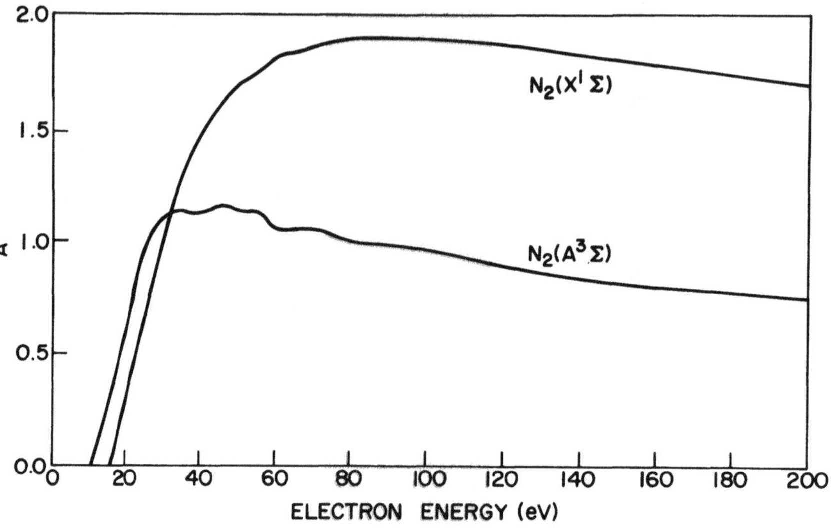

Figure 7. Cross-sections for ionization of ground and metastable state N_2 [20].

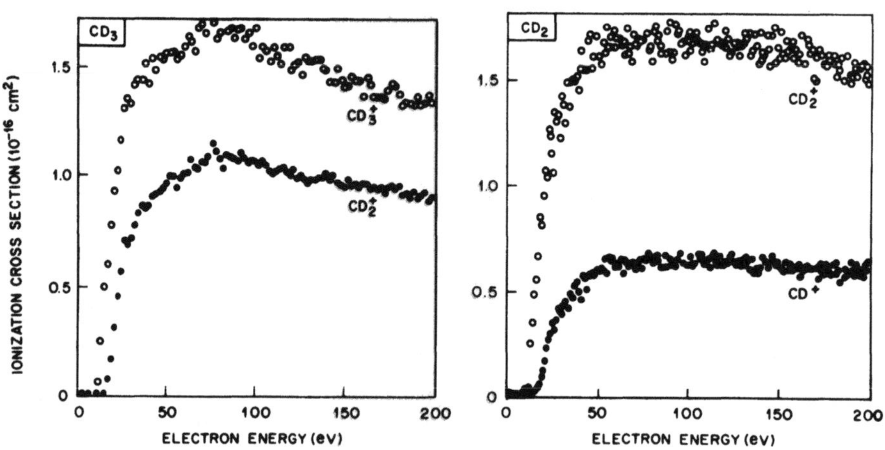

Figure 8. Cross-sections for ionization and dissociative ionization of the free radicals CD_3 and CD_2 [21].

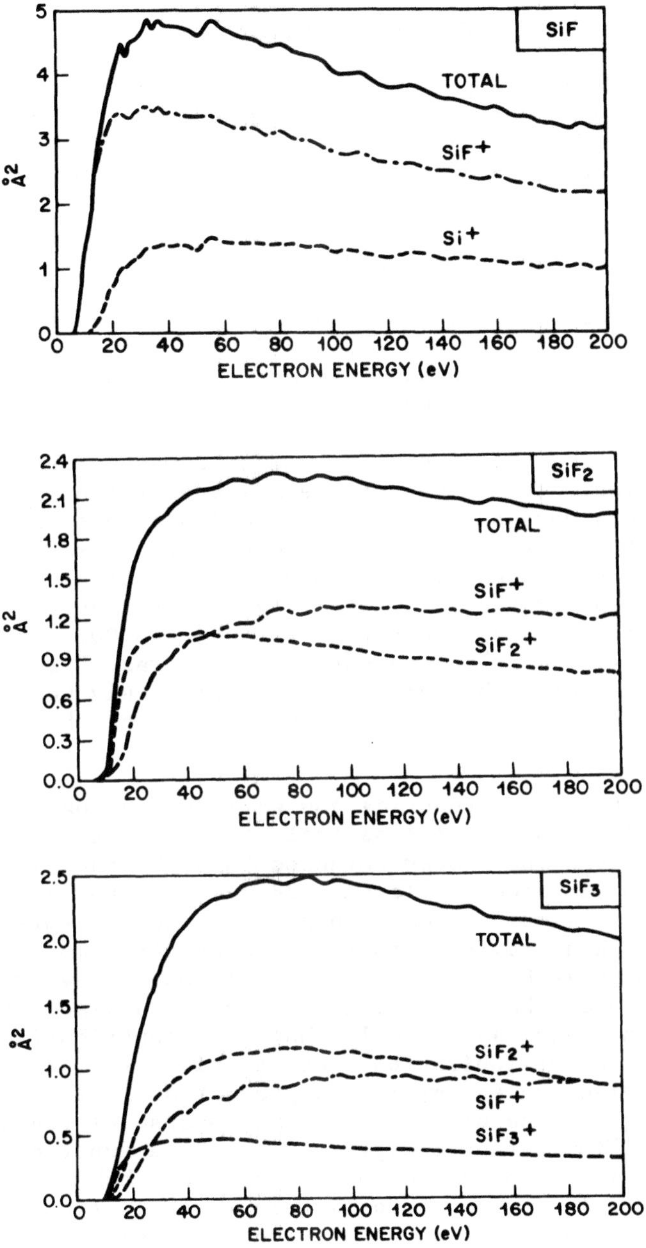

Figure 9. Preliminary cross-sections for ionization and dissociative ionization of the free radicals SiF, SiF_2, and SiF_3 [22].

Trends and Scaling

Because there are so few reliable absolute cross-sections available, and until more measurements are made, one is forced to seek guidance from observed trends, or to calculate ionization cross-sections from theory or from scaling of available data.

A number of trends are apparent in the data of Figs. 4-6. Going from left to right in the periodic table, the magnitude of the peak cross-section decreases, and the energy at which the peak occurs increases. Going down in any column of the periodic table, the peak cross-section increases, and the energy of the peak decreases. These generalizations are drawn of course from data on only a small number of atoms, and many of the individual measurements have large uncertainties. The N_2 metastable A state, which is a valence state, has a much smaller ionization cross-section than the Rydberg metastable atoms. Its cross-section is even smaller than that of its own ground state. The total ionization cross-section for the SiF radical is larger than those of the larger radicals SiF_2 and SiF_3. For CD_2 and CD_3 ionization, the principal ion is the parent. In contrast, for SiF_2 and SiF_3 ionization, the parent ion is relatively weak (except near threshold), with the loss of one F atom the principal ion.

A large number of theories and well over a dozen different semi-empirical or semi-classical formulas for ionization cross-sections have been advanced through the years, and recently have been reviewed [38,39]. Surprisingly, the relatively simple semi-empirical and semi-classical approaches have provided atomic cross-sections more accurate than quantum mechanical theories. Although they differ in detail, many of these approaches produce cross-sections which scale in the same way, namely that the contribution of each orbital n to the ionization cross-section as a function of incident electron energy E is a product of 1) the number of electrons q_n in the orbital, 2) the inverse of the square of the orbital ionization energy P_n, and 3) a function of E/P_n. This function should follow the correct behavior at the limits, namely, rise approximately linearly from threshold, reach a peak, and fall at high energy as $(\ln E)/E$.

One of the most popular theories for the limited available data is that of Lotz [40]:

$$\sigma = a\{1-b\exp[-c(E/P_1-1)]\}\sum_{n=1}^{N}(q_n/P_n^{\,2})\ln(E/P_n)/(E/P_n) \quad (2)$$

Lotz derived best fit values of the constants *a*, *b*, and *c* for a number of atoms from measurements available to him in 1967 (for more accurate fits he made *a*, *b*, and *c* functions of *n*). Without deriving values of the constants *a*, *b*, and *c*, we examine how well the rest of Eq. 2 describes all the presently available single ionization data (Table 1). Fig. 10 plots the value of the measured cross-section divided by the summation factor from Eq. 2 (including up to 3 orbitals), where values of q_n and P_n are taken from references [41] and [42]. This procedure is equivalent to plotting the function $a\{1-b\exp[-c(E/P_1-1)]\}$ for each atom. The resulting functions certainly are not identical, but the spread among them (roughly a factor of 3 above 100 eV) is much smaller than the spread among the cross-sections themselves (roughly a factor of 20), showing that the summation alone provides an initial description of the cross-sections. Peaks near

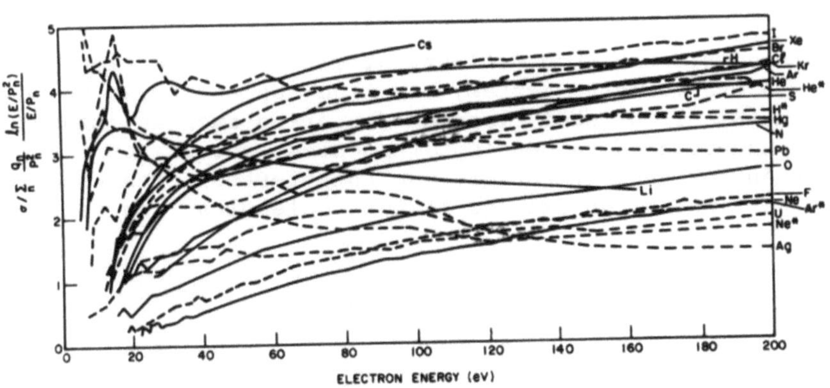

Figure 10. Ratios of measured single ionization cross-sections to the summation factor of Eq. 2. Dashed lines are for measured cross-sections less accurate than ±10%.

threshold for several atoms result from autoionization, which Lotz's theory does not even attempt to include. Adjustment of the constants *a*, *b*, and *c* would of course provide a much better fit to the data, as Lotz showed for a smaller data set. At high energy, the curves approach the value of *a*. At lower energies, the constants *b* and *c* are certainly not negligible. One of the more important conclusions we can draw from Fig. 10 is that cross-sections for the metastable atoms, roughly an order of magnitude larger than the corresponding ground state cross-sections, can be explained by the same scaling equation as the ground states. It will be interesting to analyze data on ionization of free radicals and other molecules in a similar way, but such analyses have not yet been carried out.

Conclusions

Despite their conceptual simplicity, there have been surprisingly few measurements made of absolute ionization cross-sections, especially for reactive atoms and free radicals. Accurate measurements are in fact quite difficult, as shown by the disagreement among measurements by different authors or different methods for the same atom. The fast neutral beam method, used primarily in the past 10 years, is capable of providing accurate cross-sections for a wide variety of species. Its more extensive use may open the door to many accurate measurements. In the absence of accurate measurements, one can estimate atomic cross-sections from trends along measured series, or from scaling rules based on semi-empirical or semi-classical theories such as those of Lotz. Accurate data on more species should permit the refinement of these theories, so that further cross-sections can be estimated for excited states, molecules, and radicals.

Acknowledgments

Robert Wetzel, Frank Baiocchi, and Todd Hayes are thanked for their measurements of many of the free radical and atomic ionization cross-sections. Robert Wetzel digitized the published cross-section data.

References

[1] For recent reviews, see Märk, T.D. pp 137 -197 in "Electron Impact Ionization", edited by Märk, T.D. and Dunn, G.H., Springer-Verlag, (New York-Wien) 1985.
[2] For a recent review, see Dunn, G.H., pp 277-319, in "Electron Impact Ionization", edited by Märk, T.D. and Dunn, G.H., Springer-Verlag, (Wien-New York) 1985.
[3] Tate, J.T.; Smith, P.T. Phys. Rev. 1932, 39, 270.
[4] Rapp, D.; Englander-Golden, P. J. Chem. Phys. 1965, 43, 1464.
[5] Nygaard, K.J.; Hahn, Y.B. J. Chem. Phys. 1973, 58, 3493.
[6] Zapesochnyi, I.P.; Aleksakhin, I.S. Soviet Physics JETP 1969, 28, 41.
[7] Crawford, C.K.; Wang, K.I. J. Chem. Phys. 1967, 47, 4667.
[8] Okuno, Y.; Okuno, K.; Kaneko, Y.; Kanomata, I. J. Phys. Soc. Japan 1970, 29, 164.
[9] Pavlov, S.I., Rakhovskii, V.I.; Fedorova, G.M. Soviet Physics JETP 1967, 25, 12; Pavolv, S.I.; Stotskii, G.I. Soviet Physics JETP 1970, 31, 61.
[10] Vainshtein, L.A.; Ochkur, V.I. ; Rakhovskii, V.I.; Stepanov, A.M. Soviet Physics JETP 1972, 34, 271.
[11] Karstensen, F.; Schneider, M. Z. Phys. A 1975, 273, 321; J. Phys. B 1978, 11, 167.
[12] Dettmann, J.-M.; Karstensen, F. J. Phys. B 1982, 15, 287.
[13] Peterson, J.R. Atomic Collision Processes, ed McDowell, M.R.C. 1964 (Amsterdam: North-Holland), pp 465-73.
[14] Ziegler, D.L.; Newman, J.H.; Goeller, L.N.; Smith, K.A.; Stebbings, R.F. Planet. Space Sci. 1982, 30, 1269.
[15] Dixon, A.J.; Harrison, M.F.A.; Smith, A.C.H. 8th International Conference on the Physics of Electronic and Atomic Collisions (Institute of Physics, Belgrade), 1973, abstracts p 405.
[16] Dixon, A.J.; Harrison, M.F.A.; Smith, A.C.H. J. Phys. B 1976, 9, 2617.
[17] Dixon, A.J.; von Engel, A.; Harrison, M.F.A. Proc. Roy. Soc. Lond. A 1975, 343, 333.
[18] Brook, E.; Harrison, M.F.A.; Smith, A.C.H. J. Phys. B 1978, 11, 3115.
[19] Montague, R.G.; Harrison, M.F.A.; Smith, A.C.H. J. Phys. B 1984, 17, 3295.
[20] Armentrout, P.B.; Tarr, S.M.; Dori, A.; Freund, R.S. J. Chem. Phys. 1981, 75, 2786.
[21] Baiocchi, F.A.; Wetzel, R.C.; Freund, R.S. Phys. Rev. Lett. 1984, 53, 771.

[22] Hayes, T.R.; Wetzel, R.C.; Baiocchi, F.A.; Freund, R.S. (to be published).
[23] Hayes, T.R.; Baiocchi, F.A.; Wetzel, R.C.; Freund, R.S. (to be published).
[24] de Heer, F.J.; Jansen, R.H.J.; van der Kaay, W. J. Phys. B 1979, 12, 979. de Heer, F.J.; Inokuti, M. pp 232-276, in "Electron Impact Ionization", edited by Märk, T.D. and Dunn, G.H., Springer-Verlag, (New York-Wien) 1985.
[25] Bell, K.L.; Gilbody, H.B.; Hughes, J.G.; Kingston, A.E.; Smith, F.J. J. Phys. Chem. Ref. Data 1983, 2, 891.
[26] Kieffer, L.J.; Dunn, G.H. Rev. Mod. Phys. 1966, 38, 1.
[27] Stephan, K.; Helm, H.; Märk, T.D. J. Chem. Phys. 1980, 73, 3763.
[28] Dettmann, J.-M.; Karstensen, F. J. Phys. B 1982, 15, 287.
[29] Bleakney, W. Phys. Rev. 1930, 35, 139.
[30] Halle, J.C.; Lo, H.H.; Fite, W.L. Phys. Rev. A 1981, 23, 1708.
[31] Nygaard, K.J. J. Chem. Phys. 1968, 49, 1995.
[32] Zapesochny, I.P.; Kontros, J.E.; Shimon, L.L. 9th International Conference on the Physics of Electronic and Atomic Collisions (University of Washington Press, Seattle), 1975, abstracts p. 900; Shimon, L.L.; Nepiipov, E.I.; Zapesochnyi,I.P. Sov. Phys. Tech. Phys. 1975, 20, 434.
[33] Nygaard, K.J. Phys. Lett. 1975, 51A, 171.
[34] Okuno, Y. J. Phys. Soc. Japan 1971, 31, 1189.
[35] Pavlov, S.I.; Rakhovskii, V.I.; Fedorova, G.M. Sov. Phys. JETP 1967, 25, 12.
[36] Ali, M.M.; Volovich, P.N.; Ovchinnikov, V.L.; Shimon, L.L. 14th International Conference on the Physics of Electronic and Atomic Collisions (Stanford) 1985, abstracts p. 707.
[37] Defrance, P.; Claeys, W.; Cornet, A; Poulaert, G. J. Phys. B 1981, 14, 111.
[38] Younger, S.M., in "Electron Impact Ionization ", edited by Märk, T.D. and Dunn, G.H., Springer-Verlag (Wien-New York) 1985, pp 1-23.
[39] Younger, S.M. and Märk, T.D., in "Electron Impact Ionization", edited by Märk, T.D. and Dunn, G.H., Springer-Verlag (Wien-New York) 1985, pp 24-41.
[40] Lotz, W. Z. Phys. 1967, 206, 205.
[41] Lotz, W. J. Opt. Soc. Am. 1968, 58, 915.
[42] Herman, F.; Skillman, S. "Atomic Structure Calculations" 1963, (Prentice-Hall, Englewood Cliffs, NJ).

TOTAL DISSOCIATION CROSS-SECTIONS OF FLUOROALKANES FOR ELECTRON IMPACT: THEIR USEFULNESS FOR UNDERSTANDING PLASMA-ASSISTED ETCHING ENVIRONMENTS

Harold F. Winters

Almaden Research Center, 650 Harry Road, San Jose, California

Low-pressure, low temperature gas discharges (cold plasmas) are being used in a diverse and expanding number of applications which are of great importance to industry [1]. Among these applications are plasma-assisted etching and plasma-assisted film deposition which are used extensively by most, if not all, semiconductor manufacturers. A dearth of knowledge about the important processes which occur in these chemically active plasmas clearly exists. Plasma-assisted etching is a prime illustration of this assertion. While there have been impressive advances in understanding the basic phenomena governing this technology in the past several years, significant technological progress has resulted mainly from "hard-nosed" empiricism. In effect, the science has not yet caught up with the technology, or even pointed the way. This is not an unusual situation; indeed, it is characteristic of much of modern technology development, which tends to be evolutionary. However, many of the roadblocks to successful application of nonequilibrium plasma technology could be ameliorated by expanding fundamental knowledge. In this regard, most technologically important plasma processes have been developed through inefficient, experimental parameter studies. A more desirable goal would be to partially develop future processes though mathematical modeling of the glow discharge, and in fact some effort in this direction is already available [2-4]. Unfortunately, a lack of fundamental information about cross-sections, plasma surface interactions, rate constants, synergistic effects, etc., presently make this approach of limited value.

Gases such as CF_4, CF_3H and CF_3Cl are frequently used for plasma etching technology but do not chemisorb (react) with the surfaces being etched [5]. The relatively inert nature of these gases serves to emphasize the major role of the glow discharge; i.e., to generate chemically reactive radicals by electron-impact induced dissociation of the halocarbon molecule. Therefore, a first requirement for the modeling of these reactive gas glow discharges is a fundamental knowledge about dissociative processes and dissociation cross sections - particularly near threshold. This is an area where scientists conducting basic studies of electron-molecule interactions could clearly make a contribution to technology. Despite the obvious need for this type of data, there is to our knowledge at the present time no group in the world which is attempting to make total dissociation cross section measurements.

The interactions of electrons with molecules produces ionization, electronic excitation, and dissociation. If the final state of a molecule is easily detected, which is the case for positive and negative ions, fluorescent excited states, and in some cases metastable levels, the experimental cross-sections are often available. In contrast, absolute total dissociation cross-sections are virtually unknown. This state of affairs is primarily a consequence of the experimental difficulties encountered when attempting to measure the number of neutral fragments created in electron-molecule collisions. The paucity of data is illustrated by the 1976 review article of Polak and Slovetsky [6], who in Table II of their paper list six molecules for which some published information is available. They further suggest (on p. 258 of Ref. 1) that in only one case (i.e., the dissociation of nitrogen) have all the experimental problems been satisfactorily resolved. Therefore, the accuracy of the data for many other molecules is quite uncertain.

Various techniques from the field of surface science have allowed us to develop a method for measuring the total cross-section for electron-impact-induced dissociation to an accuracy of 20 to 25%. Moreover, a reasonable effort toward improvement of the experimental technique can be expected to reduce this uncertainty to about the same magnitude as is obtained for total ionization cross-sections. Some cross-sections from our laboratories for nitrogen [7,8], methane [9], ethane [10], CF_4[11], C_2F_6 [11], CF_3H[11], and C_3F_8[11] have been published. Perrin et.al., have also published data for silane and disilane [12]. In our studies, the minimum detectable cross-section was greater than $3 \times 10^{-18} cm^2$. However, one should be able to lower this value by at least a factor of ten.

The experimental method consists of developing a situation satisfying two requirements. First, the gas of interest in a closed volume does not react with the vacuum-system surface and therefore remains in the gas phase indefinitely. Second, fragments produced from this gas by any method are removed from the gas phase with unit probability because they react chemically with a freshly evaporated metal film (often called a getter) or the surrounding walls to form a nonvolatile solid. Under this situation, the number of molecules leaving the gas phase is equal to the number of fragmentation events. In other words, the change in partial pressure within the closed volume is directly proportional to the number of gas-phase molecules dissociated. Therefore, the dissociation cross-section σ is directly proportional to (1/P)(dP/dt) where P is the pressure and t the time. Note that the cross-section measurements do not require a knowledge of the absolute pressure. This eliminates a large potential source of error.

The technique described above is particularly reliable for the fluoroalkanes such as CF_4, CF_3H, C_2F_6 and C_3F_8. Moreover, an understanding of the dissociative processes in these molecules is of interest from a theoretical point of view. The total dissociation cross-section for CF_4 has been measured for energies between threshold (≃12.5 eV) and 600 eV. The magnitude of the cross-section at its maximum is $5.5 \times 10^{-16} cm^2$. Less extensive data are available for CF_3H, C_2F_6 and C_3F_8. Their cross-sections at the maxima are $5.8 \times 10^{-16} cm^2$, $8.6 \times 10^{-16} cm^2$ and $1.18 \times 10^{-16} cm^2$, respectively. Arguments can be made which suggest that the total dissociation cross-section for each of these gases is equal to the sum of the cross-sections for excitation to all electronic and ionic states, i.e., the total cross-section for electronic excitation. From the point of view of the Bethe theory, it is concluded that the Bethe asymptotic behavior is not yet attained in the energy range of these measurements.

References

[1] For an overview see Winters, H.F.; Chang, R.P.H.; Evans, J.; Mogab, C.J.; Thornton, J.A. and Yasuda, H. Mat. Sci. Eng. 1985, 70, 53.
[2] Edelson, D. and Flamm, D.L. J. Appl. Phys. 1984, 56, 1522.
[3] Sakai, Y.; Reynolds, J.L. and Neureuther, A.R. J. Electrochem. Soc. 1984, 131, 627.
[4] Kushner, M.J. J. Appl. Phys. 1983, 54, 4958.
[5] Winters, H.F. J. Appl. Phys. 1978, 49, 5165.
[6] Polak, L.S. and Slovetsky, D.I. Int. J. Radiat. Phys. Chem. 1976, 8, 257.
[7] Winters, H.F.; Horne, D.E. and Donaldson, E.E. J. Chem. Phys. 1966, 41, 2766.
[8] Winters, H.F. J. Chem. Phys. 1966, 44,1472.
[9] Winters, H.F. J. Chem. Phys. 1975, 63, 3462.
[10] Winters, H.F. Chem. Phys. 1979, 36, 353.
[11] Winters, H.F. and Inokuti, M. Phys. Rev. A 1982, 25, 1420.

[12] Perrin, J.; Schmitt, J.P.M.; DeRoshy, G.; Drevillon, B.; Huc J. and Lloret, A. Chem. Phys. 1982, 73, 383.

ELECTRON-CLUSTER INTERACTIONS

R. G. Keesee, A. W. Castleman, Jr., and T. D. Märk*

Department of Chemistry,
The Pennsylvania State University
University Park, PA 16802

The collision of an electron with a cluster can be classified into one of three categories based on the outcome of the event: (1) scattering, (2) secondary electron ejection, and (3) attachment. Of particular emphasis in this overview are the last two categories. Current understanding of these processes is discussed and recent work undertaken to improve that understanding is surveyed.

Introduction

Investigation of the properties of atomic or molecular clusters and their interaction with electrons is directed at solving outstanding technological problems and understanding basic phenomena as well. For instance, beams of clusters have been proposed as an effective method in heating plasmas and fueling nuclear fusion devices [1]. Information on electron interactions with clusters is also applicable to designing lasers and elucidating the processes occurring in discharges and radiolysis [2]. Clusters provide a medium in which it is possible to track the changes that occur in matter as one proceeds from the gas phase to the condensed phase. Additionally, clusters and electron-cluster interactions play a role in the extent of ionization and conductivity of atmospheres [3]. In terms of fundamental science, clusters are a state of matter between the gas and condensed phase. Many problems in condensed media, such as the properties of solvated electrons and liquid structure, may be approached from the viewpoint of clusters [4].

Ionization of neutral clusters in conjunction with mass spectrometry is the most common method of detection for these species. Consequently, quantitative data on ionization of clusters are necessary in order to unambiguously interpret and relate mass spectra to the original neutral cluster distributions under study.

As the colliding partner of an electron advances from an atom to a molecule to a cluster, a greater diversity of mechanisms for energy transfer are available, due to the increased number of internal degrees of freedom of the target. An atomic target offers only electronic excitation, whereas a molecular target adds rotational and vibrational modes of excitation. A cluster, furthermore, possesses low-energy intermolecular vibrational modes and dissociation channels. The multitude of processes which may occur upon the collision of an electron with a cluster may be broadly classified into one of three categories, depending on the outcome of the event: (1) scattering, where for every incident electron, one electron departs; (2) secondary electron ejection, where additional electrons are ejected from the cluster system; and (3) attachment, where the incident electron is captured for the duration of the observation.

Scattering processes include momentum transfer, elastic recoil, rotational, vibrational, and electronic excitation. The internal excitation processes may also lead to dissociative loss of molecular species from the cluster or in some cases fragmentation of one of the molecular subunits of which the cluster is comprised. Also, in this broad definition would be included ion pair formation [5]. Secondary electron ejection may result from autoionization via electronic excitation or direct ionization. Again, dissociation or fragmentation may accompany the ionization process. Dissociation may occur by evaporation (i.e., loss of individual molecular units) or by fission into a smaller charged cluster and neutral cluster.

Following ionization, other processes must be considered. For example, structural rearrangement of the cluster in the neighborhood of the charge is likely to occur. In addition, an ionized subunit of the cluster may react with a neighboring subunit in a so-called internal reaction or ion-molecule half reaction [6,7]. With sufficient electron energies, multi-ionization may also occur and possibly result in Coulombic explosion of the cluster. Attachment processes create an excited cluster due to the electron affinity of the cluster and the initial energy of the incident electron. Relaxation may occur via radiative, dissociative, or collisional (third-body) mechanisms or autodetachment. The latter would ultimately be a scattering process. Recombination of an electron with positively charged cluster ions can also be considered to be an attachment process.

Typically, cluster beams are conveniently produced by supersonic expansions of gases from high pressures. Additionally, low concentrations of small clusters may actually be stable entities in gases at high pressures and low temperatures. A complicating factor in studying clusters as opposed to molecules is that it has not been possible to produce uncharged clusters of a specific size at a well-known number density or concentration. In an expansion, a usually not well-defined distribution of cluster sizes is produced. In dense gases, the concentration of clusters, particularly dimers, are estimated from calculations using second virial coefficients for interaction potentials of the gases. If ionized clusters are detected, dissociation processes create ambiguity in identifying the size of the parent neutral cluster. In other words, a given ionized cluster size may have contributions from more than one precursor neutral size, and the extent of ionization of a particular neutral cluster size in general cannot be determined from the intensity of the corresponding ionized cluster. Therefore, experimental studies in which clustering is extensive yield data in terms of an average cluster size for an often poorly defined size distribution. When clustering is limited to very small sizes, e.g., the dimer and trimer, large amounts of the monomer are also present. Consequently, very little quantitative information on electron-cluster interactions is available. Nevertheless, some general trends are evident. The remainder of this paper will be divided into a discussion of (1) ionization by secondary electron ejection, and (2) electron attachment.

Ionization by Secondary Electron Ejection

As mentioned in the introduction, ionization of a cluster may result in many different processes. An ionization cross-section may be defined in many ways depending on the experiment or the desired detail. A partial ionization cross-section refers to the cross-section for the production of a particular ion of a particular charge. Total charge production or ion counting cross-sections measure the gross features and are experimentally measurable for a given cluster distribution commonly characterized by an average size. For comparison with the free molecule, an effective cross-section (the total cross-section divided by the number of molecules in the cluster) is a useful concept.

When considering electron impact ionization of clusters, four "rules of thumb" are convenient to keep in mind: (1) As the energy of the incident electron increases, the variety of ions produced will increase; (2) the total ionization cross-section is approximtely proportional to the cluster size n for small n and the proportionality decreases to $n^{2/3}$ for larger n; (3) the incident electron energy at which the total ionization cross-section for a given cluster size reaches a maximum increases with increasing cluster size; and (4) the threshold energy for ionization generally decreases with increasing cluster size.

The first "rule of thumb" is rather obvious and is also true for molecules and atoms. As the electron energy is increased, more energy may be deposited into the cluster making more channels such as multi-ionization or more severe dissociation energetically favorable. The next two generalities are based on a limited number of experiments [8-11] and supported by theoretical considerations [12]. The last rule has more extensive experimental substantiation from both electron impact ionization and photoionization.

Dependence of cross-section on cluster size. The total ionization cross-section as a function of cluster size may be envisioned to be the product of three terms. The first term involves the probability that an electron strikes the cluster. This depends on the geometric cross-sectional area of the cluster and is proportional to $n^{2/3}$ for a spherical (or cubic) cluster of n molecules if the density of the cluster is assumed to be independent of cluster size. One must also assume that the cross-section of a molecule within the cluster is the same as that of the free molecule. If this assumption is not valid, then a $n^{2/3}$ dependence could still be obtained over a range of cluster sizes for which the size of the cluster does not affect the cross-section of an individual molecule in the cluster.

The next term corresponds to the probability that a colliding electron can cause an ionizing event within the cluster. This term can be expressed as $1-e^{-\ell/\lambda}$ where ℓ is the mean distance traveled inside the cluster by the incident electron and λ is the mean free path for ionization within the cluster. The mean free path λ depends only on the energy of the incident electron, whereas ℓ depends on the radial dimension of the cluster which is proportional to $n^{1/3}$. Thus, for a small cluster ($\ell \ll \lambda$), $1-e^{-\ell/\lambda}$ is proportional to $n^{1/3}$ and for a large cluster ($\ell \gg \lambda$), the term is unity.

The last term considers the probability that the ensuing secondary electron escapes the cluster. This probability is given by $1-e^{-D/\ell}$ where D is the mean distance that a secondary electron can travel in the cluster and still have sufficient energy to escape the cluster. In this case, for a small cluster ($\ell \ll D$), the probability is unity, while for a large cluster ($\ell \gg D$), the probability is proportional to $n^{-1/3}$.

The total ionization probability for a small cluster ($\ell \ll \lambda$ and D) is directly proportional to the cluster size For a larger cluster ($\ell \gg \lambda$ or D) the probability becomes proportional to $n^{2/3}$. The dimension D is expected to be smaller than λ, so the expected criterion for this proportionality is $\ell \gg D$. If the cluster is very large ($\ell \gg \lambda$ and D), the proportionality should fall to $n^{1/3}$. Bottiglioni et al. [12] point out that electrons of about 80 eV incident energy can be captured only by clusters exceeding 10^8 molecules,which is over three orders of magnitude larger than the largest average cluster sizes in the reported experiments.

From the dependence of the total charge production cross-section on cluster size (as shown in Figure 1), Henkes and coworkers determined the ionization shell thickness D to be

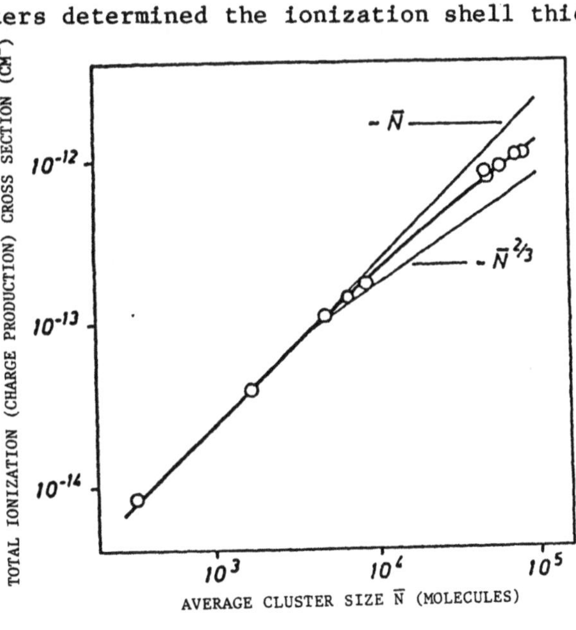

Figure 1. The ionization cross-section of clusters for 500V electron versus the mean molecular hydrogen cluster size. (From reference 11.) Reproduced with permission, copyright 1974, Elsevier.

about three molecular layers for CO_2 clusters [9] and five and one-half for the H_2 clusters [11]. In the case of the H_2 clusters, however, the effective cross-sections were found to be about half of the cross-section for the free molecule. Tay et al. [10] determined a shell thickness of one to two molecular layers for H_2 clusters. To explain the different thicknesses, Henkes and Mikosch [11] offer the explanation that since Tay et al. measured the total ion-counting cross-section (which is necessarily equal to or smaller than the total charge production cross-section), multiple charging can lead to a $n^{2/3}$ dependence.

Effects of dissociation. For the relatively large clusters examined in the above studies, the amount of dissociation which occurs upon ionization represents only a small or negligible change in the average size of the clusters. However, Gspann and Vollmar [13] observed the production of small cluster ions of about 70 atoms from ionization of large clusters (10^5–10^7 atoms) of He. They attributed these small ions to ejection from the larger clusters through electrostrictive polarization in the vicinity of newly formed ions which are near the surface of the cluster, as depicted in Figure 2. In the case of very small clusters, the extent of dissociation must be identified in order to determine cross-sections. Märk [14] has used the expected proportionality with cluster size to take into account the dissociative ionization channel of rare gas dimers and estimate the absolute partial cross-sections for the production of the dimer ion from the neutral dimer. More recently, Buck and Meyer [15] have measured the dissociation probabilities upon electron impact ionization for Ar_2 and Ar_3. At low electron energies (30 eV), 50% of the neutral dimers appear as Ar_2^+ and the remainder as Ar^+. This probability decreased to 38% for

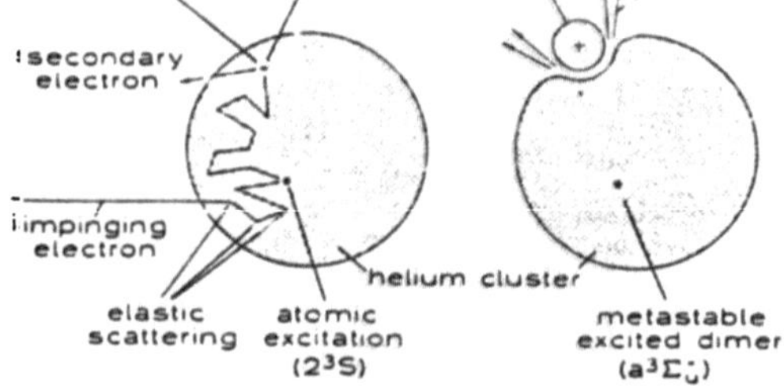

Figure 2. Diagram for small cluster ion ejection from large clusters via electrostrictive polarization. [From reference 13.]

100 eV electrons. On the other hand, no Ar_3^+ was formed in the direct ionization of Ar_3.

Several other experiments [16-19] have demonstrated the fragmentation of small clusters upon electron impact ionization, and attempts to measure relative fragmentation yields have been made in some of these experiments. Märk and Castleman [20] more completely review the work in this area.

Dependence of cross-section on electron energy. Qualitatively, the shift of incident electron energy to higher values in order to achieve the maximum cross-section (see Figure 3) can be explained on the basis that the incident electron loses energy as it passes through the cluster. In this case, the energy of the electron at the point in the cluster where the ionizing collision occurs is less than its initial energy. Assuming that the dependence of the ionization cross-secton on energy for molecules inside a cluster is the same as that of the free molecule, then the maximum of the cross-section for the cluster will occur at an incident energy which is shifted to a higher value. Therefore, the effective ionizing energy within the cluster corresponds to the electron energy at which the maximum occurs in the free molecule.

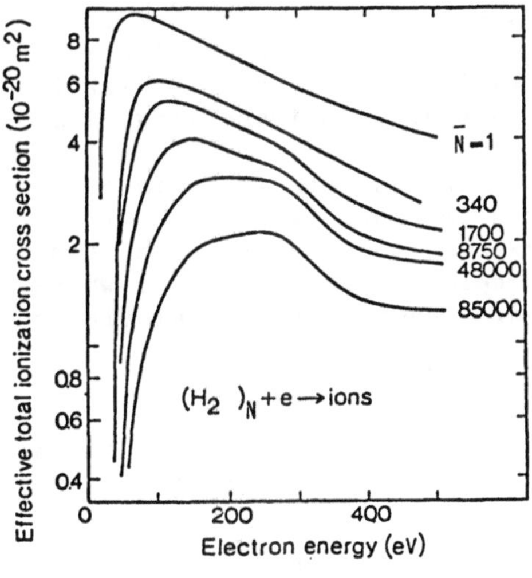

Figure 3. The effective ionization cross-section versus electron energy for clusters of hydrogen. (From reference 11.) Reproduced with permission, copyright 1974, Elsevier.

A second feature observed by Henkes and Mikosch [11] and also appearing in the calculations of Bottiglioni et al. [12] is a broadening or double peak in the energy dependence of the cross-section for larger clusters. As Henkes and Mikosch [11] point out, this feature can be interpreted in the following way. The low-energy peak occurs from ionization events toward the forward part of the cluster where the incident electrons enter. As the electron energy increases, on average ionization occurs deeper in the cluster where secondary electrons have less probability of escaping. Note that for a small cluster this probability remains at unity. Finally, at still higher energies, ionization tends to occur in the back of the cluster where the escape of secondary electrons becomes more probable again.

Threshold behavior. Another observed trend is that the threshold for ionization generally decreases with increasing cluster size. This trend is seen in the electron impact ionization of ammonia clusters $(NH_3)_n$ where ionization thresholds of 10.25, 9.40, and 9.15 eV were found for n = 1 to 3, respectively [6]. Thresholds from photoionization are much more precise, and the same trend is observed for many different systems, e.g., Na_n, K_n, Fe_n, and Cu_n [21], $(Ar)_n$ [22], $(H_2S)_n$ and $((CH_3)_2CO)_n$ [23], and $(CS_2)_n$ and $(NO)_n$ [24]. In the metallic systems, however, some fluctuations from a monotonic decrease are evident. An expected exception to the general rule is that of ionic clusters, such as $(NaCl)_n$, where removal of an electron results in a decrease in the Madelung energy, and the vertical ionization potential should incease with increasing cluster size [25].

The trend is understandable in terms of a macroscopic model [26] in which the work function W of a flat surface is corrected for curvature and the effect of removing the electron from a droplet of radius r into which the image charge is confined. Based on classical electrostatics,

$$W(r) = W_\infty + \frac{3}{8}\frac{e^2}{r} \qquad (1)$$

The decrease in ionization potential can also be understood in terms of molecular orbital theory [27]. The resulting relationship is

$$I_n = I_1 - x|J| + \frac{x|J|}{n} \qquad (2)$$

where I_n is the ionization potential of a cluster of n molecules, x is the number of nearest neighbors, and J is the transfer integral. Also, the adiabatic ionization potential is reduced in cases where local bonding effects and/or polarization stabilize the ionized cluster relative to the neutral cluster.

Multiple ionization. At sufficiently high incident electron energies, multiple ionization may occur. Often, however, small multiply charged ions may be absent when the repulsive force of the individual charges exceeds the binding forces within the cluster, as noted by Henkes [28]. The macroscopic analog, known as the Rayleigh charging limit [29], is given by

$$z_{max} = \frac{4\sqrt{\pi\sigma}}{e} r^{3/2} \tag{3}$$

where z_{max} is the maximum number of elemental charges e that can be accommodated by a droplet of radius r with surface tension σ. In some cases, very small doubly charged clusters have been observed, e.g., clusters of atomic carbon [30]. Jentsch et al. [31] have applied an empirical rule relating the neutral bond energy to the Coulombic energy, in order to explain the presence or absence of doubly charged trimers of several metals. The existence of doubly charged heteroatomic rare gas dimers has also been shown and explained by Helm et al. [32].

Henkes and Isenberg [33] produced nitrogen clusters (average size 22000 molecules) with up to five elemental charges. In their experiment, multiple electron-cluster collisions were highly probable. They also observed the appearance of small charged clusters (average molecules per charge of 135) which were not present when the probability for multiple ionization was low. Consequently, these small clusters were thought to be due to Coulombic explosion from larger highly charged clusters. Gspann and Körting [34] reported relative cross-sections as a function of electron energy for each charge up to four elemental charges for nitrogen clusters (average size 12000 molecules) and hydrogen clusters (average size 65000 molecules).

Information on the minimum size at which a multiply charged cluster is found to exist may yield insight into the bonding of the ionized cluster. Several recent studies have experimentally observed the minimum size of doubly charged clusters: 73 for Kr_n^{++} [35], 21, 31, and 53 for $(NaI)_n^{++}$, Pb_n^{++}, and Xe_n^{++} [36], 35 and 51 for doubly protonated water and ammonia clusters [37], and 45, 109, and 217 for $(CO_2)_n^{+z}$ with z = 2, 3, and 4, respectively [38]. The actual minimum size could contain one less, but it is not observable because the mass-to-charge ratio is not discernible from the singly charged species.

Electron Attachment

The study of electron attachment to clusters has uncovered phenomena which are particularly interesting in comparison to free molecules. These include the enhancement of attachment rates by clustering on species which normally attach electrons as free molecules, the collective attachment of an electron by a cluster whose individual components have negative electron affinities, and the influence of clustering on dissociative attachment processes. In contrast to secondary electron ejection, attachment is more efficient at low electron energies. As a general rule, attachment also becomes more probable with larger cluster size. Gspann and Körting [34] observed by impact of 44 eV electrons on a beam that the ratio of negatively to positively charged clusters of hydrogen and nitrogen increased with increasing cluster size for average cluster sizes in the range of 10^3–10^5 molecules.

Enhancement of attachment rates. Studies of thermal electron attachment in dense gases have revealed that attachment to van der Waals dimers plays a significant role in the overall mechanisms [39,40]. Since the relative number of van der Waals molecules is generally small, a significant enhancement in attachment rate for the clusters is necessary in these cases. The enhancement for $O_2 \cdot M$ versus O_2 has been found to be due to several effects, including (1) a reduction of the resonance energy between ground state O_2 (v=0) and the excited vibrational state O_2^- (v=4) which results from the deeper potential well for $O_2^- \cdot M$ versus $O_2 \cdot M$, (2) an increase in the resonance width and number of resonance channels due to the vibrational structure of the van der Waals molecules, (3) a lowering of molecular symmetry which allows attachment of p wave electrons in addition to d wave electrons, and (4) the presence of a built-in third-body M which makes available a relaxation channel through dissociation into O_2^- and M [41,42].

Recently, Märk et al. [43] have examined electron attachment to a beam of oxygen clusters produced in a supersonic expansion (see Figure 4). Both O_{2n}^- and O_{2n+1}^- are observed. The O_{2n+1}^- intensity peaks at around 7 eV incident electron energy, which is similar in behavior as O^- from the dissociative attachment to O_2 except that the maximum of the cross-section seems to be shifted slightly to higher energy. A similar shift is reported for condensed oxygen [44]. The O_{2n}^- ions also exhibit a peak in intensity around 7 eV and another at very low energy (near 0 eV; resolution of electron energy is 0.5 eV). The intensity of the ions at about 0 eV becomes more dominating in comparison to those at 7 eV as the size of the observed ion increases from O_2^- to O_{20}^-. The strong decrease of the 7 eV peak for O_{20}^- is thought to be in part due to the rapid decrease in the number of larger neutral clusters from which the peak is populated.

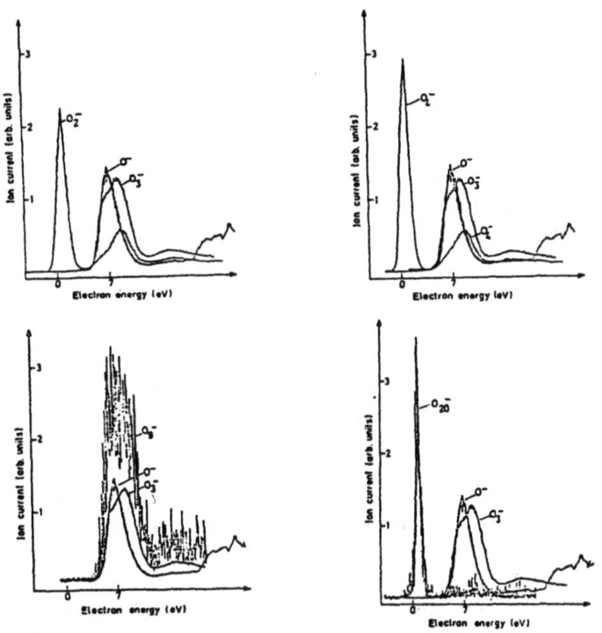

Figure 4. Ion yields from electron attachment versus electron energy for an oxygen cluster distribution produced at -140°C and 5.6 bar of oxygen expanding through a 10 μm nozzle. The O^- yield is from O_2 as is shown for comparison. [From reference 43.]

Collective effects. Another interesting aspect is the electron attachment to clusters where the free molecular units do not form stable negative ions. In many cases, however, the existence of solvated electrons, as in liquid ammonia, or of such ions in solid matrices are known. Klots and Compton [7] discovered an example of this in clusters by crossing a beam containing clusters of carbon dioxide with an electron beam. Ions of the type $(CO_2)_n^-$ for n > 2 up to the highest mass investigated (n = 6) were created, but CO_2^- is known to have an autoionization lifetime of 60-90 μs and was not observed. The maximum intensity of $(CO_2)_2^-$ was produced by electrons of about 3 eV. Evidence, based on the behavior of ion yields as a function of the stagnation pressure (i.e., the extent of clustering), indicated that $(CO_2)_2^-$ was produced from attachment to clusters larger than the trimer. In addition to the $(CO_2)_n^-$ ions, the ions of the form $O^-(CO_2)_n$ were observed where the maximum intensity shifted from 4.4 eV for O^- to 4.1 eV for CO_3^- and to slightly lower energies for higher clusters. At 8.2 eV where O^- reaches a second maximum, the other $O^-(CO_2)_n$ ions also display a peak as well. Klots [45] also reported the observation of $(H_2O)_x(CO_2)_y^-$, even for x = y = 1, where the negative ion of the free molecules do not exist.

Stamatovic et al. [46] have re-examined the CO_2 clusters and extended observations to lower energies (see also Figure 5). As in the case of O_2 clusters, attachment of near-zero energy electrons was observed and the ratio of intensities for attachment of near-zero energy electrons versus 3-4 eV electrons increased drastically with cluster size. The CO_3^- produced by the higher energy electrons and the $(CO_2)_3^-$ produced by the near-zero energy electron maximized at similar stagnation conditions. Likewise, $(CO_2)_9^-$ from about 0 eV electrons maximized at about the same stagnation pressure as $(CO_2)_3^-$ from 3 eV electrons. These results are a strong indication that the higher energy attachment process is accompanied by significantly more dissociation.

Recently, Haberland and coworkers [47,48] have produced negatively charged clusters of NH_3 and H_2O by creating electrons in the expansion region. The negatively charged cluster $(NH_3)_n^-$ was observed when n > 35. For $(H_2O)_n^-$, only n = 1 and 4 were not observed and n = 2, 6, and 7 and n > 10 were particularly prominent.

Effects on dissociative attachment. Clustering may also influence the dissociative attachment of an electron to a species in a cluster. For instance, Klots and Compton [49] observed $(CCl_4^-)(CO_2)$ which is particularly interesting

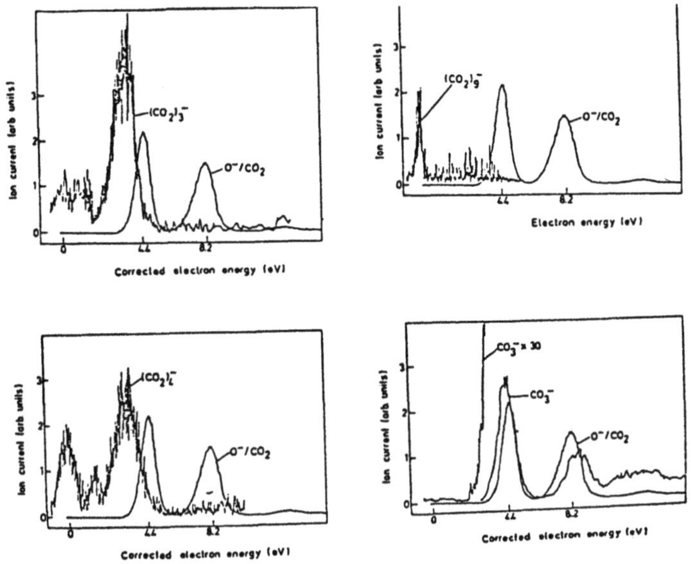

Figure 5. Ion yields from electron attachment versus electron energy for a carbon dioxide cluster beam produced by expanding gas at 5 bar and -30°C. Also shown is the O^- yield from CO_2. [From reference 46.]

since neither monomer attaches electrons to form the molecular anion,and thermal electron attachment to CCl_4 normally yields almost entirely Cl^-. The presence of clustered CO_2 molecules onto CCl_4 apparently leads to a stabilization of the negative ion by dissociating built-in CO_2 third bodies rather than fragmenting the CCl_4.

Another effect is that found in comparing the dissociative ionization of water and water clusters. Dissociative attachment to water vapor predominantly yields H^- and O^-. However, for water clusters, $OH^-(H_2O)_n$ are the most prominent ions [50]. In this case, internal ion-molecule reactions (4) and (5) occur.

$$H^-(H_2O)_n \rightarrow OH^-(H_2O)_n + H_2 \qquad (4)$$

$$O^-(H_2O)_n \rightarrow OH^-(H_2O)_n + OH \qquad (5)$$

The ion-molecule reactions of H^- and O^- with H_2O in the gas phase are well known.

Recombination with positively charged clusters. As in the case of uncharged species, clustering increases the rate of recombination. Biondi and coworkers have experimentally demonstrated this trend for several systems including $H_3O^+(H_2O)_n$ where the rate coefficient increases from $1x10^{-6}$ $cm^3\ sec^{-1}$ for n=0 to $5x10^{-6}\ cm^3\ sec^{-1}$ for n=5 [51]. Because the capture radius for clusters of this size should be similar (i.e., the charge and not the geometric size determines the collision cross-section), the electron-capture step probably involves excitation of vibrational and rotational modes in the ion cluster by the colliding electron. If the electron is captured in a high-lying Rydberg orbital, the larger number of degrees of freedom of the cluster ion-electron complex will decrease auto-ionization rates; thus the increased lifetime could allow other processes, such as dissociation into neutral products, to occur.

References

[1] Becker, E. W., Klingelhöfer, R., and Lahse, P., Z. Naturforsch. 1960, 15a, 644; Henkes, W., Phys. Lett. 1964, 12, 1322.

[2] Laudenslager, J. B., "Ion-Molecule Processes in Lasers," in Kinetics of Ion-Molecule Reactions, ed. P. Ausloos, NATO ASI 1979, Series B:40, 405; Meisels, G. G., Radiat. Phys. Chem. 1982, 20, 1.

[3] Smith, D., and Adams, N. G, Topics in Current Chemistry 1980, 89, 1.

[4] Kenney-Wallace, G. A., Acc. Chem. Res. 1978 , 11, 433.

[5] Compton, R. N., "Negative Ion States," in Proc. NATO Conf. on Photophysics and Photochemistry in the Vacuum Ultra-Violet, 1982.

[6] Stephan, K., Futrell, J. H., Peterson, K. I., Castleman, A. W., Jr., Wagner, H. E., Djuric, N., and Märk, T. D., Int. J. Mass Spect. Ion Phys., 1982, 44, 167.

[7] Klots, C. E., and Compton, R. N., J. Chem. Phys. 1978, 69, 1636.

[8] Hagena, O. F., and Henkes, W., Z. Naturforsch. 1965, 20a, 144.

[9] Falter, H., Hagena, O. F., Henkes, W., and Wedel, H., Int. J. Mass Spect. Ion Phys. 1970, 4, 145.

[10] Tay, E. S., Dawson, P. G., and Mosson, G. A. G., Proc. 6th Europ. Symp. Fusion Technol., Aachem. 1970, p. 51.

[11] Henkes, W., and Mikosch, F., Int. J. Mass Spect. Ion Phys.1974, 13, 151.

[12] Bottiglioni, F., Coutant, J., and Fois, M., Phys. Rev. 1972, A6, 1830.
[13] Gspann, J., Surf. Sci., 1981, 106, 219.
[14] Märk, T. D., Beitr. Plasmaphysik, 1982, 22, 257.
[15] Buck, U., and Meyer, H., Phys. Rev. Lett. 1984, 52, 109.
[16] Keesee, R. G., Sievert, R., and Castleman, A. W., Jr., Ber. Bunsenges. Phys. Chem., 1984, 88, 273.
[17] Geraedts, J., Stolte, S., and Reuss, J., Z. Phys. 1982, A304, 167.
[18] Worsnop, D. R., Buelow, S. J., and Herschbach, D. R., J. Phys. Chem. 1984, 88, 4506.
[19] Gough, T. E., and Miller, R. E., Chem. Phys. Lett. 1982, 87, 280.
[20] Märk, T. D., and Castleman, A. W., Jr., Adv. At. Mol. Phys. 1985, 20, 65.
[21] Peterson, K. I., Dao, P. D., Farley, R. W., and Castleman, A. W., Jr., J. Chem. Phys. 1984, 80, 1780; Hermanmn, A., Schumacher, E., and Wöste, L., J. Chem. Phys. 1978, 68, 2327; Rohlfing, E. A., Cox, D. M., and Kaldor, A., Chem. Phys. Lett. 1983, 99, 161; Powers, D. E., Hansen, S. G., Geusic, M. E., Michalpoulous, D. L., and Smalley, R. E., J. Chem. Phys. 1983, 78, 2866.
[22] Dehmer, P. M., and Pratt, S. T., J. Chem. Phys. 1982, 76, 843.
[23] Walters, E. A., and Blais, N. C., J. Chem. Phys. 1981, 75, 4208; Trott, W. M., Blais, N. C., and Walters, E. A., J. Chem. Phys. 1978, 69, 3150.
[24] Ng, C. Y., Adv. Chem. Phys. 1983, 52, 263.
[25] Martin, T. P., Phys. Rep. 1983, 95, 169.
[26] Smith, J. M., AIAA J. 1965, 3, 648.
[27] Jortner, J., Ber. Bunsenges. Phys. Chem. 1984, 88, 188.
[28] Henkes, W., Z. Naturforsch. 1962, 17a, 786.
[29] Lord Rayleigh, Phil. Mag. 1882, 14, 185.
[30] Dornenburg, E., Hintenberger, H., and Franzen, J., Z. Naturforsch. 1961, 16a, 532.
[31] Jentsch, T., Drachsel, W., and Block, J. H., Chem. Phys. Lett. 1982, 93, 144.
[32] Helm, H., Stephan, K., Mark, T D., and Huestis, D. L., J. Chem. Phys. 1981, 74, 3844.
[33] Henkes, W., and Isenberg, G., Int. J. Mass Spect. Ion Phys. 1970, 5, 249.
[34] Gspann, J., and Korting, K., J. Chem. Phys. 1973, 59, 4726.
[35] Ding, A., and Hesslich, J., Chem. Phys. Lett. 1983, 94, 54.

[36] Sattler, K., Muhlbach, J., Echt, O., Pfau, P., and Recknagel, E., Phys. Rev. Lett. 1981, 478, 160.
[37] Shukla, A. K., Moore, C., and Stace, A. J., Chem. Phys. Lett. 1984, 109, 324.
[38] Echt, O., Sattler, K., and Recknagel, E., Phys. Lett. 1982, 90A, 185.
[39] Nagra, S. S., and Armstrong, D. A., J. Phys. Chem 1977, 81, 599.
[40] Hatano, Y., and Shimamori, H., Electron Attachment in Dense Gases, in Electron and Ion Swarms, ed. L. G. Christophorou (Pergamon Press, 1981), p. 103.
[41] Huo, W. M., Fessenden, R. W., and Baurschlicher, C. W., Jr., J. Chem. Phys. 1984, 81, 5811.
[42] Toriumi, M., and Hatano, Y., J. Chem. Phys. 1985, 82, 254.
[43] Märk, T. D., Leiter, K., Ritter, W., and Stamatovic, A., Low-Energy Electron Attachment to Oxygen Clusters, Phys. Rev. Lett., submitted.
[44] Sanche, L., Phys. Rev. Lett. 1984, 53, 1638.
[45] Klots, C. E., J. Chem. Phys. 1979, 71, 4172.
[46] Stamatovic, A., Leiter, K., Ritter, W., Stephan, K., and Märk, T. D., J. Chem. Phys. 1985, in press.
[47] Haberland, H., Schindler, H.-G., and Worsnop, D. R., Ber. Bunsenges. Phys. Chem. 1984, 88, 270.
[48] Haberland, H., Ludewigt, C., Schindler, H.-G., and Worsnop, D. R., J. Chem. Phys. 1984, 81, 3742.
[49] Klots, C. E., and Compton, R. N., J. Chem. Phys. 1977, 67, 1779.
[50] Klots, C. E., and Compton, R. N., J. Chem. Phys. 1978, 69, 1644.
[51] Huang, C.-M., Whitaker, M., Biondi, M. A., and Johnsen, R., Phys. Rev. A., 1978, 18, 64.

*Leopold-Franzens University, Innsbruck, Austria

PART VII

MICROSCOPIC MODELS:

DATA NEEDS AND OUTLOOK

DISCHARGE MODELING

J.P. Boeuf* and E.E. Kunhardt**

*CNRS, École Superieure d'Electricité, Gif sur Yvette, France
**Polytechnic Institute of New York, Farmingdale, New York
USA

Introduction

Plasmas produced in low-pressure discharges are used in an increasing number of applications such as gas discharge lasers, plasma etching, plasma deposition and switching devices. The development of these technologies has led to an important need for discharge modeling; discharge modeling should be able not only to provide a better understanding of the discharge mechanisms, but also to predict and optimize the discharge regime of a given device. The availablity of high-speed computers, combined with the recent development of efficient numerical techniques, has made possible the accurate modeling of very complex systems. However, in spite of this important progress, the development of really efficient discharge modeling is still subject to the availability of basic microscopic data such as electron-molecule or molecule-molecule collision cross-sections, which exist only for a very limited number of gases.

In this paper we describe briefly the main problems which must be dealt with in modeling discharges. The need for basic data in this field will then be illustrated by the modeling of a nitrogen discharge under conditions where the coupling between the electron kinetics and the vibrational population plays a fundamental role.

Goals and Difficulties of Discharge Modeling

Given the gas components, the electrode material and geometry, a complete model of a discharge device should be able to yield the local and integral characteristics of the discharge: charged particles densities, gas composition, excited species concentrations, radiation efficiency, current-voltage characteristics, etc. We briefly describe below the problems which must be dealt with to reach this goal.

There are two important features characterizing gas discharges which are of interest in applications like gas lasers or plasma processing:

1) Due to the high electric field and the consequently high ratio of electron to gas temperatures, the plasma created by the discharge is generally far from local thermodynamic equilibrium (LTE). It is just this nonequilibrium property of a cold plasma which makes it so attractive for plasma processing or gas discharge lasers, because it makes possible the production of highly excited species while keeping the gas temperature quite low. A consequence of this nonequilibrium is that the excited species concentration cannot be deduced from the plasma temperature and must be obtained from a detailed analysis of the various elementary processes which contribute to determine these populations: electron impact excitation, dissociation, ionization, attachment, reactions between molecules or atoms, radiation emission or absorption, diffusion, etc.

Since the electron gas is the most active component of a nonequilibrium plasma, the electron distribution function (EDF) f(r,v,t) occupies a central position in any kinetic description of the chemical reactions occurring in such a plasma; the electron impact excitation rates are deduced from the distribution function.

In nonequilibrium plasmas, and for low degrees of ionization (less than 10^{-4}) the electron distribution function follows neither a Maxwellian nor a Druyvesteyn form. Rigorous determination of the EDF involves a numerical solution of the Boltzmann equation, which requires a prior knowledge of the electron-molecule collision cross-sections. In swarm physics, solving the Boltzmann equation for a given reduced electric field E/n (E is the electric field and n the gas density) and for a given set of cross-sections is the main problem. The situation is more involved in discharge modeling; for high enough electron current densities, the excited species population may become significant with respect to the gas density, so that second-kind collisions (superelastic collisions, stepwise excitation, dissociation or ionization) can play a fundamental role. In that case, the electron distribution function does not depend only on the reduced electric field E/n, but also on the excited species population and is, in turn, itself determined by the excited species population. It has been shown [1] that second-kind collisions can drastically modify the shape of the EDF and increase by orders of magnitude some excitation, dissociation or ionization rate coefficients.

Consequently, a realistic discharge model must solve the Boltzmann equation coupled with the system of differential equations describing the kinetics of vibrational and electronic excitation, dissociation and ionization processes.

2) Another important feature which must be considered in modeling discharges concerns the existence of very large spatial variations of the electric field in some regions of the discharge. In such regions the electron kinetics are not in equilibrium with the electric field [2]; the local electron transport coefficients do not depend only on the local electric field and thus cannot be deduced from swarm experiments. Since the discharge regime is controlled by the electron behaviour in these nonequilibrium regions (e.g., cathode fall of a glow discharge), a detailed analysis of these regions is necessary, involving a numerical solution of the complete Boltzmann equation (including spatial gradients terms [2] coupled with the Poisson equation). Moreover, the electron kinetics are also coupled to the gas temperature and density (the EDF depends on the E/n ratio), and a complete analysis of the energy balance in the discharge must be performed; gas heating can be due to charge exchange collisions, energy exchange between the vibrational and translational degrees of freedom of the molecules, etc. (See next section.) Evidence of the importance of gas heating in the cathode region of a glow discharge has been shown experimentally [3].

A diagram showing some aspects of discharge modeling is presented in Fig. 1. Other important phenomena such as particle surface interactions and radiation transport are not represented in this figure. Of course, depending on the specific conditions of the discharge and on the aim of the modeling, some of the couplings appearing in Fig. 1 may be nonessential, and simple models (e.g., based on a macroscopic description of the electron kinetics) can often give correct qualitative results.

Looking at Fig. 1, modeling discharges seems to be a very formidable task, and being able to predict and optimize the regime of a given discharge device appears as a rather remote goal. However, the reason for which efficient quantitative discharge modeling is still a remote goal is no longer due to mathematical or numerical difficulties: very efficient numerical techniques are now able to handle most of the problems mentioned above within a reasonable amount of computation time (see [1], [2], and the review by P. Segur in these Proceedings). There still remain some numerical problems concerning the coupling between Boltzmann and Poisson equations under steady state conditions, but this question will be solved in the very near future. The main difficulty actually stems from the lack of basic microscopic data concerning the elementary processes occurring in the discharge. Assuming that all the mathematical and numerical difficulties of the model have been overcome, one is still faced with the following problems:

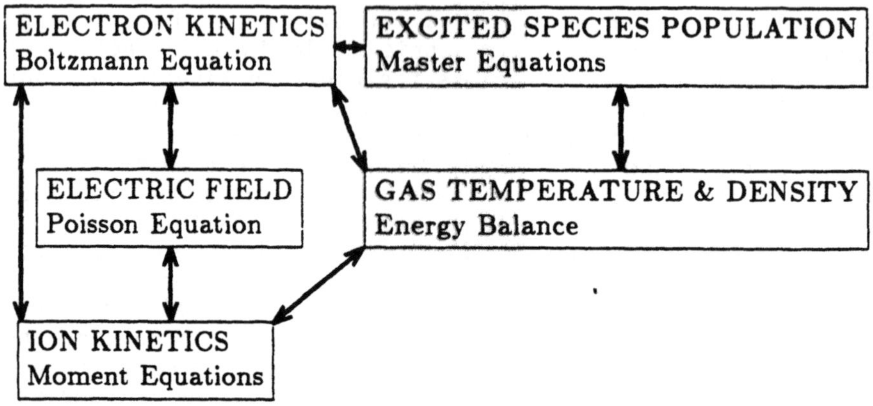

Figure 1.

1) determination of the reactions which play an essential role,
2) finding the basic data for these reactions,
3) evaluation of the sensitivity of the model to variations in the more critical (or more dubious) data [4].

This last point is essential for a proper interpretation of the results, and numerical methods able to perform systematic sensitivity analysis should be developed.

Modeling of Nonequilibrium Nitrogen Discharge

In order to illustrate some aspects of discharge modeling which are mentioned in the above section, we present a study of the coupling between the electron distribution function and the excited species population, and its consequence on the energy balance in a nitrogen discharge.

In a nitrogen discharge, the vibrational manifold forms a reservoir of energy which can be released into heat (translational energy of the molecules), electronic excitation, dissociation and ionization. The stability of the discharge depends strongly on the way this energy is released. For example, local gas heating leads to higher E/n which can lead to higher ionization, higher electron density and further heating.

The aim of this work was to analyze the effect of superelastic collisions on the EDF, to determine the vibrational population under nonequilibrium conditions, and to obtain quantitatively the fractional part of the electron energy which is transferred into heat in a nitrogen discharge.

Vibrational relaxation. Before going into the details of the model, some features of the vibrational relaxation of diatomic molecules under non LTE conditions are summarized in this section.

Under LTE conditions, the vibrational population follows a Boltzmann law:

$$N_v = N_0 \exp -\frac{\epsilon_v}{kT} \tag{1}$$

where N_v is the concentration of the v level, ε_v its energy, k the Boltzmann constant, and T the plasma temperature.

In a nonequilibrium plasma, equation (1) is no longer valid, and the vibrational population must be obtained from a detailed analysis of the elementary processes which contribute to populate or depopulate the difference vibrational levels. The main processes which must be considered in nitrogen are:

- e-V (electron-vibration) energy exchange:

$$e + N_2(v) \rightarrow e + N_2(w) \tag{i}$$

- V-V (vibration-vibration) energy exchange:

$$N_2(v) + N_2(w) \rightarrow N_2(v-1) + N_2(w+1) \tag{ii}$$

- V-T (vibration-translation) energy exchange:

$$N_2(v) + N_2 \rightarrow N_2(v-1) + N_2 \tag{iii}$$

Reactions (ii) and (iii) only account for one-quantum transitions; multi-quantum transitions are generally assumed to be negligible except at extremely high temperatures.

In a V-T reaction (iii), there is no conservation of the number of quanta and consequently no conservation of the energy among the vibrational manifold; the amount of energy

$\Delta\varepsilon_{v-T} = \varepsilon_{v-1}$ is exchanged between the vibrational and translational degrees of freedom of the molecule.

In a V-V reaction (ii), there is conservation of the number of quanta but no conservation of the vibrational energy; this is due to the fact that a diatomic molecule must be considered as an anharmonic oscillator: the energy gap $(\varepsilon_v - \varepsilon_{v-1})$ between successive levels is not constant, but decreases as v increases, according to:

$$\epsilon_v - \epsilon_{v-1} = \epsilon_{10}(1 - 2\delta v) \tag{2}$$

where δ is the anharmonicity factor (δ = 0.006217 and ε_{10} = 0.29ev in nitrogen). The consequences of the anharmonicity of the molecule are twofold:

1) Due to the energy defect $\Delta\varepsilon_{(V-V)-T} = 2\varepsilon_{10}\delta(w-v+1)$ in reaction (ii), it appears that the role of V-V processes will be to populate the high-lying vibrational levels at the expense of the lower levels (the effect of V-T processes is the opposite).

2) A small amount of energy $\Delta\varepsilon_{(V-V)-T}$ is exchanged between vibration and translation in a V-V process. Although this energy defect is much smaller than the energy exchanged in a V-T reaction, V-V processes may lead to important gas heating ((V-V)-T heating) because V-V rate coefficients are much larger than V-T rate coefficients in nitrogen (except for the highest levels).

The formation of a nonequilibrium vibrational population can be described as follows:

1) The electrons populate the first few vibrational levels. For large energy input, the "temperature" of the first level can be considerable larger than the gas temperature.

2) The energy introduced by the electrons to the first few levels is then redistributed by V-V exchange over the whole vibrational manifold. V-V processes are dominant for intermediate vibrational levels and tend to establish a population inversion (in the absence of other processes). The vibrational relaxation can be studied ana-

lytically and is characterized by the Treanor distribution [5]).

3) V-T processes dominate for the highest quantum states, prevent large population inversions and tend to establish for these levels a Boltzmann distribution at the translational temperature.

Description of the model. In this model the reduced electric field E/n (ranging from 30 to 90 Td; 1 Td=10^{21}Vm^2), electron number density n_e (10^{12}cm^{-3}), gas pressure p (100 torr) and gas temperature T_g (500 K) are kept constant. Spatial gradients are not considered. The time evolution of the electron distribution function coupled with the vibrational population and the concentration of the $N_2(A^3\Sigma_u^+)$ state is studied, assuming that all the nitrogen molecules are in the ground state $N_2(X^1\Sigma_g^+$, v = 0) at time t = 0.

The elementary processes which have been included in the model are listed below:

- electron-molecule collisions

The set of collision cross-sections derived by A.V. Phelps [6] has been used except for the vibrational excitation. This set includes elastic collisions and rotational excitation, electron excitation (13 levels), dissociation and ionization from the ground state. In the case of vibrational excitation, all the transitions between N_2(v) and N_2(w) (see reaction (i)) within the first eight vibrational levels have been considered, using the theoretical cross-sections calculated by Chandra and Temkin [7]. For higher vibrational levels, V-V or V-T processes are dominant and the effect of vibrational excitation by electron impact on the vibrational population can be neglected (the validity of this approximation may, however, become questionable for higher electron number densities).

- molecule-molecule collisions

Single quantum V-V and V-T transitions (see (ii) and (iii)) within the whole vibrational ladder (46 levels) have been taken into account, using the rate coefficients inferred from the modified SSH theory ([8] and references therein). Dissociation due to vibrational mechanisms (one-quantum V-V or V-T reactions) is considered but the recombination of atomic nitrogen is not taken into account. The $N_2(A^3\Sigma_u^+)$ concentration is obtained self-consistently, assuming that this state is populated by electron excitation and radiative decay of $N_2(B^3\Pi_g)$ and $N_2(C^3\Pi_u)$ and depopulated by the self-quenching reactions:

$$N_2(A^3\Sigma_u^+) + N_2(A^3\Sigma_u^+) \rightarrow N_2(X^1\Sigma_g^+) + N_2(B^3\Pi_g) \qquad (iv)$$

$$N_2(A^3\Sigma_u^+) + N_2(A^3\Sigma_u^+) \rightarrow N_2(X^1\Sigma_g^+) + N_2(C^3\Pi_u) \qquad (v)$$

(the rate coefficients of these reactions are taken from [9]).

The energy balance of these reactions is not known. However, in order to estimate the probable contribution of these processes to the gas heating, it has been assumed in the model that half of the remaining energy in reactions (iv) and (v) is expended in heating and the other half in vibrational excitation of the ground state.

The main approximations implied by this model are:

1) Electronic excitation, dissociation or ionization from vibrationally excited molecules are not considered. Although there are no experimental cross-sections for these processes, their effect on the EDF has been studied by Cacciatore et.al. [10] (see also the paper by M. Capitelli in these Proceedings) using Gryzinski's theoretical cross-sections.

2) The atomic nitrogen concentration is not calculated self-consistently because of the lack of data concerning the recombination of nitrogen atoms. The calculation is stopped when the atom density (which is estimated by considering only the creation processes: dissociation due to electron impact or pure vibrational mechanisms) becomes significant with respect to the molecule density.

The methods of calculation of the electron distribution function and the excited species densities are briefly described below.

The EDF is obtained from a Monte-Carlo simulation of the electron kinetics, which is equivalent to solving the time dependent Boltzmann equation,

$$(\partial_t + \vec{a}.\vec{\nabla}_v) f(\vec{v},t) = J[f](\vec{v},t) \tag{3}$$

where a is the acceleration due to the applied electric field. The collision operator J includes the electron-molecule processes mentioned above.

The excited species densities are obtained by solving the system of master equations:

$$\frac{dN_v}{dt} = \left(\frac{\partial N_v}{\partial t}\right)_{e-V} + \left(\frac{\partial N_v}{\partial t}\right)_{V-V} + \left(\frac{\partial N_v}{\partial t}\right)_{V-T} + \left(\frac{\partial N_v}{\partial t}\right)_{A-A} \tag{4}$$

$0 \leq v \leq 45,$

$$\frac{dN_A}{dt} = nn_e K_1 - N_A{}^2 K_2 \tag{5}$$

The terms in the rhs of (4) correspond to the rates of change of N_v due to e-V, V-V, V-T processes, and self-quenching of the A state (it has been assumed that half of the remaining energy in reactions (iv) and (v) goes into vibrational excitation of $N_2(X^1\Sigma_g{}^+)$). Equation (5) is the rate equation for the A state; K_1 and K_2 are the rates of production (electron impact excitation of A, B and C) and destruction (self-quenching) of the A state. More details concerning the system of master equations can be found in [11].

The coupling between the Boltzmann equation and the system of master equations appears in the vibrational part of the collision operator of (3) and in the electron source term of (4). Since the relaxation times for electron-molecule and molecule-molecule reactions are very different, it is not necessary to solve the Boltzmann equation and the system of master equations with the same time step. The EDF is calculated at times $t_0, t_1 \ldots, t_i, \ldots$ by assuming successive states of quasi-equilibrium with the excited species densities. During a time interval (t_i, t_{i+1}), the system of master equations is solved with the electron source term being kept constant and equal to its value at time t_i. The time intervals are chosen small enough to allow only small changes in the EDF [11].

Results and Discussion. The time evolution of the EDF and the vibrational population for E/n=30 Td(n_e=$10^{12}$$cm^{-3}$,Tg = 500K, p = 100 torr) are shown in Figs. 2 and 3, respectively. At time t=0, all the molecules are assumed to be in the ground state $N_2(X^1\Sigma_g^+)$. As time increases, the first few vibrational levels are populated by electron impact. As the population of these levels increases (Fig. 3), second-kind collisions become significant. As can be seen in Fig. 2, their effect is to enhance the tail of the electron distribution function. The energy provided by the electrons to the first few vibrational levels is then pumped by V-V processes towards the higher vibrational levels. V-T reactions prevent a population inversion, and the combination of V-V and V-T processes lead to the formation of a 'plateau' in the vibrational distribution, for $15 < v < 40$. For the highest vibrational levels, V-T reactions are dominant and the vibrational population decreases sharply.

It appears in Figs. 2 and 3 that the EDF and the population of the first few levels reach an equilibrium well before the higher part of the vibrational manifold. The "temperature" of the first few levels is about 4000 K at time t=0.4 ms. After longer times (which have not been in investigated in the present work), the formation of atomic nitrogen can strongly modify the EDF and the vibrational population.

The time evolution of the electron mean energy for E/n values ranging from 30 to 90 Td is shown on Figure 4. The increase of the mean energy due to superelastic collisions can be very large (up to 100% for E/n=90 Td) in (qualitative) agreement with the results of Capitelli et.al. [1].

The fractional electron power transferred to the vibrational manifold at times t=0 and t=0.5 ms is shown in Fig. 5. At time t=0 a large amount of the electron energy goes into vibrational excitation (no second kind collisions [12]). As time increases a non-negligible amount of the electron energy which had gone into vibrational excitation is recovered by the electrons through superelastic collisions. This energy later goes into electronic excitation, dissociation or ionization.

Figure 6 shows the time evolution of the fractional electron power transferred into heat by V-V and V-T processes and by self-quenching of $N_2(A^3\Sigma_u^+)$ for E/n=70 Td.

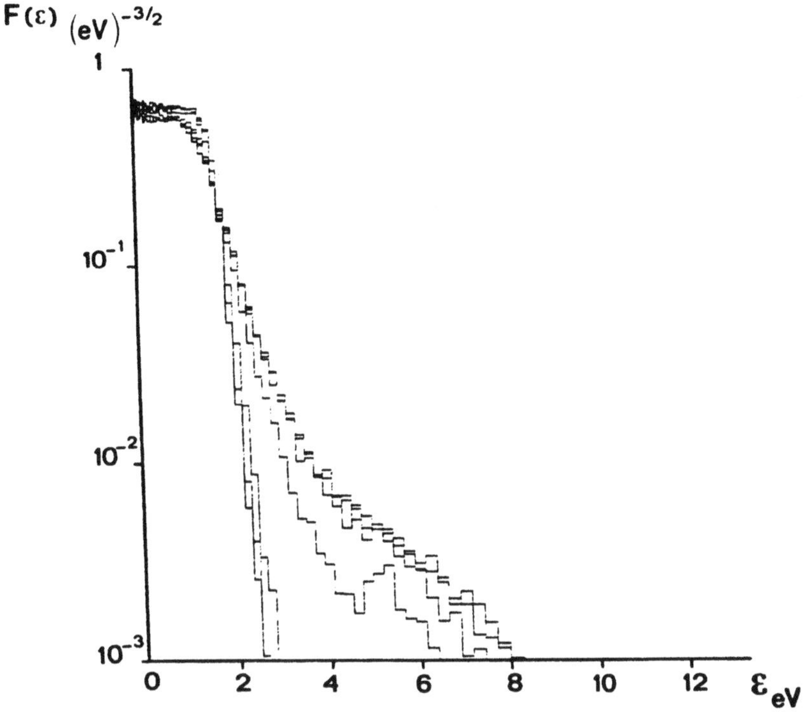

Figure 2. Variations of the EDF for E/n=30 Td at times t=0, 60, 250, 500, 600, 700 μs ($n_e = 10^{12}\,cm^{-3}$, $T_g = 500\,K$ p = 100 torr).

It appears that the contribution of the V-V processes to the gas heating ((V-V)-T or "anharmonic" heating) is large in spite of the very small amount of energy exchanged between vibration and translation in each V-V reaction. (V-V)-T heating is faster than V-T heating, but not fast enough to explain the short-time heating which is observed experimentally in nitrogen ([13] and references therein). Rotational excitation followed by R-T relaxation is the dominant process for short-time heating for E/n values below 40 Td [13]. Since the R-T relaxation time is very small, the rotational and translational degrees of freedom of the nitrogen molecule can be considered to equilibrate instantaneously. For higher E/n the contribution of rotational excitation to the gas heating decreases and the fractional energy input to gas heating via rotational excitation is less than 3% for E/n larger than 40 Td [13]. Figure 6 shows that large amounts of energy can be quickly transferred into heat via the quenching of electronically excited states.

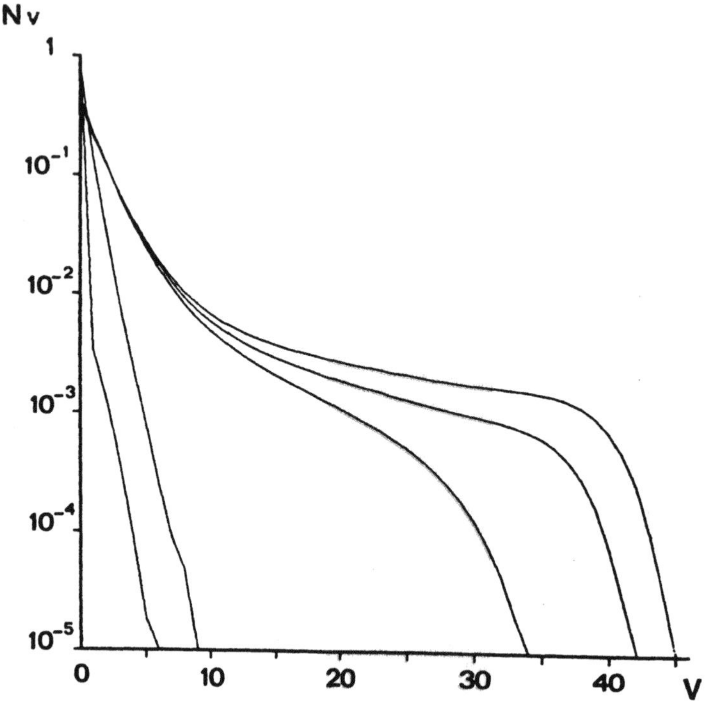

Figure 3. Reduced vibrational population at times t=2, 60, 500, 700, 1100 μs (same conditions as Fig. 2).

However, further experimental or theoretical investigations are needed to improve the model of energy balance which has been assumed for the quenching of the $A^3\Sigma_u^+$ state.

Finally, it must be stressed that a true steady-state situation has not been obtained in this work. A steady-state calculation should include a self-consistent determination of the atomic nitrogen density. Since atomic nitrogen is probably a much better quencher of vibrationally excited nitrogen molecules than molecular nitrogen, its effect on the steady-state vibrational population and energy balance can be important.

From the results presented in this section, the following conclusions can be made:

- Simulations of the time evolution of the electron distribution function and the vibrational population in a weakly ionized nitrogen plasma confirm that the coupling between the electron kinetics and the excited species

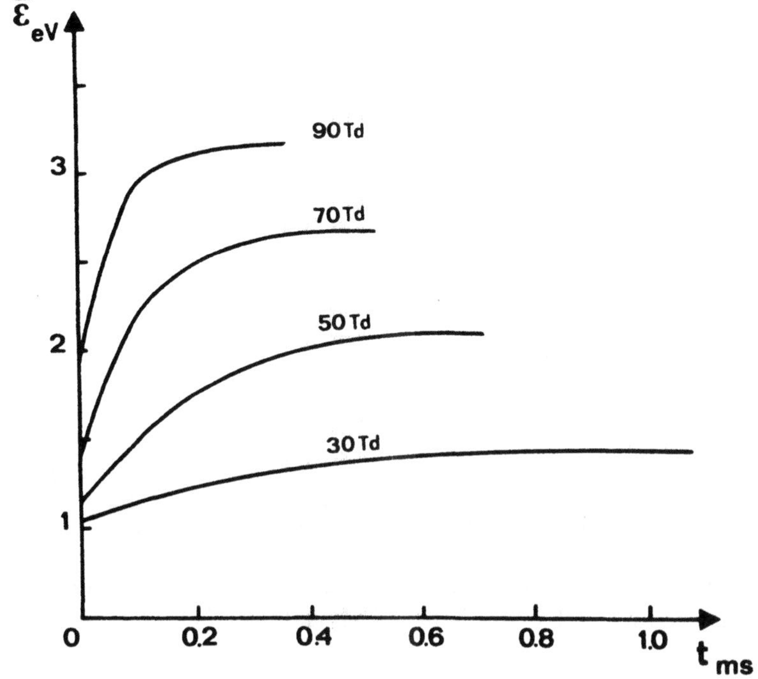

Figure 4. Time evolution of the electron mean energy for different values of the reduced electric field.

densities due to second-kind collisions can lead to an important increase of the tail of the EDF and consequently of the electron macroscopic parameters (mean energy, electronic excitation, dissociation and ionization rates).

- V-V reactions can be an important and fast mechanism for energy exchange between vibration and translational motion due to the anharmonicity of the molecule.

- The self-quenching region of the A state is a good candidate for short-time heating in nitrogen (for E/n above 40 Td).

Concluding Remarks

The example presented above has illustrated the difficulty of developing quantitative models for gas discharges. The availability of fast computers and the development of efficient numerical techniques has made possible the accurate

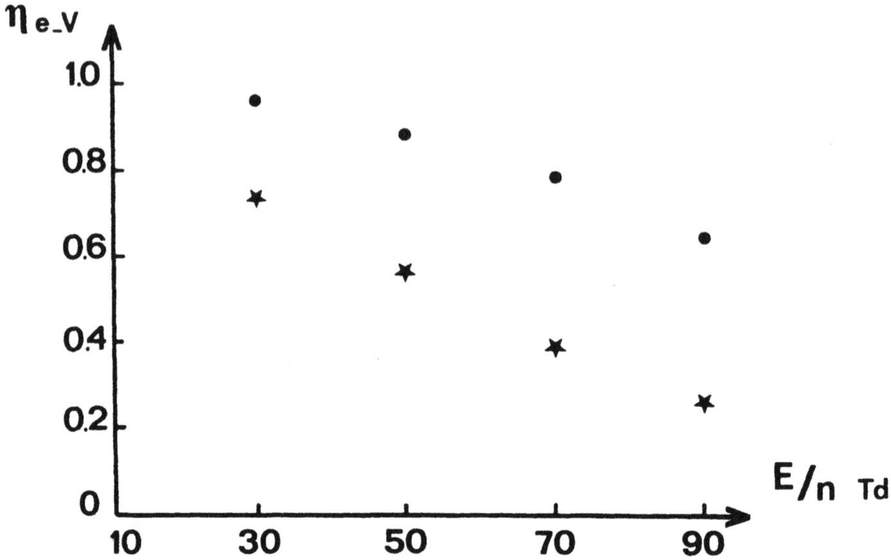

Figure 5. Fractional electron power transferred into vibrational excitation vs. reduced electric field at times t=0 (•) and t=0.5 (*) ms.

modeling of the complex and interrelated phenomena that occur within a discharge. Further progress and the development of realistic and quantitative models will depend on the availability of the appropriate basic data. It is expected that this Joint Symposium and other interactions between the disciplines of discharge and swarm physics, and electron-molecule collisions will help to develop successful discharge modeling.

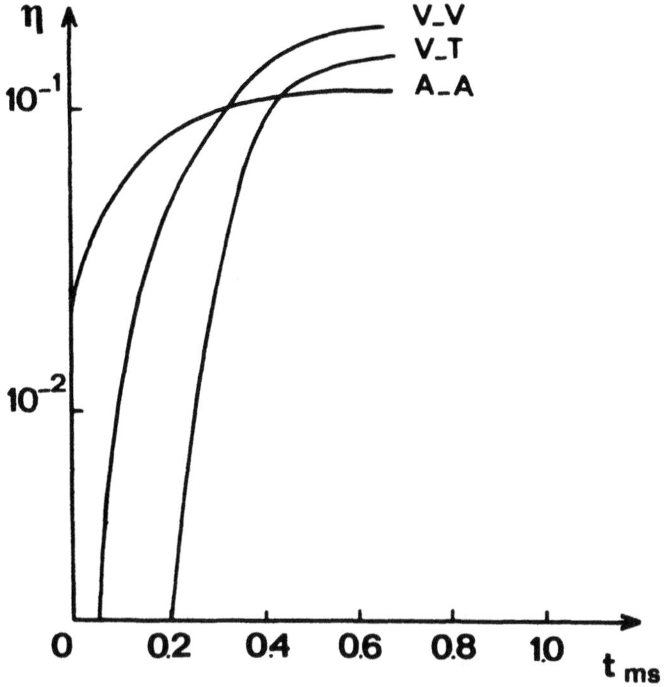

Figure 6. Time evolution of the fractional electron power transferred into heat through V-V, V-T and A-A (self-quenching of the $A^3\Sigma_u^+$ state) reactions.

Acknowledgements

This work was supported by the Office of Naval Research under Contract No. N00014-81-K-0655. We wish to express our appreciation to A.V. Phelps for his encouragement and for a number of very stimulating discussions.

References

[1] M. Capitelli, M. Dilonardo and C. Gorse, 1981, 56, 29.
[2] W.H. Long, 1979, Plasma Sheath Processes, Technical Report AFAPL-TR-79-2038; J.P. Boeuf and E. Marode, 1982, Phys. D:Appl. Phys., 15, 2069; P. Segur, M. Yousfi, J.P. Boeuf, E. Marode, A.J. Davies and J.G. Evans, 1983, in Electrical Breakdown and Discharges in Gases, ed. E.E. Kunhardt and L.H. Luessen, Plenum, New York, A 331.
[3] D.K. Doughty, E.A. DeHartog and J.E. Lawler, 1985, Appl. Phys. Lett., 46, 352. Lett., 46, 352.
[4] Group discussion on data needed from basic physics for modeling, 1984, in Gaseous Dielectrics IV, ed. L.G. Christophorou and M.O. Pace, Pergamon Press, 549.
[5] C.E. Treanor, J.W. Rich and R.G. Rehn, 1967, J. Chem. Phys., 48, 1978.
[6] A.V. Phelps, 1982, personal communication.
[7] N. Chandra and A. Temkin, 1976, NASA TN D-8347.
[8] M. Capitelli and M. Dilonardo, 1977, Chem. Phys., 24,417; Rev. Phys. Appl., 13, 115. Phys. Appl., 13, 115.
[9] G.N. Hays and H.J. Oskam, 1983, J. Chem. Phys., 59, 1507.
[10] M. Cacciatore, M. Capitelli and C. Gorse, 1982, Chem. Phys., 66, 141.
[11] J.P. Boeuf and E.E. Kunhardt, to be submitted.
[12] W.L. Nighan, 1970, Phys. Rev. A 2, 1983.
[13] A.V. Phelps and L.C Pitchford, 1985, Phys. Rev. A 31, 2932.

BOLTZMANN EQUATION IN VIBRATIONALLY AND ELECTRONICALLY EXCITED MOLECULAR PLASMAS

M. Capitelli and C. Gorse

Department of Chemistry, University of Bari (Italy)
and Research Center for Plasma Chemistry - CNR

Solution of the Boltzmann equation for the electron energy distribution function (edf) is the first step toward the understanding of many properties of molecular plasmas [1,2]. The problem is nowadays straightforward once the relevant elastic and inelastic cross-sections are known.

However, when the concentration of excited states (vibrational and electronic) becomes appreciable, complications arise because of superelastic (second-kind) and inelastic collisions involving excited states. This situation is typical of molecular plasmas (in particular N_2 and CO) of low ionization degree due to the fact that the input of vibrational energy by electrons is rapidly spread over the whole vibrational manifold of the molecule, yielding a vibrational distribution N_v more populated in the high vibrational levels than the corresponding Boltzmann one for the same ratio of N_v=0 to N_v=1. That is, a partial depletion of the ground vibrational level of the ground electronic state and a consequent filling of higher vibrational levels occurs. Solution of the Boltzmann equation under these conditions should include each vibrational level as a new species with its own cross-sections (which in general are poorly known). The same occurs for electronically excited states.

In this paper we review the attempts made in the past years to understand the role of excited levels (vibrational and electronic) in affecting the edf and related properties of molecular plasmas.

Method of Calculation

Electron energy distribution functions, vibrational distributions and the degree of dissociation in molecular plasmas can be obtained by solving simultaneously 1) the Boltzmann equation for the edf, 2) the vibrational master equation for N_v, and 3) the plasma chemistry describing the dissociation process [3,4]. In implicit form we can write

$$\delta n(\varepsilon,t)/\delta t=-(\delta J_f/\delta\varepsilon)-(\delta J_{el}/\delta\varepsilon)+In+(Sup)_v+(Sup)_e+Ion+Rot \quad (1)$$

$$dN_v/dt=(dN_v/dt)_{e-V}+(dN_v/dt)_{V-V}+(dN_v/dt)_{V-T}+(dN_v/dt)_{e-D}+\ldots (2)$$

$$dN_{v'+1}/dt=V_D(DEM)+V_D(PVM) \quad (3)$$

Equation 1 is the Boltzmann equation [5] which describes the temporal evolution of electrons with energy between ε and $\varepsilon+d\varepsilon$ subjected to the action of the electric field $(\delta J_f/\delta\varepsilon)$, elastic $(\delta J_{el}/\delta\varepsilon)$, inelastic (In), superelastic vibrational $(Sup)_v$, superelastic electronic $(Sup)_e$, ionization (Ion) and rotational (Rot) collisions. Equation 2 is the vibrational master equation describing the temporal evolution of the vth vibrational level under the influence of e-V processes, of V-V (vibration-vibration) and V-T (vibration-translation) energy transfer, of e-D (electron dissociation) and other mechanisms. Equation 3 describes the dissociation rate according to direct electronic and pure vibrational mechanisms [6]. The coupling between the three equations occurs through the superelastic vibrational collisions, the inelastic collisions involving excited vibrational and electronic levels and the chemical kinetics which modifies the plasma composition.

The net result of this coupling is that the electron energy distribution function depends not only on the reduced electric field E/N but also on the vibrational distribution N_v, on the number of atoms N_a (or reactive species) coming from the dissociation process and on the number density of electronically excited molecules N*; i.e., we can write edf = edf(E/N, N_v, N_a, N*). Different situations can be envisaged during the solution of the systems of eqs. 1-3.

First of all we consider the presence of vibrationally excited levels only in the Sup and In terms; i.e., we consider a situation in which the processes (e-V)

$$e + M_2(v) \rightleftharpoons M_2^- \rightleftharpoons e + M_2(w) \qquad a)$$

are included in the Boltzmann equation, while all the other inelastic processes start from v=0. The advantages of this situation lie in the fact that cross-sections for process (a) are today known with sufficient accuracy not only for the process $0 \to w$, but also for $v \to w$, $v \neq 0$[7]. In addition, a considerable amount of information is also available on excitation and ionization cross-sections involving the ground vibrational level of ground electronic state [8]. We refer to this situation as case A of our study.

The second situation (hereafter called case B) includes in the Boltzmann equation in addition to process (a) also inelastic and ionization processes starting from a large number of vibrational levels of the ground electronic state; i.e., we insert processes such as

$$e + M_2(v) \longrightarrow M_2^* \longrightarrow e + 2M \qquad \text{b)}$$

$$e + M_2(v) \longrightarrow e + M_2^* \qquad \text{c)}$$

$$e + M_2(v) \longrightarrow e + M_2^+ + e \qquad \text{d)}$$

$$e + M_2(v) \longrightarrow M + M^- \qquad \text{e)}$$

Case B, of course, is more realistic than case A. However, the knowledge of cross-sections involving $v \neq 0$ levels is very poor, the only exception being represented by the dissociative attachment process [9,10].

An estimation of the cross-sections for (b-d) for different diatomic molecules can be made by using the Gryzinski approximation for both bound and unbound states [11,12]. (In spite of the crudeness of this approximation, the results we have obtained are within a factor of 3-4 of the corresponding results obtained using more sophisticated quantum mechanical methods.) For internal consistency within the scope of Case B, we are forced to utilize in the Boltzmann equation Gryzinski cross sections for the v=0 processes as well, thereby obtaining results which in general do not reproduce very well the transport properties of a vibrationally cold plasma.

Figures 1-2 report the Gryzinski dissociation and ionization cross-sections of N_2 for different vibrational levels [11]. In general the threshold energy of the process decreases with increasing v while the magnitude of the cross section strongly increases with v for the dissociation, but remains practically unchanged for the ionization.

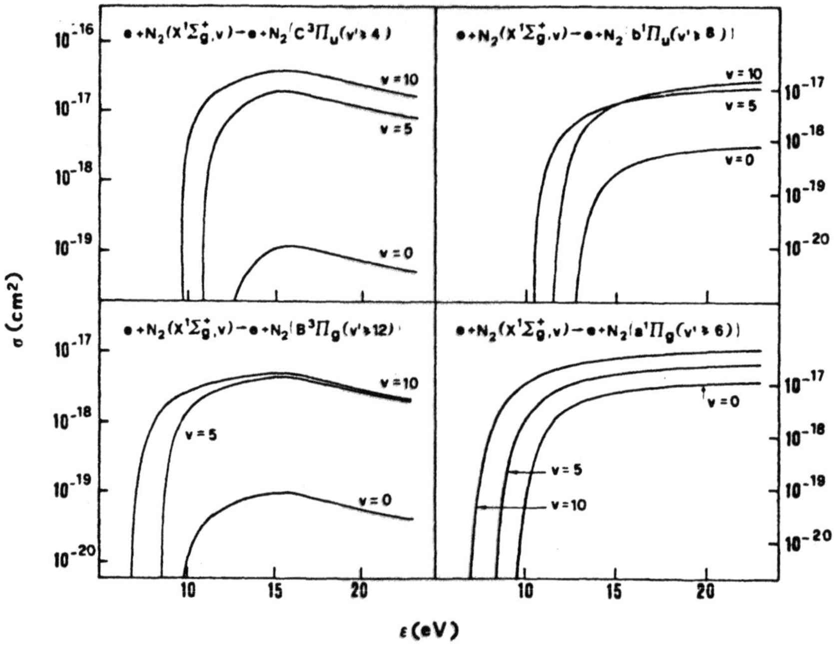

Figure 1. $N_2(X^1\Sigma_g)$ dissociation cross-sections as a function of electron energy for different vibrational levels.

Solution of the Boltzmann equation according to case B gives not only an edf consistent with a vibrationally excited molecular plasma, but also an estimation of the contribution of vibrationally excited states to the total ionization and dissociation rates.

The third case we discuss (case C) is the solution of the Boltzmann equation in the presence of both superelastic vibrational and electronic collisions. We are at the beginning of this study, but we believe that the introduction of superelastic electronic collisions; i.e., of the processes

$$e + M_2^* \longrightarrow e + M_2 \qquad f)$$

will open interesting perspectives in the study of molecular plasmas. The results we present here for this case are confined to the study of the influence of metastable CO($A^3\Pi$) on the edf in CO plasmas; i.e., we study case A with the insertion of process f.

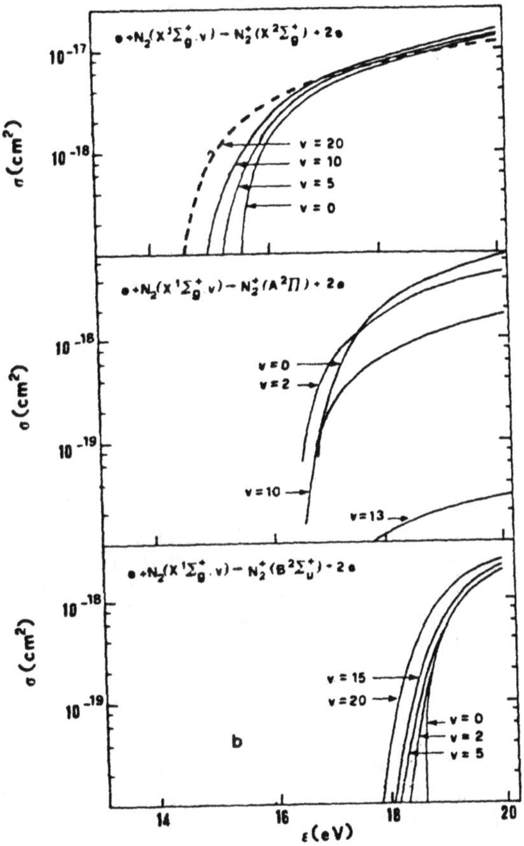

Figure 2. $N_2(X^1\Sigma_g)$ ionization cross-sections as a function of electron energy for different vibrational levels v.

Figure 3 gives a schematic representation of the different cases we will discuss in this review.

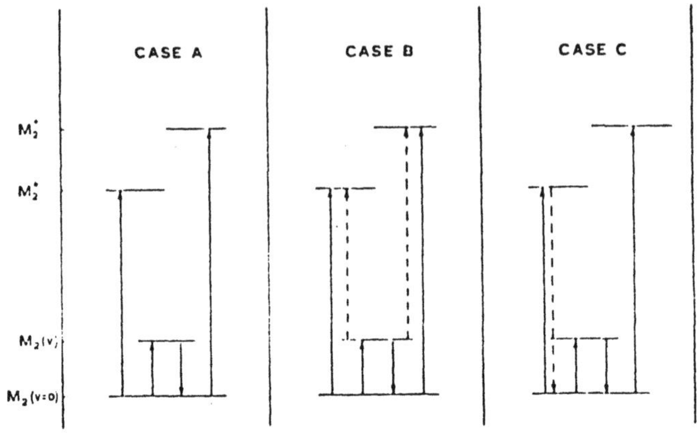

Figure 3. Schematic representation of different coupling schemes.

Results

CASE A

To illustrate case A we report in fig. 4 the electron energy distribution functions obtained by solving eqs. 1-3 for a pure CO plasma [4]. Keeping in mind that we start at t=0 with a vibrational distribution concentrated on the v=0 level, we can recognize the temporal build up of the corresponding evolution of the edf. Each curve reported in fig. 4 represents a quasistationary edf in the field of a self-consistent N_v.

These distributions, in turn, are well represented, at least over the first 10 levels, by Treanor distributions with different Θ_1 (Θ_1 is the 0-1 vibrational temperature). All the properties depending on the tail of the edf are strongly affected by the presence of superelastic vibrational collisions, as can be appreciated in fig. 5 by looking at the corresponding temporal evolution of different rate coefficients (see also ref. [4]).

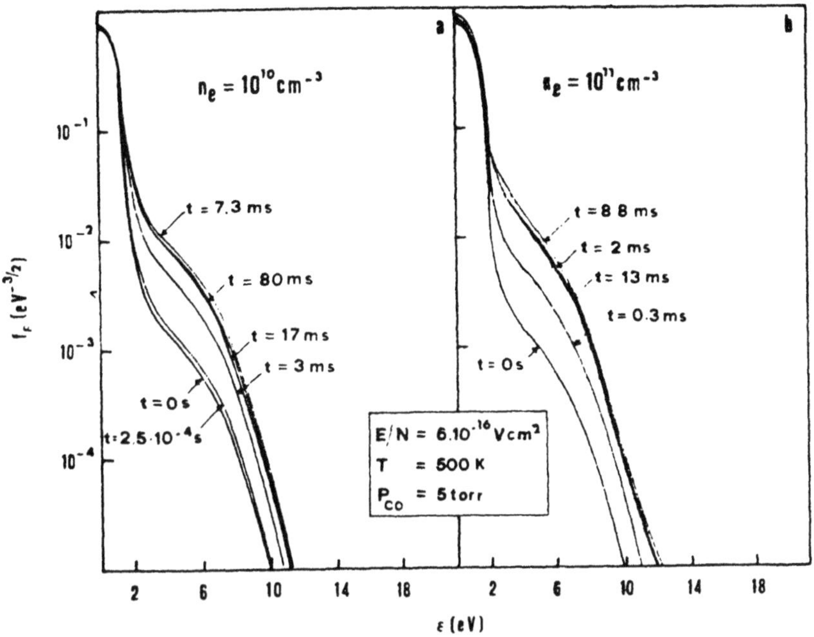

Figure 4. Temporal evolution of electron energy distribution function under discharge conditions.

Another interesting example belonging to case A is the behaviour of the edf in post discharges ($E/N = 0$, $\delta J_f/\delta\varepsilon=0$) of vibrationally excited molecules. It can be shown in this case that the edf is strongly coupled to the vibrational distribution of the diatomic molecules. As a result of this coupling, the average electron energy $\bar{\varepsilon}$ after a relaxation time of approximately $(N_1\ K_{10})^{-1}$ [N_1 is the number density of level v=1 and K_{10} is the rate coefficient for process (a) involving levels $1 \rightarrow 0$] converges toward the vibrational energy $3/2\ \Theta_1$; i.e., $\bar{\varepsilon} \simeq 3/2\ \Theta_1$. To obtain this behaviour, the following three conditions must be fulfilled; 1) a Boltzmann distribution for N_v, 2) a Maxwell distribution for the edf, and 3) the superelastic gain must compensate the inelastic losses, the other losses (elastic, rotational) being negligible [3]. Failure of any of these conditions results in the breakdown of the correlation $\bar{\varepsilon} \simeq 3/2\ \Theta_1$. For example, the

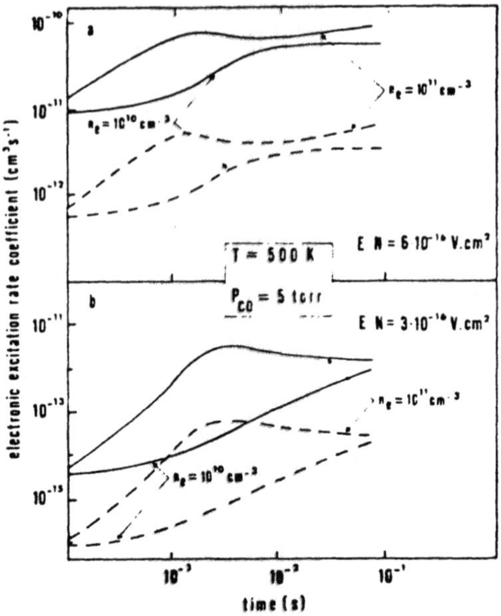

Figure 5. Temporal evolution of selected rate coefficients in CO pure plasma (—— full lines: e+CO → e+CO($a^3\Pi$) --- dashed lines: e+CO → e+CO($a^1\Pi$)

addition of elastic and rotational losses to the inelastic vibrational losses results in $\bar{\varepsilon}<3/2\ \Theta_1$, while we obtain $\bar{\varepsilon}>3/2\ \Theta_1$ if N_v is given by a Treanor distribution rather than a Boltzmann one at the same Θ_1 value.

Figure 6, which reports the quasistationary average electron energy versus the molar fraction of CO in He-CO vibrationally excited post discharges, represents the different regimes discussed above. In particular, for small molar fraction of CO, (X_{CO} < 20%), the elastic losses (from He) cannot be neglected,and we find $\bar{\varepsilon}<3/2\ \Theta_1$, while the effect of the choice of Boltzmann or Treanor distributions is clear from fig. 6 (see ref. [13] for details).

CASE B

The solution of eqs. 1-3 including both superelastic and inelastic vibrational collisions starting from v≠0 has been

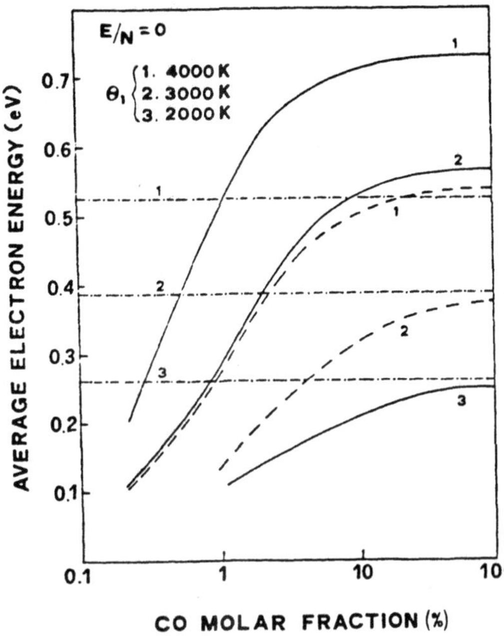

Figure 6. Average electron energy $\bar{\varepsilon}$ as a function of CO molar fraction (Full lines: Treanor's Distributions; Dashed lines: Boltzmann Distr.; Dashed dot lines: $\bar{\varepsilon} = 3/2\ \theta_1$).

carried out for H_2, HCl and N_2 plasmas [3,14,15,17]. Here we present results for N_2 for a typical experiment situation ($E/N = 7\times10^{-16}$ Vcm^2, $T_g = 490$ K, $n_e = 2\times10^{10}$ cm^{-3}, p = 1.2 Torr).

Coupling of the edf and N_v in N_2 results in a typical evolution [11] of both distributions and the related quantities. Figure 7 reports the temporal evolution of $\bar{\varepsilon}$, θ_1 and the drift velocity W_d. We note that $\bar{\varepsilon}$ and θ_1 show a similar trend (both strongly increase with time) while the drift velocity slightly decreases with time.

Let us consider now the results from the point of view of the fractional power transferred in the different dissipative channels. Figure 8 reports the different contributions as a function of the instantaneous vibrational temperature. We see that the net power going into the vibrational channel

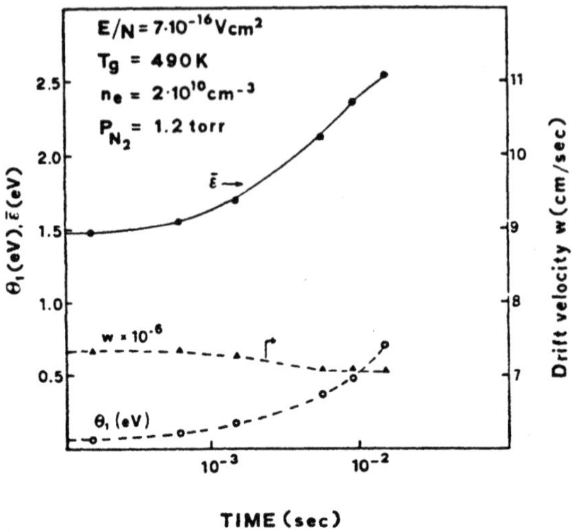

Fig. 7. Temporal evolution of $\bar{\varepsilon}$, θ_1 and drift velocity w.

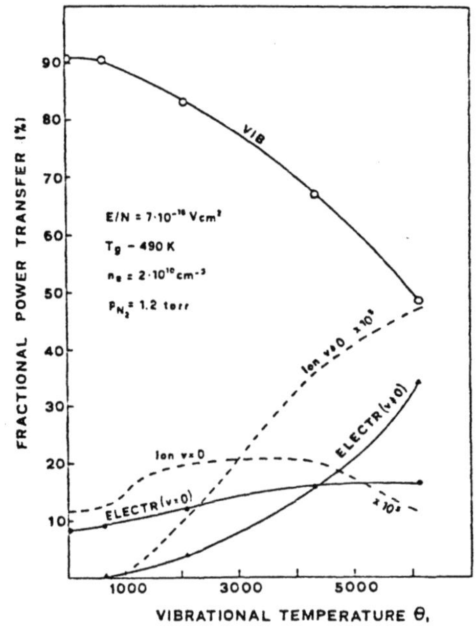

Fig. 8. Fractional power transfer as a function of θ_1.

decreases with increasing Θ_1, the reverse being true for the electronic and ionization channels. It should also be noted that past a certain Θ_1, the power gained from the electric field is preferentially dissipated through inelastic and ionization processes involving $v\neq 0$ levels. This point emphasizes the role of excited vibrational levels in the determination of the edf, and shows the shift from a plasma governed by the $v=0$ cross-sections to a plasma dominated by those involving $v\neq 0$.

The contribution of vibrationally excited states to the total dissociation rate is reported in fig. 9 as a function of Θ_1. We see that this contribution is up to an order of magnitude greater than the corresponding $v=0$ contribution.

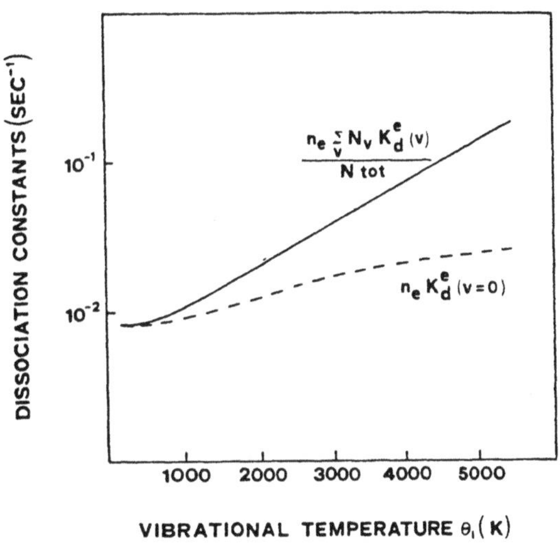

Figure 9. The influence of excited vibrational levels on the dissociation rate of N_2.

Figure 10, on the contrary, reports the vibrational distribution N_v, the dissociation rate coefficients $K_d(v)$ and the product $N_vK_d(v)$ for $\Theta_1=5400$ K. We see that under these conditions the first 20 levels contribute to the total dissociation rate of nitrogen. Similar results are obtained for the ionization process, although we must remember that other dissociation and ionization mechanisms act in nitrogen, in competition with those due to direct electron impact [11].

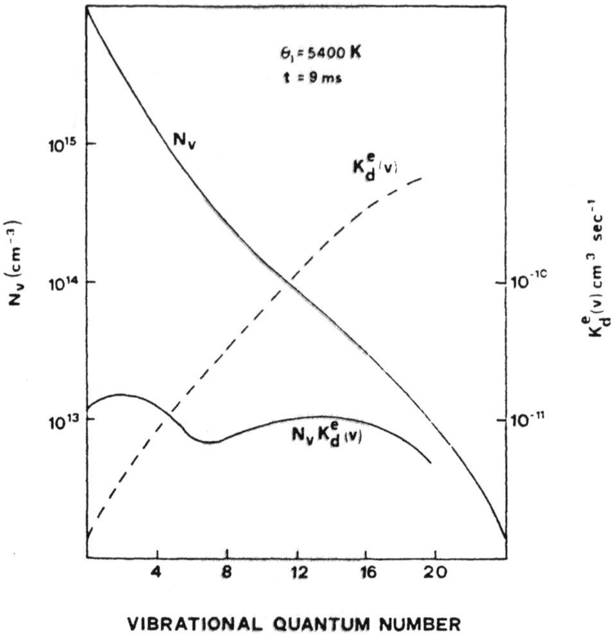

Figure 10. Vibrational distribution N_v, dissociation rate coefficients $K_d^e(v)$ and product $N_v K_d(v)$ as a function of vibrational quantum number.

Before ending this section we want to show the differences between the edf calculated according to case A and case B (see fig. 11). We note that superelastic and inelastic vibrational collisions dominate the edf up to approximately 4 eV, while the addition of electronic transitions from $v \neq 0$ has an effect on the edf for $\varepsilon > 4$ eV [11].

CASE C

Solution of the Boltzmann equation for case C has been recently considered [16] for studying the effect of the metastable state $CO(A^3\Pi)$ on the edf in CO plasmas. In this study the concentration of $CO(A^3\Pi)$ has been considered as

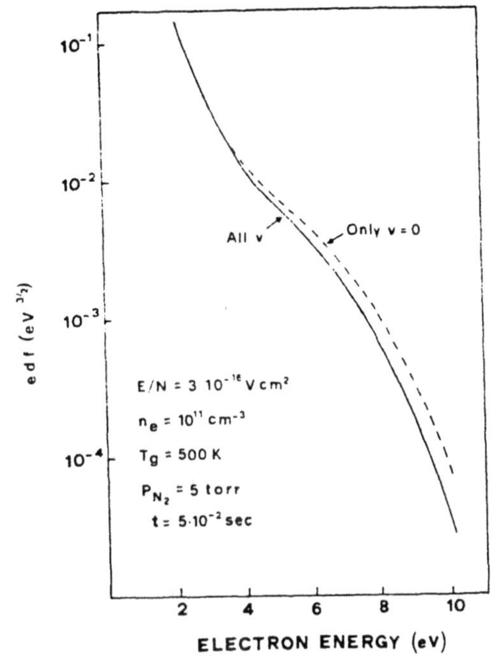

Figure 11. A comparison of edf for cases A and B.

an independent variable; i.e., we solve eq. 1 without coupling with the kinetics. Typical edf's calculated by using different concentrations of $CO(A^3\Pi)$ are reported in fig. 12. The effect of superelastic electronic collisions, which drain electrons from the energy range ε-ε^* ($\varepsilon^* = 6.2$ eV) to the energy range ε, can be understood from the appearance of plateaus in the edf. These plateaus tend to disappear with increasing E/N. The results also show that the higher the reduced electric field, the higher the concentration of metastable states necessary to affect the edf. Small concentrations of $CO(A^3\Pi)$ are sufficient to affect the edf in the post discharge regime, as can be seen in fig. 13. It should be noted that the edf's reported in figs. 12-13 contain also the effect of superelastic vibrational collisions. The corresponding N_v distribution is a Treanor distribution

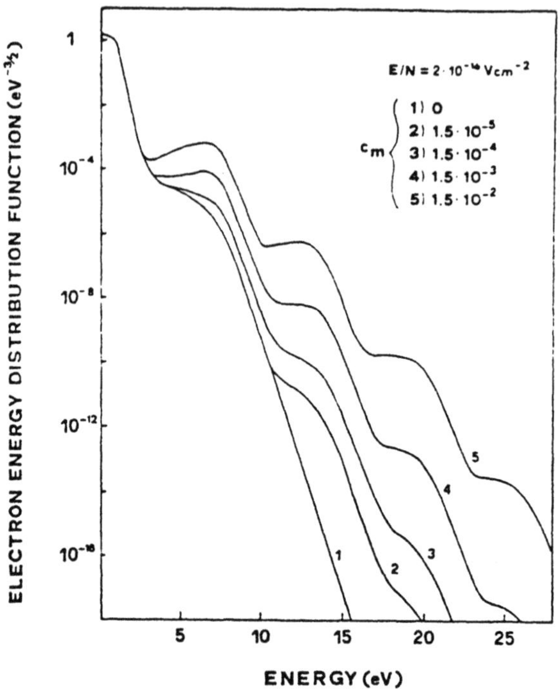

Figure 12. Influence of $CO(A^3\Pi)$ state on edf in pure CO plasmas ($c_m = N^*/N_{tot}$; N_v=Treanor's distribution at θ_1=2000K).

at Θ_1 = 2000 K. We note that the action of superelastic vibrational collisions and superelastic electronic collisions occur in different energy ranges.

Future work for understanding the effect of superelastic electronic collisions should take into account not only the appropriate kinetics for populating the different electronic states but also the cross-sections involving electronic transitions from metastable states. An important example in this context should be the insertion of the metastable $N_2(A^3\Sigma^+_u)$ state both in the Boltzmann codes as well as in the kinetics of ionization and dissociation processes at low reduced electric field ($E/N<10^{-15}$ Vcm^2).

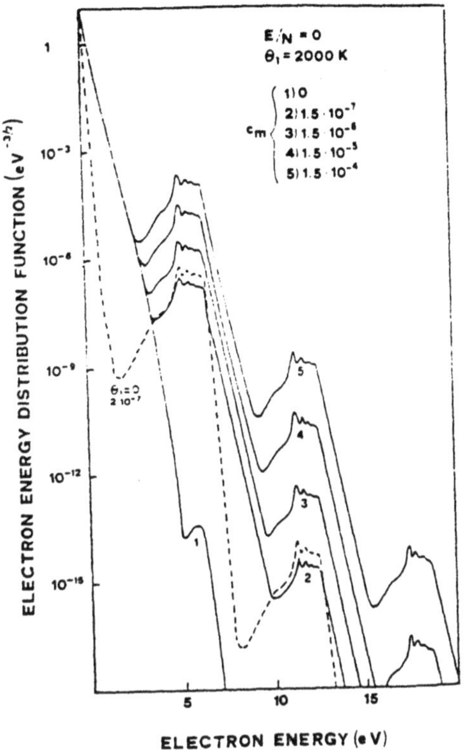

Figure 13. Influence of $CO(A^3\Pi)$ state on edf in CO post discharges ($c_m=N^*/N$; N_v as in Fig. 12).

Concluding Remarks

The results presented in this paper have shown different aspects of the solution of the Boltzmann equation in vibrationally excited molecular plasmas. As a final comment, one can ask about the accuracy of both the method utilized for solving the Boltzmann equation (the two-term expansion) as well as the cross-sections utilized in the different models. Recent work by Boeuf on vibrationally excited N_2 (case A) obtained by using the Monte-Carlo method has essentially confirmed the results obtained from the two-term expansion. The critical point remains the use of Gryzinski cross-sections in case B. We hope that this work will stimulate a theoretical effort based on more sophisticated methods for obtaining sets of electronic cross-sections involving the excited vibrational levels of N_2, CO, H_2 and other diatomics. The present results should therefore be considered as a first step toward the understanding of electron properties in vibrationally excited molecular plasmas.

Acknowledgments

The authors are very grateful to Dr. L.C. Pitchford for the oral presentation of this work at the Joint Symposium on Swarm Studies and Inelastic Electron-Molecule Collisions.

References

[1] Nighan, W.L., Phys. Rev. 1970, A2, 1989.
[2] Phelps, A.V.; Pitchford, L.C., Phys. Rev. 1985, 31A, 2932.
[3] Gorse, C.; Capitelli, M.; Ricard, A., J. Chem. Phys. 1985, 82, 1900.
[4] Gorse, C.; Capitelli, M., Chem. Phys. 1984, 85, 177.
[5] Rockwood, S.D., Phys. Rev., 1973, A8, 2348.
[6] Capitelli, M.; Molinari, E., Top. Curr. Chem., 1980, 90, 59.
[7] See for ex. Chandra, N.; Temkin, A., Phys. Rev., 1976, A13, 188; A976, A14, 507; NASA TN D-8347 (1976).
[8] See for ex. Cartwright, D.C.; Trajmar, S.; Chutjian, A.; Williams, W., Phys. Rev., 1977, 16A, 1041.
[9] Bardsley, J.M.; Wadehra, J.M., Phys. Rev., 1979, A20, 1398.
[10] Teillet-Billy, D.; Gauyacq, J.P., J. Phys. B, 1984, B17, 4041.
[11] Cacciatore, M.; Capitelli, M.; Gorse, C., Chem. Phys., 1982, 66, 141.
[12] Cacciatore, M.; Capitelli, M., Chem. Phys., 1981, 55, 67.
[13] Gorse, C.; Paniccia, F.; Ricard, A.; Capitelli, M., J. Chem. Phys., in press.
[14] Garscadden, A.; Bailey, W.F.; Engel, J.W., 5th ISPC (Edinburgh 1981) V. 1 p. 4.
[15] Capitelli, M; Dilonardo, M.; Gorse, C., Chem. Phys. 1979, 43, 403.
[16] Gorse, C.; Paniccia, F.; Bretagne, J.; Capitelli, M., J. Appl. Phys. (to be published, 1986).
[17] Boeuf, J.P., These d'Etat Université Paris Sud Orsay 1985.

APPLICATIONS OF CROSS-SECTIONS FOR ELECTRON-MOLECULE COLLISION PROCESSES

David C. Cartwright

Los Alamos National Laboratory
P. O. Box 1663, MS E527
Los Alamos, New Mexico 87545

Electron-impact processes with atoms and molecules play a central role in a wide variety of naturally occurring and laboratory-produced phenomena and, as a result, have been the subject of considerable research, particularly in the last ten years. The importance of electron-impact excitation processes involving ground-state molecules, and molecules in metastable excited states, has been recognized for many years in the study of ionospheric [1] and auroral [2] processes in planetary atmospheres. The discovery of high-power (electron-beam and discharge-pumped) gas lasers has resulted in an increased interest in electron collision processes because of the important role they play in creating the population inversion [3]. This renewed interest in electron scattering has resulted in the measurement of many of the electron-impact excitation cross-sections of importance during the past ten years. The need for cross-section information involving targets that are presently difficult to study experimentally (eg, molecular free radicals, etc), and recent results from electron-photon coincidence measurements [4] which probe the details of the scattering process, have further stimulated theoretical studies. In spite of this renewed activity, however, much of the electron-impact excitation cross-section information needed to model a variety of plasma and planetary processes is still not known and the absence of these basic cross-section data hinders progress toward a more thorough understanding of various phenomena.

The Earth's Ionosphere and Aurora

Electron impact processes play a major role in two naturally occurring atmospheric phenomena. In the earth's ionosphere, processes associated with production and loss of free electrons (and the related ion-chemistry) determine the characteristics of the secondary electron distribution and the ionospheric emissions. Auroral emissions are produced as a result of energy deposition by charged particles, primarily electrons, entering the earth's polar atmospheres along magnetic field lines. These two atmospheric phenomena involve electrons of somewhat different energies, but require essentially the same cross-section information.

Free electrons in the ionosphere are produced by photoionization of atmospheric constituents, and the subsequent cascade-ionizations of these "primary" electrons as they move through the atmosphere. These electrons lose their energy by ionization, electronic, vibrational and rotational excitation of the atmospheric constituents, and collisions with other electrons, until they thermalize and ultimately recombine with the positive ions.

The full range of electron impact processes, elastic scattering through ionization, is possible because the flux of electrons ranges from 0 to greater than 100 eV. Molecular nitrogen, however, is the only atmospheric constituent for which extensive results have been reported for the differential and integral cross-sections for electronic excitation and, even for N_2, there remain a number of atmospherically important issues for which the available data are in major disagreement. Two of the leading issues involving electronic excitation in N_2 has to do with the magnitudes and shapes of the integral cross-sections from threshold to about 25 eV and with the absolute cross-sections for dissociation of N_2 into $N(^4S)$, $N(^2D)$, and $N(^2P)$ fragments. As illustrated in Figure 1, Zipf and McLaughlin [5] have reported integral "apparent" cross-sections for excitation of the dipole-allowed states that are a factor of two larger than those determined [6] from electron energy-loss spectra and the origin of this difference has not yet been explained. This difference needs to be resolved because nitrogen atoms, in their ground and excited states, play key roles in the formation of NO and similar constituents in the ionosphere, and dissociation of N_2 by electron impact appears to be the major mechanism for producing N atoms.

The data situation for O_2, also a major molecular constituent in the ionosphere, is currently worse than for N_2 because there are relatively few absolute differential and integral cross-sections for excitation of specific states in O_2. One of the major reasons for this deficiency is because the very strong perturbations present in the excited

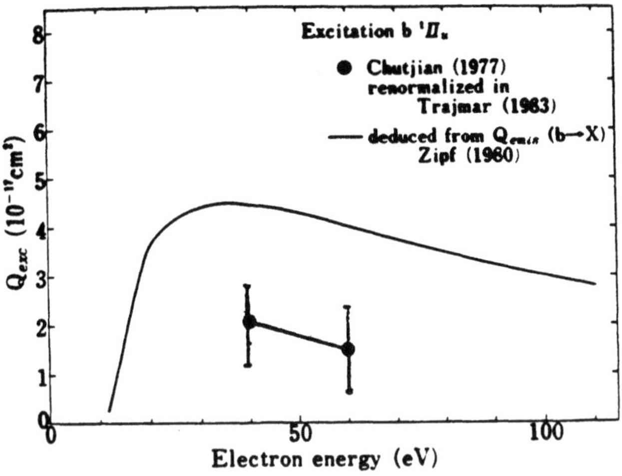

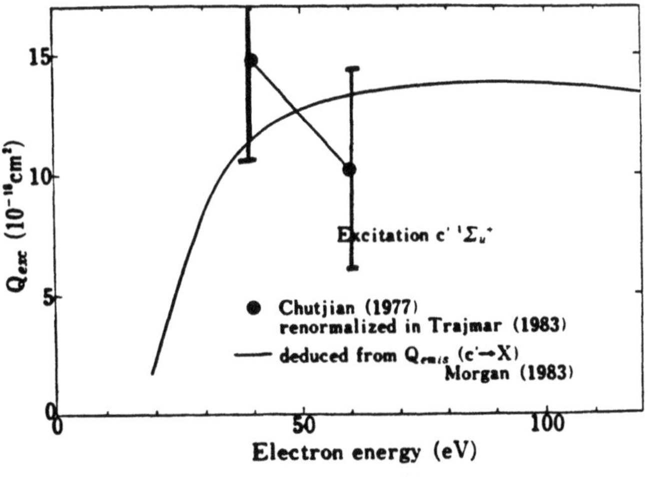

Figure 1. Comparison of recent measurements of the integral cross-sections for excitation of the b $^1\pi_u$ state (upper panel) and c' $^1\Sigma_u^+$ state (lower panel) of N_2. The solid curves are from Zipf and Gorman [11] and from Morgan and Mentall [12]. The data points are from Chutjian and Cartwright [6]. (Composite figures from Itikawa et al 1985 [13]).

states of O_2 so complicate the electron energy-loss spectra in O_2 that a straightforward analysis of the data is not possible. In the atmospheric sciences, the primary interest in electron impact excitation of O_2 is as a source of excited-state O-atoms, because electron-impact excitation of essentially every electronic state above the b $^1\Sigma_g^+$ state (at 1.6 eV) results in dissociation into O-atom fragments. Progress has been made recently in understanding the photodissociation of O_2 via the atmospherically important Schumann-Runge continuum [7] and the magnitudes of the electron impact cross-sections for excitation of composite (ie, more than one) electronic state in the energy-loss spectra [8]. However, differential and integral cross-sections for excitation of individual electronic states, and the correlation to excited state fragments, is still not available. Providing this information will require a particularly close collaboration between theorists and experimentalists because the very strong perturbations in the electronic state of O_2 preclude application of existing theoretical techniques on the one hand, and make straightforward interpretation of the energy-loss data impossible on the other.

As a final comment on molecules for which electron impact excitation data are useful in the atmospheric sciences, we mention the molecule NO. This molecule, although not a dominant neutral species in the earth's atmosphere, is of particular interest because of the key role it plays in the ion chemistry in the ionosphere and the aurora [2]. The excited electronic states of this molecule are also strongly perturbed and, again, very little data are currently available on the direct electronic excitation and/or the excited state and kinetic energy distribution of the dissociation atomic fragments. As in the case of O_2, determination of the cross-sections for electronic excitation of NO is a fertile area for collaborative theoretical and experimental efforts because of strong perturbations in the excited state spectrum.

Gas discharges/excimer lasers. There is considerable renewed interest in the general properties of gas discharges, primarily because they serve as efficient means to drive gas lasers. Gas breakdown and the transport of moderate-energy electrons through gas media are not yet very well understood and remain the subject of considerable research. The origin of the current difficulties with modeling these phenomena can be traced to be a combination of (1) the mathematical and computational problems in dealing with a non-LTE plasma, and (2) the general absence of the cross-section data necessary to model the charged-particle interactions in the medium.

Excimers are the class of exciplex molecules formed from identical atoms or atomic groups (eg, rare-gas excimers).

Exciplex molecules are defined as those which are bound in their excited electronic states but dissociate readily in their ground electronic state. The general characteristics of, and detailed processes occurring in, exciplex lasers have been discussed in a number of excellent review articles [9]. Because of the importance in a wide variety of applications, the rare-gas KrF excimer system will be used as the example to illustrate the relevant mechanisms. Figure 2 illustrates the potential-energy curves for KrF. The upper lasing level is the B state and is dipole connected to the weakly bound X ground electronic state.

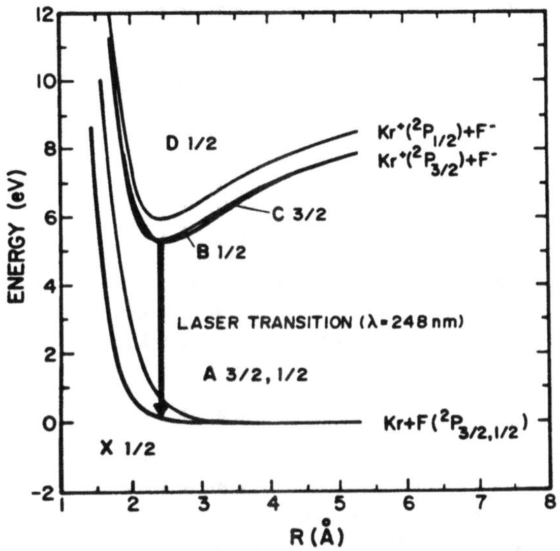

Figure 2. Electronic structure characteristics of the lasing KrF monohalide molecule. The lowest electronic state is repulsive, or only slightly bound, and therefore the ground-state (X) molecule is thermally unstable against dissociation into atomic fragments. The upper laser level (B) is an ionic electronic state that dissociates into the indicated ionic fragments.

The KrF excimer laser is usually produced from an atmospheric-pressure mixture of Ar-Kr-F_2 having various percentage proportions of the component gases. Usually, the constituents that form the lasing molecule are diluents in a carrier gas.

The production and loss of the active exciplex media involve a large number of electron and heavy-particle collision processes. A thorough analysis of the basic collision processes that take place in an electron-beam stabilized electrical discharge in atmospheric Ar–Kr–F_2 mixtures has been provided by Nighan [10]. Figure 3 is an adaptation

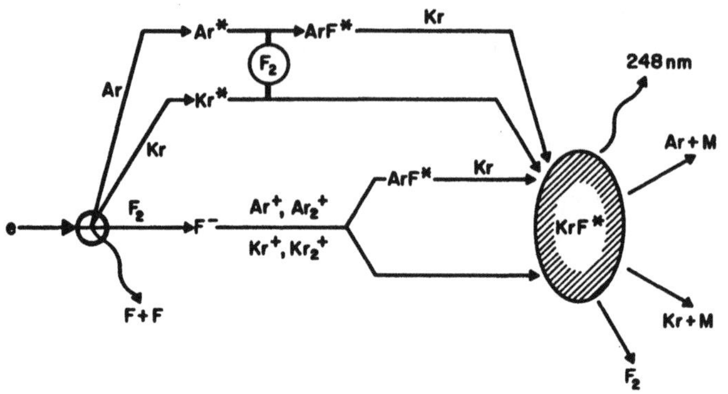

Figure 3. Schematic diagram of the primary reactions for formation of KrF by electron-impact excitation of atmospheric-pressure mixtures of Ar–Kr–F_2 (Adapted from Nighan [10]).

of his schematic summary of the primary collision processes contributing to KrF molecule which, as shown in Figure 3, occurs via either ion-molecule reactions or reactions involving metastable rare-gas atoms. Although no single mechanism completely dominates the production of KrF, Nighan's analysis shows that two- and three-body positive and negative ion recombination is most important for electron-beam excitation (lowest path in Figure 3), while reactions between F_2 and metastable rare-gas atoms are most important in the case of gas-discharge pumping (upper two paths in Figure 3).

A key factor in achieving high efficiency for these lasers is the efficient production of rare-gas atoms in metastable states. Analysis shows [10] that there is a substantial electron energy loss to electron-impact dissociation of F_2. Vibrational excitation of F_2 is not an important energy loss mechanism (because the energy lost per collision is relatively small) but is very important in the formation of KrF because of its possible effect on dissociative attachment. Accurate assessments of these loss mechanisms are, unfortunately, not yet possible because there are currently very few data available on the electronic and vibrational excitation of F_2, and these deficiencies hamper the assessment of the role of electron-F_2 processes in the KrF laser.

As shown schematically in Figure 3, F^- plays a central role in the formation of KrF but the role of excitation to electronic states of F_2 that dissociate into $F^+ + F^-$ could not be determined because the relevant cross-section data are not available. Theoretical work suggests that ionic-character states that dissociate into $F^+ + F^-$ are abundant in F_2 and, consequently, the electron-impact excitation of these states should be studied to determine if this source of F^- is important in the KrF laser. In the short term, theoretical techniques may be the best way to obtain this information.

Table 1 is a summary of the key physics issues that need to be addressed in order to better understand the KrF laser. One of the most important is the determination of electron impact cross-sections involving molecules in excited electronic states.

Table 1. **UNRESOLVED PHYSICS ISSUES RELATED TO KrF**

- *ELECTRON CROSS-SECTION FOR GROUND AND EXCITED STATE SPECIES*
- *TEMPERATURE DEPENDENCE OF RATE CONSTANTS*
- *DETAILED VIBRATIONAL STATE-TO-STATE KINETICS*
- *PHOTON ABSORPTION BY Kr_2F AND OTHER SPECIES*

Table 2 is a summary of the kinds of electron-impact cross-section data that are needed to better understand certain aspects of atmospheric and discharge processes.

Table 2. *ELECTRON-MOLECULE (ATOM) COLLISION PROCESSES PLAY A MAJOR ROLE IN UNDERSTANDING A NUMBER OF PHENOMENA*

BUT MANY MORE CROSS-SECTION DATA ARE NEEDED

ATMOSPHERES	*PARTIALLY IONIZED PLASMAS*
• N_2 ($E_0 \leq 30$ eV)	• EXCITED STATES → EXCITATION & IONIZATION
• N_2 (DIPOLE STATES, $E_0 \leq 100$ eV)	• DIELECTRONIC RECOMBINATION
• O_2 (SR CONTINUUM)	• TRANSIENT SPECIES
• NO (ALL STATES, ALL E_0)	
• O (TOTAL SCATTERING)	
• EXCITED STATES	

References

[1] Whitten, R.C.; Poppoff, I.G. "Fundamentals of Aeronomy", Wiley, New York, 1971.
[2] Vallance-Jones, A. "Aurora", Reidel, 1974.
[3] Willett, C.S. "Introduction to Gas Lasers: Population Inversion Mechanisms", Pergamon, New York, 1974.
[4] Blum, K; Kleinpoppen, H. Phys. Rep. 1979, 52, 204-261.
[5] Zipf, E.C.; McLaughlin, R.W. Planet Space Sci., 1978, 26, 449-462.
[6] Chutjian, A.; Cartwright, D.C.; Trajmar, S. Phys. Rev., 1977, A16, 1052-1060.
[7] Lee, L.C.; Slanger, T.G.; Black, G.; Sharpless, R.L. J. Chem. Phys., 1977, 67, 5602.
[8] Wakiya, K. J. Phys. B: At. Molec. Phys., 1978 11, 3913, ibid 3931.
[9] Huestis, D. "Gas Lasers" (E.W. McDaniel and W.L. Nigan, eds.), Chap. 1 Academic Press, New York, 1982.
[10] Nighan, W.L. IEEE J. Quent. Elect. QE-14, 714-726, 1978.
[11] Zipf, E.C.; Gorman, M.R. J. Chem, Phys. 1980, 73, 813.
[12] Morgan, H.D.; Mentall, J.E. J. Chem. Phys. 1983, 78, 1747.
[13] Itakawa, Y.; Hayashi, M.; Ichimura, A.; Onda, K.; Sakimota, K.; Takayanagi, K.; Nakamura, M.; Nishimura, H.; Takijanagi, T. ISAS Research Note 291, June 1985.

CURRENT APPLICATIONS AND OPPORTUNITIES IN SWARM STUDIES

Alan Garscadden

Aero Propulsion Laboratory
Air Force Wright Aeronautical Laboratories
Wright-Patterson AFB OH 45433

The assessment of the applications and prospects of the electron swarm scientific area is a daunting task. Addressing the topics at all implies on the one hand an appreciation of the scientific and/or technical utility of some available results,and on the other hand an encouragement, even impatience, that other interesting areas should be investigated. The evaluations and predictions are highly subjective and hence some will be wrong, but it is the scientific method to attempt the task and to stimulate responses and criticism. The development of swarm studies has suffered because of a general lack of funding to small-scale classical atomic physics experiments and a certain amount of inertia that has prevented novel approaches or new problem areas being explored by the method. It is imperative to be more imaginative in experimental objectives and also to be more aggressive in advocacy of the method and its utility. In contemporary terms we might sub-title this critique the swarm defense initiative!

The field has provided excellent quality low-energy drift velocity and characteristic energy data for the rare gases, the atmospheric gases and some other relatively common gases [1,2,3]. The complementary studies using electron-atom and electron-molecule crossed beams have demonstrated continued progress to lower energies and high resolution. Recent experiments have been able to attain energy resolution of 20 meV and beam energies down to 1 eV. The thresholds and shapes of the low energy cross-sections are better defined by these collision experiments. The absolute amplitude calibration of the cross-sections receive confirmation from precision swarm experiments, although it may be noted that with precision capacitance manometers and more sophistication in effusion calculations, accurate absolute data is becoming available from refined cross-beam measurements. Hence, it is appropriate to attempt to define the future role of swarm studies in collision research. The present advantage of the swarm method has been its capability to explore the very low-energy regime below a few electron volts. This uniquely permits some vibrational excitation and rotational excitation cross-sections to be derived. However, after 70 years or so of investigations,what directions should the research now take?

The evaluation of any topical area involves assessments whether: (i) it is exploring critical scientific tests: (ii) there are still important advances to be made in relevant theoretical techniques: (iii) there are continuing significant advances in experimental techniques: (iv) there is production of applied scientific data, and: (v) generally there is evidence of some tangible return on the required investments, scientific and technical.

On the theoretical side there has been reasonable progress on the understanding and handling of the various moment methods [4] and reconciliation of these methods with Monte-Carlo calculations [5], both of which benefit from the availability of time on large, fast computers. These are important but not exceptional theoretical contributions. The methods addressed have included solutions of the integral form of the Boltzmann transport equation [6,7]; expansion of the Boltzmann equation into a series of spherical harmonics and consideration of terms of order higher than the second [8]; expansion into a series of Sonine orthogonal polynomials [9,10] and development of a method initiated by Kitamori, et. al., [11] that uses a finite difference method on the integro-differential Boltzmann equation. This last method is useful because it permits the analysis of the temporal development of the swarm (if needed) and the influences of ionization and attachment. Segur, et. al., [12] have emphasized the benefits of iterative methods especially for deconvolution of cross-sections, and they derived a modified and fast form of the S_n Method to calculate swarm parameters for arbitrary conditions. An analytical method for assumed strictly forward and backward scattering including electron nonconservation was obtained by Blevin, et. al., [13]. A similar approach was used by Long [14] to examine the cathode sheath region of a discharge.

Experiments have improved slowly, taking advantage of advances in vacuum techniques, pressure measurements, laser-activated photocathodes and fast electronics for signal detection and averaging. A precision of better than 1% or so in the original data has been achieved in a few locations, notably in Canberra. These studies are valuable, and the methods developed are being used to achieve better precision in the transport data for the traditional gases that are investigated. However, this appears to be a saturation approach, as it does not address unexplored areas or breach a scientific frontier. Perhaps some new conditions or situations can be explored that satisfy the above criteria.

Applications of Swarm Data. There are several main avenues that appear interesting to investigate. Some of these are motivated by the technologies where such transport data should be useful. The gas discharge lamp industry is using new gas mixtures to enhance efficiency and to provide more acceptable color. The plasma chemistry reactors (low pressure and high pressure) are using a great number of gases, some of which have not been studied by the swarm methods or collision methods at all. Much of the microelectronics industrial processing relies on plasma-assisted methods ranging from "ashing" [15] to sophisticated contouring of integrated circuits by plasma etching [16]. Therefore, because of the promise of catalytic interactions between swarm studies and these other fields, I venture to suggest some new directions as follows:

(i) Studies in gases and gas mixtures that are important to thin film deposition and etching.

(ii) Measurements in gases under conditions of internal excitation, including dissociated products, atoms and radicals.

(iii) Measurements in selected parameterized gas mixtures so as to vary the electron-energy distribution approximately independent of the ratio of the applied electric field to total gas density.

(iv) Use of a drift experiment to simulate interesting discharge field conditions.

(v) The development of a structured information-theoretical approach that allows comparison with (iv) and conventional uniform field experiments and subsequent derivation of cross-sections with defined accuracy.

(vi) Investigation of electron transport parameters with simultaneous measurements of ion species (positive and negative), excitation-, attachment- and ionization rates, especially the species referenced in (i). In these experiments there is also the opportunity to develop standards, i.e., if we are measuring excitation cross-section (whose upper states have known transition probabilities) under defined boundary conditions of electron current, pressure and geometry, then it should be possible to define in absolute terms the radiant output at specified, well-defined wavelengths.

The present largest application areas of swarm data are in plasma chemistry and gas discharge lamp physics. Note that I deliberately de-emphasized gas discharge laser physics: while laser physics has benefited from the data bank provided by swarm studies, lasers are no longer a growing area in low-energy electron kinetics. Most gas discharge lasers are in quite mature stages of development and much of the progress is empirical engineering not necessarily related to homogeneous electron kinetics. There are important exceptions, however the wavelength regions that do introduce new collision processes are now the far infrared and the near ultraviolet and shorter wavelengths which are not really addressed by traditional swarm experiments. There appear to be more requirements in the laser-materials interactions area if the swarm experiments are tailored appropriately (especially (ii) and (vi) above).

Opportunities for Swarm Studies

I. The success of integrated circuit technology has been achieved in part due to the advances in deposition and in surface etching and contouring on the micron and sub-micron scales. One of the most notable advances in silicon technology is the development of high-performance metal-oxide-semiconductor field effect transistors (MOSFETS) based on the Si-SiO_2 interface. These devices require reliable and reproducible interfaces that have low densities of trapped charge and low interface state densities (where low means less than $10^{11} cm^{-2} eV^{-1}$). The growth of thin films or layers was previously achieved by pyrolysis or 'wet' chemical methods. However, deposition at high temperatures causes thermal strain leading to the generation of defect states and to degradation of previous processing steps [17]. Also III-V materials such as GaAs decompose above 350°C causing loss of stoichiometry at the surface. Chemical etching methods are usually isotropic and are not favored for finer resolutions. Plasma-enhanced deposition and etching have been found to be more economic and more extrinisic in their applications. Glow discharge plasma deposited hydrogenated amorphous silicon is being used for solar cells, thin film transistors and imaging devices. The high resistivity (10^{12} ohm-cm), high absorption coefficient (10^5 cm^{-1}) and sensitive photoconductive response make it attractive for low-cost vidicons and photoreproduction.

The techniques of chemical vapor deposition and pyrolytic deposition of silicon are being challenged by the plasma-enhanced deposition from silane, disilane and other higher silanes. The electron kinetics of silane discharges were derived [18] from electron swarm data and spectroscopic results. The first electron swarm measurements in pure silane were made as long ago as 1968 by Pollock [19] and they have proved to be very valuable. The electron drift velocity versus E/N displays a local maximum. From the results and spectral data the cross-sections of SiH_4 were obtained (Figure 1). More recent data by Andrews, et. al., [20], in gas mixtures with silane at low concentrations have permitted comparisons with theory without concerns about the use of the "two-term" approximation. M. Hayashi has applied similar analysis techniques to assemble a large number of cross-sections (these proceedings) for gases of interest in amorphous silicon deposition and reactive etching.

The other technical application of great interest is plasma etching. Industry has derived a large number of extremely accurate and useful plasma techniques to etch anisotropically many materials [16]. The results of Christophorou's group for the fluorocarbons [21], while derived for other purposes, have been very useful in initiating kinetic studies in some of the etching gases. However, many electron swarm studies do not include the measurement of the positive and negative ions, especially under conditions approaching those used in the semiconductor etching operations. A very important process for many applications is electron impact-induced dissociation. Crossed beam studies have, of course, measured dissociative ionization, dissociative attachment and dissociative excitation. These are the easy measurements! The dissociation into unexcited (or low-energy excited levels) is much harder to measure. Fortunately Winters [22,23] has made some very important contributions by measuring dissociation from pressure changes and selective absorption. The total dissociation cross-sections of CF_4, CF_3H, C_2F_6, C_3F_8, CH_4, C_2H_6 and N_2 have been measured by interacting an electron beam with the gases of interest in closed volumes. It was observed that the fluoromethanes and probably the fluoroethanes have similar cross-sections near their maxima. New work has been initiated by Cosby and Helm [24] using fast neutral beams. The target molecules will be created at high energy by near-resonant charge transfer of the corresponding positive ion. Dissociation of the fast (10^7 cms^{-1}) neutral molecule will give two or more fast neutral fragments that are correlated in space and time through the fragmentation energy and angular distributions of the dissociation. The fragments have enough energy to permit detection without an auxiliary ionizer. This removes the ambiguity caused by new species introduced by the ionizer in a conventional quadrupole mass analyzer detector.

The advantages of a consistent attack on a range of similar gases is seen from the work of Winters, in that some generalizations become possible; thus it appears that the total dissociation cross-section for these gases is equal to the sum of the cross-sections for all electronic and ionization states. Some of the gases where some data exists but where more work is needed are given in Table 1. It is true that many of these gases are quite difficult to handle, however procedures for these gases and many others have been established by the semiconductor industry [25]. There are many other gases that are of interest to thin films and plasma reactor technology. Some of these gases are listed in Table 2. It is a long list and many of these are gases that have not been investigated at all by either crossed beam studies

Table 1. Plasma Reactor Gases where some Data Base Exists

SiH_4, Si_2H_6, $SiCl_4$
Cl_2, HCl, HBr
NH_3, CCl_4, O_2
CH_4, C_2H_6, C_3H_8, CF_4
F_2, NF_3, SF_6, C_2F_4

Table 2. Species where Electron Transport and Collision Cross-Section Data are Needed

SiF_4, GeH_4, Ge_2H_6
F, Cl, CF_2, CF_3
CCl, CClF, CCl_2, CCl_3
Si, Si_2, Si_3
NH, NH_2, NF_2, N_2F_4, N_2F_2
SiF, SiF_2, SiF_3, HSiF
SiCl, $SiCl_2$, $SiCl_3$
Ga, Ga_2, As, AsH, AsH_2, AsH_3
BCl_3, BCl_2, BCl
B_2H_6, PH_3, H, O, N

or by swarm studies. These gases should be viewed as research opportunities. There is some justified skepticism that basic research-type swarm studies will provide useful results if they proceed at their traditional pace. The empirical studies of the semiconductor and optics industries often have achieved dramatic success and also developed rules for work with various substrates and specific device processes. However, research opportunities still do exist due to the increasing complexity of devices and to the requirements of higher density of components. This leads to emphasis on efficient throughput at each process step or procedure, and a requirement for actual understanding of the etch or deposition process. The provision of a data base should assist scaling of the reactors to different geometries, excitation conditions and materials. However, it should be noted that it is still an area where theoretical elegance would be sacrificed for some timely estimates of the momentum transfer, vibrational, dissociation and ionization cross-sections of these complex molecules. Thus, ab initio studies are not really encouraged, because it would take too long to treat the many species of interest, but approximate methods that permit some generalizations would be very useful. Some new studies on semi-empirical formulae for ionization cross-sections have been made recently by Deutsch and Schmidt [26].

II. The second area where there exist opportunities for swarm data is under conditions of dissociation or internal excitation. The lifetime of many free radicals and of metastable states is sufficiently long (Table 3), that with modest flow rates, measurements could be performed within a gas

Table 3. Metastable Species that Might Be Explored with Electron Drift Tube

Species	State	Excitation Energy (eV)	Approx Radiative Lifetime (secs)
H_2	$^1\Sigma_g^+$ (v=1)	0.52	10^6
N_2	$^1\Sigma_g^+$ (v=1)	0.29	10^5
N_2	$A^3\Sigma_u^+$	6.17	1.3
O_2	$a^1\Delta$	0.98	3×10^3
O_2	$b^1\Sigma_g^+$	1.63	10

residence time or a diffusion time. Some changes in geometry or compromises in quality of the data may have to be accepted but the "non-equilibrium drift tube" (NEDT) appears to offer interesting possibilities. The reason that these studies are suggested is that the many applications of such drift velocities, characteristic energies, etc. are for gases or gas mixtures that are excited or have been discharged with energy densities up to joules/liter-atmosphere. This level of excitation is not uncommon in high-power lasers or in pulsed power switches. Thus, it is a moot question that the present drift tube data for many infrared laser gas mixtures is particularly relevant as these gases have, of necessity, high concentrations of excited states and sometimes atoms and radicals. Unless one performs detailed chemical kinetics for dissociation and energy transfer kinetics at appropriate energy densities with the detailed balancing including the effects on the electron energy distribution function, completely different results will be obtained from the drift tube data as compared to a discharge experiment. Therefore, the present approach requires great effort in cross-section measurements coupled with significant skill and time in solving consistently the Boltzmann transport equation and the chemical and electron kinetics. This method should still be pursued, but it is appropriate also to use the high-pressure, e-beam ionized drift tube as an analog computer. The definition of the excitation conditions is given by active or passive spectroscopy such as CARS (coherent anti-stokes Raman spectroscopy) LIF (laser-induced fluorescence) and multi-channel emission spectroscopy. In gas mixtures containing nitrogen or carbon monoxide, it is not unusual to have a vibrational 'temperature' higher than 2000°K. For other gases, dissociation occurs readily and the gas mixture is a strong function of excitation conditions. It is surprising that electron swarm measurements in vibrationally excited nitrogen or in atomic hydrogen and other atomic species have not been attempted.

Regarding swarm measurements under high temperature thermal equilibrium conditions, the situation has actually deteriorated since the last swarm seminar, as the planned studies of J. Lucas at the University of Liverpool have been cancelled. This means that there is no center addressing electron transport in the gases of high temperature materials or materials that have low vapor pressures at room temperature. It might be claimed that the transport properties can be calculated for both equilibrium and non-equilibrium conditions. This is true in principle, but it is by no means a trivial task and it requires an even broader data base: vibrational-vibrational energy transfer rates, vibrational-translational transfer rates, cumulative excitation rates, excitation rates and transition probabilities. There exist technical situations, especially in space, where efficient thermal management requires that radiators must be at high temperatures. Hence, under these contraints switches, lasers and the like may operate at high temperatures. It can, therefore, be argued that careful drift measurements, at high temperatures, perhaps with external excitation (laser, e-beam or x-ray), are economical in that the direct measurement of the transport properties of interest is achieved.

If the relaxation time of the excited states is shorter than the residence time of the flowing gas through the drift tube, laser pumping of selected levels at the saturation intensity could still permit measurement of the influence of excited states. This proposal would appear to be more attractive for infrared pumping, as there would be difficult side issues of photoelectric emission and charging with visible and shorter wavelength lasers. To avoid excessive pressures or path lengths, a probable requirement for the method will be that optical pumping of the ground state is possible.

Many of the same arguments presented for gases of interest for switching also apply to gases that are used in high and low-pressure discharge lamps. Especially in the high-pressure lamps, the gases are at high temperatures and at high current densities, so that measurements should be devised to provide data relevant to these situations. The gases that are being added often include the metallic halides and other rare earth compounds. Some of the species whose cross-sections are of interest to the new lamps for old community are given in Table 4. While the impetus for these studies is to provide a return on investment, it is anticipated that the studies will lead to interesting physics results per se. Christophorou, et. al., (these proceedings) have found that even at relatively low temperatures, there can be a precipitous effect of temperature on dissociative and non-dissociative electron attachment. Similar non-equilibrium excitations have been advanced for improved hydrogen negative ion sources [27].

Table 4. Species of Interest to Discharge Lamp Development

Metal Halides, especially Iodides
Sn, Na, Hg, Sc, Th, including dimers
I, Br, In, Cs, Rb, K, Ba

III. The third area that I wish to emphasize is that of selected parameterized gas mixtures that permit variation of the characteristic energy at fixed E/N by varying the relative concentrations, and at fixed gas mixture for spatial E/N variations of interest. The necessity for such measurements is to be able to separate the different cross-sections at low energies. Thus, it is possible to separate the influence of momentum transfer and inelastic vibrational excitation in silane by adding a buffer gas such as helium that has no inelastic losses in the energy region of interest but will dominate the elastic losses. The buffer gas contribution to the elastic losses is considered to be known and well defined. If measurements are made on only one particular gas or gas mixture, then the theoretical comparison is made with a curve representing the variation of the drift velocity or characteristic energy versus E/N. By making the measurement over a wide range of gas mixtures where the buffer gas contributes little to momentum scattering at low concentrations, the comparison between theory and experiment is made over a surface (coordinates E/N and gas mixture). Argon, krypton, and xenon are the usual Ramsauer buffer gases, although it is noted that silane and methane demonstrate Ramsauer behavior. Helium is also useful in that by comparison it will dominate elastic scattering and provides a different surface. There is then obtained a mapping of the cross-section of the minority gas as E/N is increased. The sensitivities of the maximum of the differential negative conductivity and of the relative increase in drift velocity are quite dramatic, and permit a comparison of the inelastic cross-section (Qinel) and elastic cross-section (Qel) at very low energies. There is a caution here in the comparison of theoretical and experimental cross-sections. When Qel is less than Qinel the theory requires higher order terms in the polynomial expansion, or an integral equation approach to the solution of the Boltzmann transport equation. However, it must be remembered that the derived cross-sections are exactly appropriate only for the method or code whereby they were derived. If one obtains cross-sections for a process using the two-term approximation, then it is not appropriate to use the same cross-section for higher order term methods, and vice versa. The literature citations have not always been careful regarding this point.

IV. It is also appropriate in the same vein to consider using the swarm method as an analog computer for given gases or gas mixtures of interest and a selected spatial variation of E/N that is known to be of interest. New spectroscopic methods using laser-induced optogalvanic spectroscopy of high Rydberg states permit fairly accurate measurements of the electric field profiles in the plasma sheath. The difficulty with the actual discharge measurements is that the discharge determines all the conditions and at higher currents other effects such as metastable generation, cumulative excitation, photoelectric emission and gas heating complicate the interpretation of experimental data and its comparison with theory. It is therefore suggested that, rather than working a complete self-consistent field analysis for such a situation, with consideration of cross-section and excitation anisotropies and non-equilibrium, one can set up the configuration and fields of the drift tube to simulate the measured electric fields. It is therefore possible to measure directly the effects of such non-equilibrium conditions on ionization and attachment rates. The sensitivity of the results to the assumed field profiles can be tested and in this "simulation" all of the cross-section anisotropies and electron energy distribution function anisotropies are included. A variable drift space is included to simulate the negative glow and an anode field can also be imposed. This type of apparatus would appear to be able to give at least approximate and quick results for electronegative gases of interest to thin film etching. The electronegative character of the gases usually does not appear until low E/N values are attained. In fact, at high E/N values many of these "electronegative" gases have larger ionization rates than the buffer gas. The transition point between electronegative character and electropositive character for interesting non-equilibrium conditions, even time-averaged radio-frequency fields, should be accessible in the proposed configuration.

V. The fifth topic advocated is consideration of how the swarm method can be improved with respect to deconvolution of the cross-section data. The objective would be to develop standard methods of evaluation of the transport properties into cross-sections with quantitative assessment of the accuracy obtained. Independent of which theoretical method that is used, the procedure at most laboratories is to elect a trial set of cross-sections derived from theory, other experiments or prior experience, and then to adjust this set to improve agreement with experiment. When there are only a few cross-sections, the procedure works reasonably well. However, if there is a manifold of, say, vibrational cross-sections, then assistance from cross-beam experiments and from spectroscopy is mandatory. Now that I am advocating measurements in gases where there is not always an auxiliary data bank, it is especially appropriate and necessary to examine the refinement of the deconvolution techniques. Formally, the experiment is a non-linear transducer changing the input signals of applied fields into current responses that are determined by the self-consistent electron energy distribution controlled by the gases and their cross-section. The measured electron transport values are constraints that specify mean values of certain functions or moments of the distribution function. Then for the two-dimensional set of distribution functions and correlated cross-sections that satisfy the constraints, there should be a decider in terms of the surprisal analysis or entropy maximization [28]. The constraints can be imposed both from the swarm measurements and from other types of experiments. Then the most probable distribution is that distribution which satisfies the moment constraints and maximizes the entropy of the distribution. The maximization is derived by stationarity using the Lagrange multiplier techniques (It should be noted that these procedures have been derived in different ways by workers in various fields, and the techniques are also known as autoregression modeling). It would be extremely helpful to the swarm experimentalist if a standard or even partial standard method of analysis were available so that the influence of the constraints, sometimes physical measurements, sometimes assumptions, could be analyzed quantitatively. For better comparisons with Monte Carlo calculations, two-dimensional detector arrays should be utilized as anodes in the drift experiments. This improvement would permit more sophisticated statistical methods to be applied to the measurement of transverse diffusion and forward scattering.

VI. The concept of standards applies also to the geometry, signal generation, number of primary electrons, signal detection and procedures used. These standards will be very important for measurements in reactive gases and/or transient species, because the original data may be geometry and pressure dependent. The other standard that appears mandatory for the more complex gases appears to be the identification and measurement of positive and negative ion species. These measurements provide additional constraints that the electron energy distribution function has to satisfy, and hence, define it better. The mass spectrometric measurements are probably more useful in gases that show resonant or dissociative attachment at low energies or in studies on gases that have been previously excited. The easier moments to measure experimentally are probably electron impact excitation rates. While some studies have been performed along this direction, they have been the 'easier' measurement of visible radiation [29] excited by the swarm electrons in the pure rare gases (He, Ne, and Ar). However, infrared detectors have become quite efficient [30]. Also, infrared detector linear arrays are available, so that multichannel spectral analysis and rapid measurements of spatial growth can be performed. The importance of infrared measurements is that they permit moments of the tail of the electron energy distribution function to be obtained at relatively low $\frac{E}{N}$ values. The geometry and the gas density of the swarm tube can be defined very accurately. The electron current can also be measured very accurately. Therefore, if excitation cross-sections and transition probabilities for selected excited states are known, say from theory, it should be possible to define an absolute standard for a wide range of wavelengths. This capability would seem also to be important in the ultraviolet region at high E/N values.

Conclusions. The swarm method has reached a maturity in its development, as evidenced in the books of Gilardini [1] and of Huxley and Crompton [2]. The scope of the investigations has (naturally) tended to become more concentrated on some of the finer details of cross-sections of particular molecules. The method itself is being challenged by other techniques such as crossed beams and electron traps. All of this is healthy. In this assessment, it has been argued that there are still many interesting opportunities for swarm studies, with the caveat that some of these topics require a timely response in order to influence the progress of perhaps the most important changes in solid state technology of the next 15 years. There are other interesting areas where swarm data is relevant that were not assessed. These include the atmospheric and space sciences, and the combustion and plume areas. An increased interest in the transport properties of sodium in various gas mixtures is anticipated now that it has been determined [31] to be a major constituent of the atmospheres of Mercury and of Io, the satellite of Jupiter. The electron attachment processes in combustion products is of interest to define the electron density in plumes and to reconcile microwave and infrared signatures. In optics and microelectrons, there is an interest in the plasma-enhanced deposition of transition metals and transition metal silicides. This interest introduces the halides and the carbonyls as the source gases [32]. The buffer gases are usually hydrogen and silane, for the depositions of the metals and the metal silicides respectively. There may be other applications in pollution control and plasma-induced decomposition of toxic compounds [15]. Thus, Mucha [25] has emphasized that plasma technology is still in its infancy, and that the control of plasma parameters to achieve desirable results (in his case, etching selectivity and profile control) is more an art than a science. To unravel the electron kinetics of these gas mixtures offers a challenge and an opportunity to swarm practitioners, theoretical and experimental.

References

[1] Gilardini, A.; "Low-Energy Electron Collisions in Gases: Swarm and Plasma Methods Applied to Their Study", Wiley, New York 1972.

[2] Huxley, L. G. H.; Crompton, R. W.; "The Diffusion and Drift of Electrons in Gases", Wiley, New York 1974.

[3] Dutton, J.; J. Phys and Chem. Ref. Data, 4, 577 1975.

[4] Lin, S. L.; Robson. R. E.; Mason, E. A.; J. Chem. Phys., 71, 3483 1979.

[5] Reid, I. D.; Aust. J. Phys., 32, 331 1979.

[6] Rees, H. D.; J. Phys. Chem. Solids, 30, 643 1969.

[7] Kleban, P.; Davis, H. T.; J. Chem. Phys., 68, 2999 1978.

[8] Pitchford, L. C.; O'Neil, S. V.; Rumble, J. R.; Phys. Rev., 23, 294 1981.

[9] Grad, H.; Handbuch der Physik (S. Flugge, ed.) 12, 205, Springer-Verlag, Berlin, 1956.

[10] Kunar, K.; Robson, R. E.; Aust. J. Phys. 26, 157 1973.

[11] Kitamori, K.; Tagishira, H.; Sakai, Y.; J. Phys. D, 13, 535 1980.

[12] Segur, P.; Bordage, M-C; Balaguer, J-P; Yousfi, M.; J. Comp. Physics, 50, 116 1983.

Segur, P.; Yousfi, M.; Bordage, M-C; J. Phys. D: 17, 2199, 1984.

[13] Blevin, H. A.; Fletcher, J.; Hunter, S. R.; Phys. Rev., 31, 2215 1985.

[14] Long, W. H. Jr.; "Plasma Sheath Studies", USAF Technical Report AFAPL TR-79-2038.

[15] Boenig, H. V.; "Plasma Science and Technology," Cornell University Press, Ithaca, 1982.

[16] Flamm, D. L.; "Plasma Chemistry and Plasma Processing, 1, 37, 1981.

[17] Pande, K. P.; Nair, V.K.R.; Gutierrez, D.; J. Appl. Phys., 54, 5436, 1983.

[18] Garscadden, A.; Duke, G. L.; Bailey, W. F.; Appl. Phys. Lett., 43, 1012, 1983.

[19] Pollock, W. J.; Faraday Discuss. Chem. Soc., 2, 2919, 1968.

[20] Andrews, M. A.; Kirkendall, K.; Garscadden, A.; to be published.

[21] Christophorou, L. G.; "Electron and Ion Swarms," 261, edited by L. G. Christophorou, Pergamon Press, New York, 1981.

[22] Winters, H. G.; Inokuti, M.; Phys. Rev. A25, 1420, 1982.

[23] Winters, H. F.; J. Chem. Phys., 63, 3462, 1975.

[24] Cosby, P. C.; Helm, H.; private communication, 1985.

[25] Mucha, J. A.; Solid State Technology, 28, 123, 1985.

[26] Deutsch, H.; Schmidt, M.; Beit, aus der Plasmaphys., 25, 475, 1985.

[27] Bacal, M.; Bruneteau, A. M.; Graham, W. G.; Hamilton, G. W.; Nachman, M.; J. Appl. Phys., 52, 1247, 1981.

[28] Levine, R. D.; Tribus, M.; editors; "The Maximum Entropy Formalism", MIT Press, Cambridge, 1978.

[29] Fletcher, J.; "Non-Equilibrium in Low-Pressure Rare Gas Discharges"; J. Phys. D: Appl. Phys., 18 1985.

[30] Morgan, B. L.; editor; "Photo-Electronic Image Devices" (Advances in Electronics and Electron Physics 64B), Academic Press, London, 1985.

[31] Potter, A.; Morgan, M.; Science, 229, 651, 1985.

[32] Tang, C. C.; Chu, J. K.; Hess, D. W.; Solid State Technology, 26, 125, 1983.

MIX
Papier aus verantwortungsvollen Quellen
Paper from responsible sources
FSC® C105338

If you have any concerns about our products,
you can contact us on
ProductSafety@springernature.com

In case Publisher is established outside the EU,
the EU authorized representative is:
Springer Nature Customer Service Center GmbH
Europaplatz 3, 69115 Heidelberg, Germany

Printed by Libri Plureos GmbH
in Hamburg, Germany